Lecture Notes in Artificial Intelligence 11676

Subseries of Lecture Notes in Computer Science

Series Editors

Randy Goebel
University of Alberta, Edmonton, Canada
Yuzuru Tanaka
Hokkaido University, Sapporo, Japan
Wolfgang Wahlster
DFKI and Saarland University, Saarbrücken, Germany

Founding Editor

Jörg Siekmann
DFKI and Saarland University, Saarbrücken, Germany

More information about this series at http://www.springer.com/series/1244

Vicenç Torra · Yasuo Narukawa ·
Gabriella Pasi · Marco Viviani (Eds.)

Modeling Decisions for Artificial Intelligence

16th International Conference, MDAI 2019
Milan, Italy, September 4–6, 2019
Proceedings

Springer

Editors
Vicenç Torra (iD)
Maynooth University
Maynooth, Ireland

Yasuo Narukawa
Tamagawa University
Machida, Tokyo, Japan

Gabriella Pasi (iD)
University of Milano-Bicocca
Milan, Italy

Marco Viviani (iD)
University of Milano-Bicocca
Milan, Italy

ISSN 0302-9743   ISSN 1611-3349   (electronic)
Lecture Notes in Artificial Intelligence
ISBN 978-3-030-26772-8   ISBN 978-3-030-26773-5   (eBook)
https://doi.org/10.1007/978-3-030-26773-5

LNCS Sublibrary: SL7 – Artificial Intelligence

This Springer imprint is published by the registered company Springer Nature Switzerland AG
The registered company address is: Gewerbestrasse 11, 6330 Cham, Switzerland

Preface

This volume contains papers presented at the 16th International Conference on Modeling Decisions for Artificial Intelligence (MDAI 2019), held in Milan, Italy, during September 4–6, 2019. This conference followed MDAI 2004 (Barcelona), MDAI 2005 (Tsukuba), MDAI 2006 (Tarragona), MDAI 2007 (Kitakyushu), MDAI 2008 (Sabadell), MDAI 2009 (Awaji Island), MDAI 2010 (Perpinyà), MDAI 2011 (Changsha), MDAI 2012 (Girona), MDAI 2013 (Barcelona), MDAI 2014 (Tokyo), MDAI 2015 (Skövde), MDAI 2016 (Sant Julià de Lòria), MDAI 2017 (Kitakyushu), and MDAI 2018 (Mallorca).

The aim of this conference was to provide a forum for researchers to discuss different facets of decision processes in a broad sense. This includes model building and all kinds of mathematical tools for data aggregation, information fusion, and decision-making; tools to help make decisions related to data science problems (including, e.g., statistical and machine learning algorithms as well as data visualization tools); and algorithms for data privacy and transparency-aware methods so that data processing procedures and the decisions made from them are fair, transparent, and avoid unnecessary disclosure of sensitive information.

The MDAI conference included tracks on the topics of (a) data science, (b) data privacy, (c) aggregation functions, (d) human decision-making, (e) graphs and (social) networks, and (f) recommendation and search.

The organizers received 50 papers from 15 different countries, 30 of which are published in this volume. Each submission received at least two reviews from the Program Committee and a few external reviewers. We would like to express our gratitude to them for their work. This volume also includes some of the plenary talks.

The conference was supported by the Information Retrieval Laboratory (IR Lab), the Department of Informatics, Systems, and Communication (DISCO), the University of Milano-Bicocca, the European Society for Fuzzy Logic and Technology (EUSFLAT), the Catalan Association for Artificial Intelligence (ACIA), the Japan Society for Fuzzy Theory and Intelligent Informatics (SOFT), the UNESCO Chair in Data Privacy, and Axioms – MDPI.

June 2019

Vicenç Torra
Yasuo Narukawa
Gabriella Pasi
Marco Viviani

Organization

General Chairs

Gabriella Pasi — University of Milano-Bicocca, Milan, Italy
Marco Viviani — University of Milano-Bicocca, Milan, Italy

Program Chairs

Vicenç Torra — Maynooth University, Maynooth, Ireland
Yasuo Narukawa — Tamagawa University, Tokyo, Japan

Advisory Board

Didier Dubois — Institut de Recherche en Informatique de Toulouse, CNRS, France
Jozo Dujmović — San Francisco State University, CA, USA
Lluis Godo — IIIA-CSIC, Spain
Kaoru Hirota — Beijing Institute of Technology; JSPS Beijing Office, China
Janusz Kacprzyk — Systems Research Institute, Polish Academy of Sciences, Poland
Sadaaki Miyamoto — University of Tsukuba, Japan
Michio Sugeno — Tokyo Institute of Technology, Japan
Ronald R. Yager — Machine Intelligence Institute, Iona Collegue, NY, USA

Program Committee

Eva Armengol — IIIA-CSIC, Spain
Edurne Barrenechea — Universidad Pública de Navarra, Spain
Gloria Bordogna — Consiglio Nazionale delle Ricerche, Italy
Humberto Bustince — Universidad Pública de Navarra, Spain
Francisco Chiclana — De Montfort University, UK
Susana Díaz — Universidad de Oviedo, Spain
Josep Domingo-Ferrer — Universitat Rovira i Virgili, Spain
Yasunori Endo — University of Tsukuba, Japan
Zoe Falomir — Universität Bremen, Germany
Katsushige Fujimoto — Fukushima University, Japan
Michel Grabisch — Université Paris I Panthéon-Sorbonne, France
Enrique Herrera-Viedma — Universidad de Granada, Spain
Aoi Honda — Kyushu Institute of Technology, Japan
Van-Nam Huynh — JAIST, Japan

Masahiro Inuiguchi	Osaka University, Japan
Simon James	Deakin University, Australia
Aránzazu Jurío	Universidad Pública de Navarra, Spain
Yuchi Kanzawa	Shibaura Institute of Technology, Japan
Petr Krajča	Palacky University Olomouc, Czech Republic
Marie-Jeanne Lesot	Université Pierre et Marie Curie (Paris VI), France
Xinwang Liu	Southeast University, China
Jun Long	National University of Defense Technology, China
Jean-Luc Marichal	University of Luxembourg, Luxembourg
Radko Mesiar	Slovak University of Technology, Slovakia
Andrea Mesiarová-Zemánková	Slovak Academy of Sciences, Slovakia
Toshiaki Murofushi	Tokyo Institute of Technology, Japan
Guillermo Navarro-Arribas	Universitat Autònoma de Barcelona, Spain
Jordi Nin	Esade, Universitat Ramon Llull, Spain
Pierangela Samarati	Università degli Studi di Milano, Italy
Sandra Sandri	Instituto Nacional de Pesquisas Espaciais, Brazil
H. Joe Steinhauer	University of Skövde, Sweden
László Szilágyi	Sapientia-Hungarian Science University of Transylvania, Hungary
Aida Valls	Universitat Rovira i Virgili, Spain
Zeshui Xu	Southeast University, China
Yuji Yoshida	University of Kitakyushu, Japan

Local Organizing Committee Chair

Fabio Stella	University of Milano-Bicocca, Milan, Italy

Local Organizing Committee

Elias Bassani	University of Milano-Bicocca, Milan, Italy
Magdalena Nikola	University of Milano-Bicocca, Milan, Italy
Emanuele Cavenaghi	University of Milano-Bicocca, Milan, Italy
Francesco Stranieri	University of Milano-Bicocca, Milan, Italy

Additional Reviewers

Jimmy Devillet
Fadi Hassan
Alberto Blanco
Carles Anglès
Sergio Martinez
Michael Bamiloshin
Jordi Casas
Julian Salas

Supporting Institutions

The Information Retrieval Laboratory (IR Lab)
The Department of Informatics, Systems, and Communication (DISCO)
The University of Milano-Bicocca
The European Society for Fuzzy Logic and Technology (EUSFLAT)
The Catalan Association for Artificial Intelligence (ACIA)
The Japan Society for Fuzzy Theory and Intelligent Informatics (SOFT)
The UNESCO Chair in Data Privacy
Axioms – MDPI (who supported a Best Paper Award)

Abstracts of Invited Talks

Incomplete Knowledge in Computational Social Choice

Jérôme Lang

Laboratoire d'Analyse et Modélisation des Systèmes pour l'Aide à la Décision
(LAMSADE)
Université Paris-Dauphine,
Place du Maréchal de Lattre de Tassigny,
75775 Paris Cedex 16

Abstract. Social choice theory studies the aggregation of individual preferences towards a collective choice. Computational social choice emerged in the late 1980s, and mostly uses computational paradigms and techniques to provide a better analysis of social choice mechanisms (especially in the fields of voting and of fair division of resources), and to construct new ones. Among the subfields of artificial intelligence that are concerned by this interaction, knowledge representation plays an important role (other subfields being machine learning, reasoning with uncertainty, search, and constraint programming). The reasons for which it plays an important role include: representing preferences and reasoning about them; computing collective decisions with incomplete knowledge of agents' preferences; the role of knowledge in strategic behavior; and using logic for automated theorem proving in social choice.

As Simple as Possible But Not Simpler in Multiple Criteria Decision Analysis: The Robust Stochastic Level Dependent Choquet Integral Approach[1]

Salvatore Greco

Department of Economics and Business, University of Catania, Corso Italia 55,
95024 Catania, Italy
salgreco@unict.it

Abstract. The level dependent Choquet integral has been proposed to take into account multiple criteria decision making problems in which the importance of criteria, the sign and the magnitude of their interactions may depend on the level of the alternatives' evaluations. This integral is based on a level dependent capacity, which is a family of single capacities associated to each level of evaluation for the considered criteria. Since, in general, there is not only one but many level dependent capacities compatible with the preference expressed by the Decision Maker, we propose to take into account all of them by using the Robust Ordinal Regression (ROR) and the Stochastic Multicriteria Acceptability Analysis (SMAA). On one hand, ROR defines a necessary preference relation (if an alternative a is at least as good as an alternative b for all compatible level dependent capacities), and a possible preference relation (if a is at least as good as b for at least one compatible level dependent capacity). On the other hand, considering a random sampling of compatible level dependent capacities, SMAA gives the probability that each alternative reaches a certain position in the ranking of the alternatives as well as the probability that an alternative is preferred to another. A real decision problem related to ranking of universities is provided to illustrate the proposed methodology.

References

1. Angilella, S., Greco, S., Matarazzo, B.: Non-additive robust ordinal regression: a multiple criteria decision model based on the Choquet integral. Eur. J. Oper. Res. **201**(1), 277–288 (2010)
2. Angilella, S., Corrente, S., Greco, S.: Stochastic multiobjective acceptability analysis for the Choquet integral preference model and the scale construction problem. Eur. J. Oper. Res. **240** (1), 172–182 (2015)
3. Arcidiacono, S.G., Corrente, S., Greco, S.: As simple as possible but not simpler in multiple criteria decision aiding: the ROR-SMAA level dependent choquet integral approach. arXiv preprint, arXiv:1905.07941 (2019)

---

[1] Joint work with Sally Giuseppe Arcidiacono and Salvatore Corrente

4. Giarlotta, A., Greco, S.: Necessary and possible preference structures. J. Math. Econ. **49**(2), 163–172 (2013)
5. Greco, S., Matarazzo, B., Giove, S.: The Choquet integral with respect to a level dependent capacity. Fuzzy Sets Syst. **175**(1), 1–35 (2011)
6. Greco, S., Mousseau, V., Słowiński, R.: Ordinal regression revisited: multiple criteria ranking using a set of additive value functions. Eur. J. Oper. Res. **191**(2), 416–436 (2008)

Contents

Data Privacy and Security

Aggregation Operators and Decision Making

General Chebyshev Type Inequality
for Seminormed Fuzzy Integral

Michał Boczek[1], Anton Hovana[2], and Ondrej Hutník[2(✉)]

[1] Institute of Mathematics, Lodz University of Technology,
ul. Wólczańska 215, 90-924 Lodz, Poland
michal.boczek.1@p.lodz.pl
[2] Institute of Mathematics, Faculty of Science,
Pavol Jozef Šafárik University in Košice, Jesenná 5, 040 01 Košice, Slovakia
anton.hovana@student.upjs.sk, ondrej.hutnik@upjs.sk

Abstract. We give a necessary and sufficient condition for Chebyshev type inequality in the class of seminormed fuzzy integrals with respect to m-positively dependent functions. Reviewing the literature many known results are generalized. We also present the Chebyshev type inequality for Shilkret integral for independent random variables.

Keywords: Aggregation functions · Semicopula ·
Seminormed integral · Capacity · Chebyshev inequality ·
Positively dependent functions

1 Introduction

In the last decade, a huge amount of papers appeared in the connection of inequalities and non-additive integrals. In this paper, we restrict our attention to the Chebyshev type inequality for the class of integrals called the seminormed fuzzy integrals. This class includes the prominent Sugeno and Shilkret integral. For the definitions, we refer to Sect. 2.

The classical Chebyshev integral inequality reads as follows, see [9]: *If $(X, \mathcal{A})$ is a measurable space, then $\mathcal{A}$-measurable functions $f, g \colon X \to \mathbb{R}$ satisfy the inequality*

$$\int fg \, d\mathsf{P} \geqslant \int f \, d\mathsf{P} \int g \, d\mathsf{P} \tag{1}$$

for any probability measure P if and only if f and g are comonotone.

It is well-known that the classical integral inequalities (including the Chebyshev one) need not hold in general when considering non-additive measure and

This work was supported by the Slovak Research and Development Agency under the contract No. APVV-16-0337 and also cofinanced by bilateral call Slovak-Poland grant scheme No. SK-PL-18-0032 with the Polish National Agency for Academic Exchange PPN/BIL/2018/1/00049/U/00001.

V. Torra et al. (Eds.): MDAI 2019, LNAI 11676, pp. 3–16, 2019.
https://doi.org/10.1007/978-3-030-26773-5_1

integral setting. In fact, when replacing the probability measure P by a non-additive measure m, and the additive (Lebesgue) integral in (1) by the semi-normed fuzzy integral $\mathbf{I_S}$, in most cases we have to require additional assumptions on semicopula S or non-additive measure m in order to the Chebyshev inequality hold even for comonotone functions. For the reader's convenience, we briefly describe the present state of art of the Chebyshev type inequalities for $\mathbf{I_S}$.

A Short Historical Overview. The first paper dealing with the Chebyshev type inequality for seminormed fuzzy integral was published in 2007 by Flores-Franulič and Román-Flores [12] in the particular case for the Sugeno integral. Their results were generalized in [18] by proving the Chebyshev type inequality for an arbitrary continuous monotone measure-based Sugeno integral and for any two monotone functions (in the same sense). Using the concept of comonotonicity, in 2009, Mesiar and Ouyang [16] generalized the results for a non-decreasing continuous binary operation $\star$. In that paper, the condition $\star \leqslant \wedge$, with $\wedge$ being the pointwise minimum, appeared for the first time and then this condition was frequently applied in further papers, see [3–5,20,23]. Later on, Ouyang and Mesiar [19] proved the result on Chebyshev type inequality for seminormed fuzzy integral in general. They showed that the inequality is true for all comonotone functions whenever

$$S(a \star b, c) \geqslant (S(a, c) \star b) \vee (a \star S(b, c))$$

holds for all $a, b, c \in [0, 1]$. This is equivalent to $\star \leqslant \wedge$ (see Corollary 1) in the case of Sugeno integral, i.e., for the minimum semicopula S = M. The result from [19] was further extended by other researchers, see e.g. [1,2,7,8]. In 2014, Kaluszka et al. [14] provided not only a sufficient condition for validity of the Chebyshev type inequalities, but also a necessary one for a wider class of functions than comonotone ones. This is the first time in the literature of various generalizations of the seminormed fuzzy integral inequalities when the necessary condition is given. It is also worth noting that already in 2011 certain equivalent conditions for validity of the Chebyshev inequality were given by Girotto and Holzer [13] providing a characterization of comonotonicity by the Chebyshev inequality for the Sugeno integral and all monotone measures, analogically with the above mentioned work of Armstrong [9].

Aim and Structure of the Paper. In Sect. 2, we review some basic concepts and notations including the class of m-positively dependent functions with examples. The introduction of m-positively dependent functions in the consideration of the Chebyshev type inequality allows us to obtain the equivalent conditions in Sect. 3. Furthermore, in Sect. 4, we discuss the existing sufficient conditions in the literature and provide relationships among them. These results together with Theorem 1 enable us to cover many known results in the literature related to the Chebyshev type inequalities for $\mathbf{I_S}$.

2 Preliminaries

A *semicopula* S: $[0,1]^2 \to [0,1]$ is a non-decreasing function, i.e., $S(a,c) \leqslant S(b,d)$ whenever $a \leqslant b$ and $c \leqslant d$, such that $S(x,1) = S(1,x) = x$ for all x. From it follows that $S(x,0) = 0 = S(0,x)$ for any x and $S(x,y) \leqslant x \wedge y$ for all x,y, where $a \wedge b = \min(a,b)$. Typical examples of semicopulas are the functions $M(a,b) = a \wedge b$, $\Pi(a,b) = ab$ and $W(a,b) = (a+b-1) \vee 0$, where $a \vee b = \max(a,b)$. The set of all semicopulas S will be denoted by $\mathfrak{S}$.

Furthermore, we deal with a measurable space $(X, \mathcal{A})$, where $\mathcal{A}$ is a σ-algebra of subsets of a non-empty set X. For the given measurable space $(X, \mathcal{A})$ we denote the set of all $\mathcal{A}$-measurable functions $f \colon X \to [0,k]$, $k \in (0,1]$, by $\mathcal{F}^k_{(X,\mathcal{A})}$. We also consider the set $\mathcal{M}_{(X,\mathcal{A})}$ of all *monotone measures* (or, *capacities*), i.e., set functions $m \colon \mathcal{A} \to [0,1]$ satisfying $m(A) \leqslant m(B)$ whenever $A, B \in \mathcal{A}$ such that $A \subset B$, and the boundary conditions $m(\emptyset) = 0$ and $m(X) = 1$. We denote the range of m by $m(\mathcal{A})$. We say that a monotone measure m is *continuous* if $m\left(\bigcup_{n=1}^{\infty} A_n\right) = \lim_{n \to \infty} m(A_n)$ and $m\left(\bigcap_{n=1}^{\infty} B_n\right) = \lim_{n \to \infty} m(B_n)$ for $A_1 \subset A_2 \subset \ldots$ and $B_1 \supset B_2 \supset \ldots$.

Let $S \in \mathfrak{S}$. A functional $\mathbf{I}_S \colon \mathcal{M}_{(X,\mathcal{A})} \times \mathcal{F}^1_{(X,\mathcal{A})} \to [0,1]$ defined by

$$\mathbf{I}_S(m,f) := \sup_{t \in [0,1]} S(t, m(\{f \geqslant t\})),$$

where $\{f \geqslant t\} = \{x \in X \colon f(x) \geqslant t\}$, is called the *seminormed fuzzy integral* [10,11,21]. It is clear that $\mathbf{I}_S(m, f \mathbb{1}_A) = \mathbf{I}_{S,A}(m,f)$ for $A \in \mathcal{A}$, where $\mathbb{1}_A$ denotes the indicator function of a subset $A \in \mathcal{A}$ and

$$\mathbf{I}_{S,A}(m,f) := \sup_{t \in [0,1]} S(t, m(\{f|_A \geqslant t\})).$$

Here, $h|_C$ is the restriction of an $\mathcal{A}$-measurable function h to a set $C \subset X$. Obviously, $\{h|_C \geqslant t\} = \{x \in C \colon h(x) \geqslant t\} = C \cap \{h \geqslant t\}$. Replacing semicopula S with W, Π and M, we get the opposite-Sugeno integral, Shilkret integral and Sugeno integral of $[0,1]$-valued functions, respectively.

Definition 1. *Let* $f, g \in \mathcal{F}^k_{(X,\mathcal{A})}$, $m \in \mathcal{M}_{(X,\mathcal{A})}$ *and* $A, B \in \mathcal{A}$ *with* $k \in (0,1]$. *Functions* $f|_A$ *and* $g|_B$ *are called m-positively dependent with respect to an operator* $\triangle \colon m(\mathcal{A}) \times m(\mathcal{A}) \to m(\mathcal{A})$, *if*

$$m(\{f|_A \geqslant a\} \cap \{g|_B \geqslant b\}) \geqslant m(\{f|_A \geqslant a\}) \triangle m(\{g|_B \geqslant b\})$$

holds for all $a, b \in [0,k]$.

This concept was introduced by Kaluszka et al. in [14]. It includes all comonotone functions[1] on A being m-positively dependent with respect to $\triangle = \wedge$ with $B = A$ and any capacity, and all independent random variables with respect to probability measure when $\triangle = \cdot$ and $A = B = X$.

---

[1] Functions $f, g \colon X \to \mathbb{R}$ are called *comonotone* on $D \subset X$ if $(f(x) - f(y))(g(x) - g(y)) \geqslant 0$ for all $x, y \in D$. In the case $D = X$ we will simply say "f, g are comonotone".

3   Main Results

Recall that a non-decreasing operator $\circ \colon [0,1]^2 \to [0,1]$ is *left-continuous* if it is left-continuous with respect to each coordinate.

Theorem 1. *Suppose that* $k \in (0,1]$, $m \in \mathcal{M}_{(X,\mathcal{A})}$, $S_i \in \mathfrak{S}$ *for* $i = 1,2,3$ *and* $\star, \Diamond \colon [0,1]^2 \to [0,1]$ *are non-decreasing and* $\Diamond$ *be left-continuous. Let* $\varphi_i \colon [0,1] \to [0,1]$ *be functions such that* φ_1 *is non-decreasing and* φ_j *are increasing and right-continuous,* $\psi_i \colon [0, \varphi_i(1)] \to [0,1]$ *be non-decreasing and* ψ_j *be left-continuous, where* $i = 1,2,3$ *and* $j = 2,3$.

(a) Assume that the inequality

$$\psi_1\big(S_1(\varphi_1(a \star b), c \,\triangle\, d)\big) \geqslant \psi_2\big(S_2(\varphi_2(a), c)\big) \,\Diamond\, \psi_3\big(S_3(\varphi_3(b), d)\big) \tag{2}$$

holds for all $a,b \in [0,k]$ *and* $c,d \in m(\mathcal{A})$. *If* $f|_A, g|_B \in \mathcal{F}^k_{(X,\mathcal{A})}$ *are* m-*positively dependent with respect to* $\triangle \colon m(\mathcal{A}) \times m(\mathcal{A}) \to m(\mathcal{A})$, *then*

$$\psi_1\big(\mathbf{I}_{S_1, A \cap B}(m, \varphi_1(f \star g))\big) \geqslant \psi_2\big(\mathbf{I}_{S_2, A}(m, \varphi_2(f))\big) \,\Diamond\, \psi_3\big(\mathbf{I}_{S_3, B}(m, \varphi_3(g))\big). \tag{3}$$

(b) Suppose that $a \,\triangle\, 0 = 0 \,\triangle\, a = 0$ *for all* $a \in m(\mathcal{A})$ *and the condition* (Z_1) *holds, i.e., for all* $c,d \in m(\mathcal{A})$ *there exists sets* $C, D \in \mathcal{A}$ *such that* $c = m(C)$, $d = m(D)$ *and* $m(C \cap D) = m(C) \,\triangle\, m(D)$. *If* (3) *is true for all* m-*positively dependent functions* $f|_A$ *and* $g|_B$ *with respect to* $\triangle$, *then* (2) *holds for all* $a,b \in [0,k]$ *and* $c,d \in m(\mathcal{A})$.

Proof. (a) Observe that ψ's are well defined in (2) as $S_j(\varphi_j(a), c) \in [0, \varphi_j(1)]$ for all $a \in [0,k]$ and $c \in m(\mathcal{A})$, where $j = 2,3$ and $S_1(\varphi_1(a \star b), c \,\triangle\, d) \in [0, \varphi_1(1)]$ for all $a,b \in [0,k]$ and $c,d \in m(\mathcal{A})$. From monotonicity of $\star$, the inequality $f(x) \star g(x) \geqslant a \star b$ holds for $x \in \{f \geqslant a\} \cap \{g \geqslant b\}$. Then from m-positive dependency of $f|_A$ and $g|_B$ and monotonicity of m, we get

$$m(A \cap B \cap \{f \star g \geqslant a \star b\}) \geqslant m(A \cap \{f \geqslant a\}) \,\triangle\, m(B \cap \{g \geqslant b\})$$

for any $a,b \in [0,k]$. As φ_1 is non-decreasing and φ_j are increasing for $j = 2,3$, we have

$$m\big(A \cap B \cap \{\varphi_1(f \star g) \geqslant \varphi_1(a \star b)\}\big) \geqslant c_a \,\triangle\, d_b$$

for all $a,b \in [0,k]$ with $c_a := m(A \cap \{\varphi_2(f) \geqslant \varphi_2(a)\})$ and $d_b := m(B \cap \{\varphi_3(g) \geqslant \varphi_3(b)\})$. From the assumption on the monotonicity of S_1 and ψ_1, as well as definition of $\mathbf{I}_{S_1}$ we obtain

$$\psi_1\big(\mathbf{I}_{S_1, A \cap B}(m, \varphi_1(f \star g))\big) \geqslant \psi_1\big(S_1(\varphi_1(a \star b), c_a \,\triangle\, d_b)\big)$$

for any $a,b \in [0,k]$. From (2) it follows that

$$\psi_1\big(\mathbf{I}_{S_1, A \cap B}(m, \varphi_1(f \star g))\big) \geqslant \psi_2\big(S_2(\varphi_2(a), c_a)\big) \,\Diamond\, \psi_3\big(S_3(\varphi_3(b), d_b)\big) \tag{4}$$

for all $a, b \in [0, k]$. Let $H(x) = x \lozenge y$ for a fixed $y \in [0, 1]$ and all $x \in [0, 1]$ as well as $h(a) = S_2(\varphi_2(a), c_a)$ for all $a \in [0, k]$. The function $H(\psi_2) \colon [0, \varphi_2(1)] \to [0, 1]$ is non-decreasing and left-continuous as $\lozenge$ and ψ_2 are non-decreasing and left-continuous, thus

$$\sup_{a \in [0,k]} H\big(\psi_2(h(a))\big) = H\big(\psi_2(\sup_{a \in [0,k]} h(a))\big). \tag{5}$$

Moreover, $\sup_{a \in [0,k]} h(a) = \sup_{a \in [0,1]} h(a)$ due to $S_2(x, 0) = 0$. Putting $t = \varphi_2(a)$, we get

$$\sup_{a \in [0,k]} h(a) = \sup_{t \in \varphi_2([0,1])} S_2\big(t, m(A \cap \{\varphi_2(f) \geqslant t\})\big),$$

where $\varphi_2([0, 1])$ is the image of φ_2. By the right-continuity of φ_2 and properties of S_2, we have

$$\sup_{a \in [0,k]} h(a) = \sup_{t \in [0,1]} S_2\big(t, m(A \cap \{\varphi_2(f) \geqslant t\})\big). \tag{6}$$

From (4)–(6) we conclude

$$\psi_1\big(\mathbf{I}_{S_1, A \cap B}(m, \varphi_1(f \star g))\big) \geqslant \psi_2\big(\mathbf{I}_{S_2, A}(m, \varphi_2(f))\big) \lozenge \psi_3\big(S_3(\varphi_3(b), d_b)\big)$$

for all $b \in [0, k]$. Proceeding similarly with the supremum in $b \in [0, k]$, we obtain (3).

(b) Fix $a, b \in [0, k]$ and $c, d \in m(\mathcal{A})$. Define $f = a\mathbb{1}_X$ and $g = b\mathbb{1}_X$, and consider $A, B \in \mathcal{A}$ satisfying the condition (Z_1). Then $f|_A$ and $g|_B$ are m-positively dependent for $\triangle$. Thus,

$$\mathbf{I}_{S_1, A \cap B}(m, \varphi_1(f \star g)) = S_1(\varphi_1(a \star b), m(A \cap B)) = S_1(\varphi_1(a \star b), m(A) \triangle m(B)),$$
$$\mathbf{I}_{S_2, A}(m, \varphi_2(f)) = S_2(\varphi_2(a), m(A)),$$
$$\mathbf{I}_{S_3, B}(m, \varphi_3(g)) = S_3(\varphi_3(b), m(B)).$$

Applying (3) we obtain (2) for all $a, b \in [0, k]$ and $c, d \in m(\mathcal{A})$. $\square$

4   On Existing Sufficient Conditions

The most frequently used sufficient conditions appearing in the literature in the context of Chebyshev type inequality for comonotone functions are:

(C_1) $\psi_1\big(S_1(\varphi_1(a \star b), c \wedge d)\big) \geqslant \psi_2\big(S_2(\varphi_2(a), c)\big) \star \psi_3\big(S_3(\varphi_3(b), d)\big)$ for all $a, b \in [0, k]$ and $c, d \in D$;

(C_2) $\psi_1\big(S_1(\varphi_1(a \star b), c)\big) \geqslant \big[\psi_2\big(S_2(\varphi_2(a), c)\big) \star \psi_3\big(S_3(\varphi_3(b))\big)\big] \vee \big[\psi_2(\varphi_2(a)) \star \psi_3\big(S_3(\varphi_3(b), c)\big)\big]$ for all $a, b \in [0, k]$ and $c \in D$;

(C_3) $\star \leqslant \wedge$, i.e., $a \star b \leqslant a \wedge b$ for all $a, b \in [0, 1]$, where D is some subset of $[0, 1]$. Note that (C_1) is a special case of (2) with $\triangle = \wedge$ and $\Diamond = \star$. The inequality in (C_2) with $\psi_i = \varphi_i^{-1}$ for $i = 1, 2, 3$, is the result of research stated in [2]. Lastly, the condition (C_3) appears in the literature on non-additive integral inequalities exclusively in connection with the Sugeno integral. We describe relations among the conditions. In what follows, let id denote the identity function on the interval $[0, 1]$.

Proposition 1. *Fix $D \subset [0, 1]$ such that $1 \in D$. Let $\star \colon [0, 1]^2 \to [0, 1]$ be a non-decreasing operation and $k \in (0, 1]$. Assume that $\varphi_i \colon [0, 1] \to [0, 1]$ and $\psi_i \colon [0, \varphi_i(1)] \to [0, 1]$ are non-decreasing for $i = 1, 2, 3$. Then*

(a) $(C_1) \Leftrightarrow (C_2)$
(b) $(C_3) \Rightarrow (C_1)$ if $k = 1$, $S_i = M$, $\varphi_1(1) = \varphi_j(1)$, $\psi_1 \geqslant \psi_j$ and $\psi_j(\varphi_j) \leqslant id \leqslant \psi_1(\varphi_1)$, where $i = 1, 2, 3$ and $j = 2, 3$.

Proof. To shorten ongoing expressions, we put $L(a, b, c) := \psi_1\big(S_1(\varphi_1(a \star b), c)\big)$ and $P(a, b, c, d) := \psi_2(\varphi_2(a) \wedge c) \star \psi_3(\varphi_3(b) \wedge d)$.

(a) "$(C_1) \Rightarrow (C_2)$" Putting $d = 1 \in D$ in (C_1) and then $c = 1$ in (C_1), we get (C_2).
"$(C_2) \Rightarrow (C_1)$" By assumptions, we get

$$L(a, b, c) \geqslant \psi_2\big(S_2(\varphi_2(a), c)\big) \star \psi_3(\varphi_3(b)) = \psi_2\big(S_2(\varphi_2(a), c)\big) \star \psi_3\big(S_3(\varphi_3(b), 1)\big)$$
$$\geqslant \psi_2\big(S_2(\varphi_2(a), c)\big) \star \psi_3\big(S_3(\varphi_3(b), d)\big)$$

for any $a, b \in [0, k]$ and $c, d \in D$. Also, the inequality

$$L(a, b, d) \geqslant \psi_2(\varphi_2(a)) \star \psi_3\big(S_3(\varphi_3(b), d)\big)$$

for all $a, b \in [0, k]$ and $d \in D$ implies

$$L_d \geqslant \psi_2\big(S_2(\varphi_2(a), c)\big) \star \psi_3\big(S_3(\varphi_3(b), d)\big)$$

for any $a, b \in [0, k]$ and $c, d \in D$. Combining these inequalities we get (C_1).
(b) From the obvious inequalities and the assumptions on ψ_i and φ_i, we have

$$P(a, b, c, d) \leqslant \psi_2(\varphi_2(a)) \star \psi_3(\varphi_3(b)) \leqslant a \star b \leqslant \psi_1(\varphi_1(a \star b)) \qquad (7)$$

for all $a, b \in [0, 1]$ and $c, d \in D$. Consider four cases:

(i) Firstly, suppose $c \in D \cap [0, \varphi_2(1)]$ and $d \in D \cap [0, \varphi_3(1)]$. Applying $\psi_j \leqslant \psi_1$ and $\star \leqslant \wedge$, we get

$$P(a, b, c, d) \leqslant \psi_2(c) \star \psi_3(d) \leqslant \psi_1(c) \wedge \psi_1(d)$$

for any $a, b \in [0, 1]$. By (7) and monotonicity of ψ_1, we obtain

$$P(a, b, c, d) \leqslant \psi_1(\varphi_1(a \star b) \wedge c \wedge d)$$

for all $a, b \in [0, 1]$.

(ii) Let $c \in D \cap (\varphi_2(1), 1]$ and $d \in D \cap (\varphi_3(1), 1]$. From (7) and the condition on $\star$, we conclude that

$$P(a, b, c, d) \leqslant \psi_1(\varphi_1(a \star b)) = \psi_1(\varphi_1(a \star b) \wedge \varphi_1(1))$$

for any $a, b \in [0, 1]$. By $\varphi_1(1) = \varphi_j(1)$ for $j = 2, 3$,

$$P(a, b, c, d) \leqslant \psi_1(\varphi_1(a \star b) \wedge c \wedge d)$$

for all $a, b \in [0, 1]$.

(iii) Suppose that $c \in D \cap (\varphi_2(1), 1]$ and $d \in D \cap [0, \varphi_3(1)]$. Then

$$P(a, b, c, d) \leqslant \psi_2(\varphi_2(a)) \wedge \psi_3(d) \leqslant \psi_1(\varphi_1(a)) \wedge \psi_1(d) \qquad (8)$$

for any $a, b \in [0, 1]$. Combining (7) and (8), we get

$$P(a, b, c, d) \leqslant \psi_1(\varphi_1(a \star b) \wedge \varphi_1(a) \wedge d) \leqslant \psi_1(\varphi_1(a \star b) \wedge c \wedge d)$$

for all $a, b \in [0, 1]$.

(iv) The case $c \in D \cap [0, \varphi_2(1)]$ and $d \in D \cap (\varphi_3(1), 1]$ is similar to (iii). Finally, by (i)–(iv) we get (C_1). $\qquad \square$

Now, we will provide other connections among the conditions (C_i).

Proposition 2. *Let $D = [0, 1]$, $k = 1$ and $\star: [0, 1]^2 \to [0, 1]$ be a non-decreasing operation. Assume that $\varphi_i, \psi_i: [0, 1] \to [0, 1]$ are non-decreasing with $\varphi_i(1) = 1$ and $\psi_1([0, 1]) = [0, 1]$ for $i = 1, 2, 3$. Then*

(a) $(C_1) \Rightarrow (C_3)$ if $\psi_j \geqslant \psi_1$ for $j = 2, 3$.
(b) $(C_2) \Rightarrow (C_3)$ if either
 (i) $a \star 1 \leqslant a$ for all $a \in [0, 1]$, $\psi_2(\varphi_2) \geqslant id \geqslant \psi_1(\varphi_1)$ and $\psi_3 \geqslant \psi_1$, or
 (ii) $1 \star b \leqslant b$ for all $b \in [0, 1]$, $\psi_3(\varphi_3) \geqslant id \geqslant \psi_1(\varphi_1)$ and $\psi_2 \geqslant \psi_1$.

Proof. (a) Putting $a = b = 1$ in (C_1) we obtain

$$\psi_1\big(S_1(\varphi_1(1 \star 1), c \wedge d)\big) \geqslant \psi_2(c) \star \psi_3(d) \geqslant \psi_1(c) \star \psi_1(d)$$

for all $c, d \in [0, 1]$. Since $M \geqslant S_1$ and ψ_1 is non-decreasing, we get $\psi_1(c) \wedge \psi_1(d) \geqslant \psi_1(c) \star \psi_1(d)$ for any $c, d \in [0, 1]$. To obtain (C_3) observe that $\psi_1([0, 1]) = [0, 1]$.

(b) Putting $b = 1$ in (C_2) we get

$$\psi_1\big(S_1(\varphi_1(a \star 1), c)\big) \geqslant \psi_2(\varphi_2(a)) \star \psi_3(c)$$

for all $a, c \in [0, 1]$. By the fact that $a \geqslant a \star 1$ and $M \geqslant S_1$, we conclude that

$$\psi_1(\varphi_1(a)) \wedge \psi_1(c) \geqslant \psi_2(\varphi_2(a)) \star \psi_3(c)$$

for any $a, c \in [0, 1]$. Using $\psi_2(\varphi_2) \geqslant id \geqslant \psi_1(\varphi_1)$ and $\psi_3 \geqslant \psi_1$ we have $a \wedge \psi_1(c) \geqslant a \star \psi_1(c)$ which finishes the proof of the first part of (b).

Now, let $1 \star b \leqslant b$ for all $b \in [0, 1]$. Setting $a = 1$ in (C_2) and proceeding similarly as above, we get (C_3). $\qquad \square$

From Propositions 1 and 2 (a) we get the conditions under which all (C_i)'s are equivalent.

Corollary 1. *Let* $\star\colon [0,1]^2 \to [0,1]$ *be a non-decreasing operation. Assume that* $\varphi\colon [0,1] \to [0,1]$ *are increasing with* $\varphi([0,1]) = [0,1]$. *Then*

$$(C_1) \Leftrightarrow (C_2) \Leftrightarrow (C_3)$$

with $D = [0,1]$, $k = 1$, $\mathrm{S}_i = \mathrm{M}$, $\varphi_i = \varphi$ *and* $\psi_i = \varphi^{-1}$, *where* $i = 1,2,3$.

5   Special Cases

We describe in detail the relationship of many results from the literature with Theorem 1. At first, we cover Theorem 2.3 with $Y = [0,1]$ and $\circ_i \in \mathfrak{S}$ in [14].

Corollary 2. *Let* $k \in (0,1]$, $\mathrm{S}_i \in \mathfrak{S}$, $m \in \mathcal{M}_{(X,\mathcal{A})}$ *and* $\star, \Diamond\colon [0,1]^2 \to [0,1]$ *be non-decreasing such that* $\Diamond$ *is left-continuous for* $i = 1,2,3$. *Assume that* $\Delta\colon m(\mathcal{A}) \times m(\mathcal{A}) \to m(\mathcal{A})$ *satisfies the boundary conditions* $0 \,\Delta\, a = a \,\Delta\, 0 = 0$ *for all* a. *Let* $\varphi_i\colon [0,1] \to [0,1]$ *be increasing and continuous such that* $\varphi_i(0) = 0$ *for all* $i = 1,2,3$, *and the condition* (Z_1) *is fulfilled. The Chebyshev type inequality*

$$\varphi_1^{-1}\big(\mathbf{I}_{\mathrm{S}_1, A \cap B}(m, \varphi_1(f \star g))\big) \geqslant \varphi_2^{-1}\big(\mathbf{I}_{\mathrm{S}_2, A}(m, \varphi_2(f))\big) \Diamond \varphi_3^{-1}\big(\mathbf{I}_{\mathrm{S}_3, B}(m, \varphi_3(g))\big)$$

is satisfied for arbitrary m-*positively dependent functions* $f|_A, g|_B \in \mathcal{F}_{(X,\mathcal{A})}^k$ *if and only if the inequality*

$$\varphi_1^{-1}\big(\mathrm{S}_1(\varphi_1(a \star b), c \,\Delta\, d)\big) \geqslant \varphi_2^{-1}\big(\mathrm{S}_2(\varphi_2(a), c)\big) \Diamond \varphi_3^{-1}\big(\mathrm{S}_3(\varphi_3(b), d)\big)$$

is true for all $a, b \in [0,k]$ *and* $c, d \in m(\mathcal{A})$.

Proof. Putting $\psi_i = \varphi_i^{-1}$ in Theorem 1 we get the statement. Note that ψ's are well defined. $\qquad\square$

Remark 1. The condition $\varphi_i(0) = 0$ can be abandoned being a consequence of the inequality (12). Indeed, consider $\varphi_2^{-1}\big(\mathrm{S}(\varphi_2(a), c)\big)$ for all a, c. Setting $c = 0$, we get $\varphi_2^{-1}\big(\mathrm{S}(\varphi_2(a), 0)\big) = \varphi_2^{-1}(0)$, and therefore $0 \in \varphi_2([0,1])$, so $\varphi_2(0) = 0$. Proceeding similarly, one can check that $\varphi_i(0) = 0$ for $i = 1, 3$. Adding the assumptions on $\varphi_i(0) = 0$ simplifies the use of Corollary 2.

Since the previous result still refers to the wider class of functions than the most usually used in the literature, we give a consequence of Theorem 1 (a) which will be useful in order to compare results of the Chebyshev type inequalities for the seminormed fuzzy integral and comonotone functions with other known results.

Corollary 3. *Assume that* $k \in (0,1]$, $m \in \mathcal{M}_{(X,\mathcal{A})}$ *and* $\star: [0,1]^2 \to [0,1]$ *is non-decreasing and left-continuous. Let* $\varphi_i: [0,1] \to [0,1]$ *be increasing and right-continuous, and* $\psi_i: [0, \varphi_i(1)] \to [0,1]$ *be non-decreasing and left-continuous for* $i = 1,2,3$. *If for* $S \in \mathfrak{S}$ *the inequality*

$$\psi_1\big(S(\varphi_1(a \star b), c)\big) \geqslant \big[\psi_2\big(S(\varphi_2(a), c)\big) \star \psi_3(\varphi_3(b))\big] \vee \big[\psi_2(\varphi_2(a)) \star \psi_3\big(S(\varphi_3(b), c)\big)\big] \tag{9}$$

is true for any $a, b \in [0, k]$ *and* $c \in m(\mathcal{A})$, *then for all comonotone functions* $f, g \in \mathcal{F}^k_{(X,\mathcal{A})}$ *on* $A \in \mathcal{A}$ *it holds*

$$\psi_1\big(\mathbf{I}_{S,A}(m, \varphi_1(f \star g))\big) \geqslant \psi_2\big(\mathbf{I}_{S,A}(m, \varphi_2(f))\big) \star \psi_3\big(\mathbf{I}_{S,A}(m, \varphi_3(g))\big).$$

Proof. Put $S_i = S$, $A = B$, $\Diamond = \star$ and $\triangle = \wedge$ in Theorem 1 (a). Then the inequality (9) for all $a, b \in [0, k]$ and $c \in m(\mathcal{A})$ is equivalent to (2) for all $a, b \in [0, k]$ and $c, d \in m(\mathcal{A})$ from Proposition 1 (a) with $D = m(\mathcal{A})$. Then Theorem 1 (a) provides the statement, since comonotone functions $f|_A$ and $g|_A$ are m-positively dependent with respect to $\wedge$. $\qquad\square$

Now, we provide the power-mean type Chebyshev inequality for $\mathbf{I}_S$. Hereafter, we use the convention $S^p(x, y) = (S(x, y))^p$.

Corollary 4. *[1, Theorem 3.1] Let* $m \in \mathcal{M}_{(X,\mathcal{A})}$ *and* $\star: [0,1]^2 \to [0,1]$ *be a left-continuous and non-decreasing operation. If* $S \in \mathfrak{S}$ *satisfies*

$$S^\lambda((a \star b)^\alpha, c) \geqslant (S^\nu(a^\beta, c) \star b) \vee (a \star S^\tau(b^\gamma, c)) \tag{10}$$

for all $a, b, c \in [0,1]$, *where* $\alpha, \beta, \gamma, \lambda, \nu, \tau \in (0, \infty)$ *such that* $\gamma\tau \geqslant 1$ *and* $\beta\nu \geqslant 1$, *then*

$$\big(\mathbf{I}_{S,A}(m, (f \star g)^\alpha)\big)^\lambda \geqslant \big(\mathbf{I}_{S,A}(m, f^\beta)\big)^\nu \star \big(\mathbf{I}_{S,A}(m, g^\gamma)\big)^\tau$$

is valid for any comonotone functions $f, g \in \mathcal{F}^1_{(X,\mathcal{A})}$ *on* $A \in \mathcal{A}$.

Proof. From (10) and assumptions we get

$$S^\lambda((a \star b)^\alpha, c) \geqslant (S^\nu(a^\beta, c) \star b^{\gamma\tau}) \vee (a^{\beta\nu} \star S^\tau(b^\gamma, c)) \tag{11}$$

for all $a, b, c \in [0,1]$. Put in Corollary 3 the following functions: $\psi_1(x) = x^\lambda$, $\psi_2(x) = x^\nu$, $\psi_3(x) = x^\tau$, $\varphi_1(x) = x^\alpha$, $\varphi_2(x) = x^\beta$, $\varphi_3(x) = x^\gamma$ for all $x \in [0,1]$. Then the inequality (11) takes the form (9) and Corollary 3 with $k = 1$ gives the conclusion. $\qquad\square$

Next, we present the result proved in [7] which generalizes other known results in the literature for the seminormed integral $\mathbf{I}_S$.

Corollary 5. *[7, Corollary 3.10] Let* $k \in (0,1]$, $m \in \mathcal{M}_{(X,\mathcal{A})}$, $S \in \mathfrak{S}$, *operation* $\star: [0,1]^2 \to [0,1]$ *be non-decreasing and left-continuous,* $\eta: [0, k] \to [0,1]$ *be non-decreasing and left-continuous, and* $\varphi_i: [0,1] \to [0,1]$ *be increasing and continuous functions such that* $\varphi_i(0) = 0$ *for* $i = 1,2,3$. *If*

$$\varphi_1^{-1}\big(S(\varphi_1(\eta(a) \star \eta(b)), c)\big) \geqslant \big[\eta(\varphi_2^{-1}(S(\varphi_2(a), c))) \star \eta(b)\big] \vee \big[\eta(a) \star \eta(\varphi_3^{-1}(S(\varphi_3(b), c)))\big] \tag{12}$$

for all $a, b \in [0, k]$ and $c \in [0, 1]$, then

$$\varphi_1^{-1}\Big(\mathbf{I}_{\mathrm{S},A}\big(m, \varphi_1(\eta(f) \star \eta(g))\big)\Big) \geqslant \eta\Big(\varphi_2^{-1}\big(\mathbf{I}_{\mathrm{S},A}(m, \varphi_2(f))\big)\Big) \star \eta\Big(\varphi_3^{-1}\big(\mathbf{I}_{\mathrm{S},A}(m, \varphi_3(g))\big)\Big) \tag{13}$$

is fulfilled for any comonotone functions $f, g \in \mathcal{F}_{(X,\mathcal{A})}^k$ on $A \in \mathcal{A}$.

Proof. Define $a \star_\eta b = \eta(a) \star \eta(b)$ for all $(a, b) \in [0, k]^2$. It is clear that $\star_\eta \colon [0, k]^2 \to [0, 1]$ is non-decreasing and left-continuous. Then the inequality (12) takes the form

$$\varphi_1^{-1}\big(\mathrm{S}(\varphi_1(a \star_\eta b), c)\big) \geqslant \big[\varphi_2^{-1}\big(\mathrm{S}(\varphi_2(a), c)\big) \star_\eta b\big] \vee \big[a \star_\eta \varphi_3^{-1}\big(\mathrm{S}(\varphi_3(b), c)\big)\big] \tag{14}$$

for all $a, b \in [0, k]$ and $c \in [0, 1]$. Observe that (14) is the same as (9) (when exchanging $\star$ for $\star_\eta$) with functions $\psi_i = \varphi_i^{-1}$ for $i = 1, 2, 3$. Then Corollary 3 gives the integral inequality

$$\varphi_1^{-1}\big(\mathbf{I}_{\mathrm{S},A}(m, \varphi_1(f \star_\eta g))\big) \geqslant \varphi_2^{-1}\big(\mathbf{I}_{\mathrm{S},A}(m, \varphi_2(f))\big) \star_\eta \varphi_3^{-1}\big(\mathbf{I}_{\mathrm{S},A}(m, \varphi_3(g))\big),$$

which is the same as (13) by definition of $\star_\eta$. □

Putting $\varphi_i = id$ for all $i = 1, 2, 3$ in Corollary 5, we get [8, Theorem 2 with $n = 2$]. Moreover, [2, Theorem 4.1] is the special case of Corollary 5 with $k = 1$ and $\eta = id$. In addition, if $\varphi_i = \varphi$, where φ is a continuous and increasing function such that $\varphi(0) = 0$, we get [6, Corollary 3.5] and for $\varphi = id$, we obtain [19, Theorem 3.1].

Using Theorem 1 for $\mathrm{S}_i = \mathrm{M}$ or $\mathrm{S}_i = \Pi$ we can get the Chebyshev type inequality for the Sugeno and Shilkret integral, respectively. In the special cases of $\triangle$, the condition (2) can be replaced by another one which is easier to verify. Consequently, the integral inequality is still true.

5.1   Sugeno Integral on $[0, 1]$

In what follows, we say that a binary operation $\star$ satisfies the condition (M_1), if $\star \colon [0, 1]^2 \to [0, 1]$ is non-decreasing, left-continuous and $\star \leqslant \wedge$.

Corollary 6. *Let $m \in \mathcal{M}_{(X,\mathcal{A})}$ and $\star$ satisfy (M_1). Let $\varphi_i \colon [0, 1] \to [0, 1]$ be increasing and right-continuous and $\psi_i \colon [0, \varphi_i(1)] \to [0, 1]$ be non-decreasing and left-continuous, where $i = 1, 2, 3$. Assume that $\varphi_1(1) = \varphi_j(1)$, $\psi_1 \geqslant \psi_j$ and $\psi_j(\varphi_j) \leqslant id \leqslant \psi_1(\varphi_1)$ for $j = 2, 3$. Then the Chebyshev type inequality*

$$\psi_1\big(\mathbf{I}_{\mathrm{M},A}(m, \varphi_1(f \star g))\big) \geqslant \psi_2\big(\mathbf{I}_{\mathrm{M},A}(m, \varphi_2(f))\big) \star \psi_3\big(\mathbf{I}_{\mathrm{M},A}(m, \varphi_3(g))\big)$$

holds for all comonotone functions $f, g \in \mathcal{F}_{(X,\mathcal{A})}^1$ on $A \in \mathcal{A}$.

Proof. Put $k = 1$, $\mathrm{S}_i = \triangle = \wedge$, $A = B$ and $\Diamond = \star$ in Theorem 1 (a). Then the assumptions yield the inequality (2) for all $a, b \in [0, 1]$ and $c, d \in m(\mathcal{A})$ from Proposition 1 (b) with $D = m(\mathcal{A})$. Then Theorem 1 (a) gives the statement. □

Corollary 7. *Let $m \in \mathcal{M}_{(X,\mathcal{A})}$ and $\star$ satisfy (M_1). Let $\eta\colon [0,1] \to [0,1]$ be left-continuous and non-decreasing such that $\eta \leqslant id$, and $\varphi_j\colon [0,1] \to [0,1]$ be continuous and increasing such that $\varphi_j(0) = 0$ and $\varphi_2(1) = \varphi_3(1)$, where $j = 2,3$, and $\varphi_1 = \varphi_2 \wedge \varphi_3$. Then*

$$\varphi_1^{-1}\Big(\mathbf{I}_{\mathrm{M},A}\big(m, \varphi_1(\eta(f) \star \eta(g))\big)\Big) \geqslant \eta\Big(\varphi_2^{-1}\big(\mathbf{I}_{\mathrm{M},A}(m, \varphi_2(f))\big)\Big) \star \eta\Big(\varphi_3^{-1}\big(\mathbf{I}_{\mathrm{M},A}(m, \varphi_3(g))\big)\Big) \tag{15}$$

is true for any comonotone functions $f, g \in \mathcal{F}^1_{(X,\mathcal{A})}$ on $A \in \mathcal{A}$.

Proof. Define $a \star_\eta b = \eta(a) \star \eta(b)$ for all $a, b \in [0,1]$. Then $\star_\eta$ is left-continuous and non-decreasing such that $\star_\eta \leqslant \wedge$. By Corollary 6 with $\psi_i = \varphi_i^{-1}$ for $i = 1,2,3$ and $\star_\eta$ instead of $\star$, we obtain

$$\varphi_1^{-1}\big(\mathbf{I}_{\mathrm{M},A}(m, \varphi_1(f \star_\eta g))\big) \geqslant \varphi_2^{-1}\big(\mathbf{I}_{\mathrm{M},A}(m, \varphi_2(f))\big) \star_\eta \varphi_3^{-1}\big(\mathbf{I}_{\mathrm{M},A}(m, \varphi_3(g))\big),$$

which is the same as (15). □

The assumption on $\varphi_i(0) = 0$ cannot be omitted in Corollary 7. To see this, it is enough to put $\eta = id$, $\varphi_i(x) = 0.5(x+1)\mathbb{1}_{[0,1]}(x)$ for $i = 1,2,3$, $f = g = 0.5\mathbb{1}_X$, $\star = \cdot$ and $m(A) = 0.4$ in Corollary 7. Putting $\varphi_i = id$ in Corollary 7 we get the Chebyshev type inequality presented in [4, Theorem 3.1 with $(k,n) = (1,2)$].

It can be shown that the boundary conditions of functions φ_i in Corollary 7 can be replaced by continuity of $m \in \mathcal{M}_{(X,\mathcal{A})}$. To do so, we firstly derive the helpful result.

Lemma 1. *Let $m \in \mathcal{M}_{(X,\mathcal{A})}$ be continuous and $V_1, V_2\colon [0,1] \to [0,1]$ be continuous and increasing such that $V_1 \leqslant V_2$ and $V_2(0) = 0$. Then*

$$V_1^{-1}\big(\mathbf{I}_{\mathrm{M},A}(m, V_1(f))\big) \geqslant V_2^{-1}\big(\mathbf{I}_{\mathrm{M},A}(m, V_2(f))\big)$$

is satisfied for each $f \in \mathcal{F}^1_{(X,\mathcal{A})}$ and $A \in \mathcal{A}$.

Proof. Since $\mathbf{I}_{\mathrm{M},A}(m, V_i(f)) \in [0, V_i(1)]$, there exists $a_i \in [0,1]$ such that $V_i^{-1}\big(\mathbf{I}_{\mathrm{M},A}(m, V_i(f))\big) = a_i$ for $i = 1,2$, as $V_i(0) = 0$. From monotonicity of V_i and [22, Lemma 9.5 (4)] we obtain

$$m(A \cap \{f \geqslant a_i\}) = m(A \cap \{V_i(f) \geqslant V_i(a_i)\}) \geqslant V_i(a_i)$$

for $i = 1,2$. Since $V_1 \leqslant V_2$, then

$$a_1 = V_1^{-1}\big(\mathbf{I}_{\mathrm{M},A}(m, V_1(f))\big) \geqslant V_1^{-1}\big(V_1(a_2) \wedge m(A \cap \{f \geqslant a_2\})\big)$$
$$\geqslant V_1^{-1}\big(V_1(a_2) \wedge V_2(a_2)\big) = V_1^{-1}\big(V_1(a_2)\big) = a_2$$

and the proof is complete. □

Now, we cover the result [5, Theorem 4.11 with $n = 2$] for the Sugeno integral of $[0,1]$-valued functions.

Corollary 8. *Let $m \in \mathcal{M}_{(X,\mathcal{A})}$ be continuous and $\star$ satisfy (M_1). Let $\eta\colon [0,1] \to [0,1]$ be left-continuous and non-decreasing such that $\eta \leqslant id$, and $\varphi_i\colon [0,1] \to [0,1]$ be continuous and increasing such that $\varphi_i(0) = 0$, $i = 1,2,3$, and $\varphi_1 \leqslant \varphi_2 \wedge \varphi_3$. Then the Chebyshev type inequality* (15) *is valid for all comonotone functions $f, g \in \mathcal{F}^1_{(X,\mathcal{A})}$ on $A \in \mathcal{A}$.*

Proof. Let $\tilde{\varphi}_j\colon [0,1] \to [0,1]$ be continuous and increasing such that $\tilde{\varphi}_j(0) = 0$, $\tilde{\varphi}_2(1) = \tilde{\varphi}_3(1)$, where $j = 2,3$, and $\tilde{\varphi}_1 = \tilde{\varphi}_2 \wedge \tilde{\varphi}_3$. Moreover assume that $\tilde{\varphi}_j \leqslant \varphi_j$ and $\varphi_1 \leqslant \tilde{\varphi}_1$. By Lemma 1 we get

$$\varphi_1^{-1}\big(\mathbf{I}_{\mathrm{M},A}(m, \varphi_1(h))\big) \geqslant \tilde{\varphi}_1^{-1}\big(\mathbf{I}_{\mathrm{M},A}(m, \tilde{\varphi}_1(h))\big), \qquad (16)$$

$$\tilde{\varphi}_j^{-1}\big(\mathbf{I}_{\mathrm{M},A}(m, \tilde{\varphi}_j(h))\big) \geqslant \varphi_j^{-1}\big(\mathbf{I}_{\mathrm{M},A}(m, \varphi_j(h))\big) \qquad (17)$$

with $j = 2,3$, for all $h \in \mathcal{F}^1_{(X,\mathcal{A})}$. By Corollary 7 and inequalities (16)–(17) we get the statement. □

Corollary 8 generalizes many results in the literature for the Sugeno integral of $[0,1]$-valued functions, e.g. [5, Theorem 4.11], [23, Theorem 4.1], [3, Theorem 3.1], [16, Theorem 3.1], [20, Theorem 3.1], [18, Theorems 2.6 and 2.7], [15, Theorems 2.1 and 2.2] and [12, Theorems 3 and 4].

5.2   Shilkret Integral on $[0,1]$

Similarly as in the previous subsection we replace the condition (2) for $S_i = \Pi$ with another one which is easier to check and still provides a sufficient condition guaranteeing the validity of the Chebyshev type inequality for the Shilkret integral.

Theorem 2. *Suppose that $m \in \mathcal{M}_{(X,\mathcal{A})}$ and $\star\colon [0,1]^2 \to [0,1]$ is a non-decreasing left-continuous binary operation. Let $\varphi_i\colon [0,1] \to [0,1]$ be increasing and right-continuous, $\psi_j\colon [0,1] \to [0,1]$ be non-decreasing and left-continuous such that $\psi_j \vee \varphi_j \leqslant id \leqslant \varphi_1$, where $i = 1,2,3$ and $j = 2,3$. If $\cdot \leqslant \star$, then*

$$\sqrt{\mathbf{I}_{\Pi,A\cap B}(m, \varphi_1(f \star g))} \geqslant \psi_2\big(\mathbf{I}_{\Pi,A}(m, \varphi_2(f))\big) \star \psi_3\big(\mathbf{I}_{\Pi,B}(m, \varphi_3(g))\big) \qquad (18)$$

is satisfied for arbitrary m-positively dependent functions $f|_A, g|_B \in \mathcal{F}^1_{(X,\mathcal{A})}$ with respect to $\cdot$.

Proof. Set $S_i = \Delta = \cdot$, $\lozenge = \star$ and $\psi_1(x) = \sqrt{x}$ in Theorem 1 (a). Note that

$$\varphi_1(a \star b) \geqslant a \star b \geqslant \varphi_2(a) \star \varphi_3(b) \geqslant \psi_2(\varphi_2(a)c) \star \psi_3(\varphi_3(b)d) \qquad (19)$$

for any $a, b, c, d \in [0,1]$. Due to $\star \leqslant \cdot$, we have

$$cd \geqslant c \star d \geqslant \psi_2(\varphi_2(a)c) \star \psi_3(\varphi_3(b)d) \qquad (20)$$

for any a, b, c, d. Multiply (19) and (20) we obtain

$$\sqrt{\varphi_1(a \star b)cd} \geqslant \psi_2(\varphi_2(a)c) \star \psi_3(\varphi_3(b)d).$$

for all a, b, c, d. Then Theorem 1 (a) finishes the proof. □

Corollary 9. *Let* $(X, \mathcal{A}, \mathsf{P})$ *be a probability space. Then the Chebyshev type inequality*

$$\mathbf{I}_\Pi(\mathsf{P}, fg) \geqslant \left(\mathbf{I}_\Pi(\mathsf{P}, f)\right)^2 \cdot \left(\mathbf{I}_\Pi(\mathsf{P}, g)\right)^2$$

is true for any independent random variables $f, g \in \mathcal{F}^1_{(X, \mathcal{A})}$.

6 Concluding Remarks

We have presented a necessary and sufficient condition guaranteeing the validity of the Chebyshev type inequality

$$\psi_1\left(\mathbf{I}_{\mathsf{S}_1, A \cap B}(m, \varphi_1(f \star g))\right) \geqslant \psi_2\left(\mathbf{I}_{\mathsf{S}_2, A}(m, \varphi_2(f))\right) \Diamond \psi_3\left(\mathbf{I}_{\mathsf{S}_3, B}(m, \varphi_3(g))\right)$$

for any m-positively dependent functions $f|_A$ and $g|_B$. It has enabled us to generalize many known results from the literature related to the Chebyshev type inequalities for $\mathbf{I}_\mathsf{S}$ for the class of comonotone functions. Using the relationship between the conditions appearing in the literature we were able to present the Chebyshev type inequality for the Sugeno integral and Shilkret integral with easier conditions to be checked.

Since the considered seminormed integrals are aggregation functions [17], we expect applications of our results everywhere where some bounds on the aggregation process is needed, such as information aggregation, or decision making.

References

1. Agahi, H., Eslami, E.: A general inequality of Chebyshev type for semi(co)normed fuzzy integrals. Soft Comput. **15**(4), 771–780 (2011)
2. Agahi, H., Eslami, E., Mohammadpour, A., Vaezpour, S.M., Yaghoobi, M.A.: On Non-additive Probabilistic Inequalities of Hölder-type. Results Math. **61**, 179–194 (2012)
3. Agahi, H., Mesiar, R., Ouyang, Y.: New general extensions of Chebyshev type inequalities for Sugeno integrals. Int. J. Approx. Reason. **51**, 135–140 (2009)
4. Agahi, H., Mesiar, R., Ouyang, Y.: Further development of Chebyshev type inequalities for Sugeno integral and T-(S-)evaluators. Kybernetika **46**, 83–95 (2010)
5. Agahi, H., Mesiar, R., Ouyang, Y.: On some advanced type inequalities for Sugeno integral and T-(S-)evaluators. Inform. Sci. **190**, 64–75 (2012)
6. Agahi, H., Mesiar, R., Ouyang, Y., Pap, E., Štrboja, M.: General Chebyshev type inequalities for universal integral. Inform. Sci. **187**, 171–178 (2012)
7. Agahi, H., Mohammadpour, A., Vaezpour, S.M.: Predictive tools in data mining and k-means clustering: universal inequalities. Results Math. **63**, 779–803 (2013)
8. Agahi, H., Yaghoobi, M.A.: On an extended Chebyshev type inequality for semi(co)normed fuzzy integrals. Int. J. Uncertain. Fuzziness Knowl. Based Syst. **19**, 781–798 (2011)
9. Armstrong, T.E.: Chebyshev inequalities and comonotonicity. Real Anal. Exchange **19**, 266–268 (1993/94)
10. Boczek, M., Kaluszka, M.: On comonotone commuting and weak subadditivity properties of seminormed fuzzy integrals. Fuzzy Set Syst. **304**, 35–44 (2016)

11. Borzová-Molnárová, J., Halčinová, L., Hutník, O.: The smallest semicopula-based universal integrals I: properties and characterizations. Fuzzy Set Syst. **271**, 1–17 (2015)
12. Flores-Franulič, A., Román-Flores, H.: A Chebyshev type inequality for fuzzy integrals. Appl. Math. Comput. **190**, 1178–1184 (2007)
13. Girotto, B., Holzer, S.: A Chebyshev type inequality for Sugeno inegral and comonotonicity. Int. J. Approx. Reason. **52**, 444–448 (2011)
14. Kaluszka, M., Okolewski, A., Boczek, M.: On Chebyshev type inequalities for generalized Sugeno integrals. Fuzzy Set Syst. **244**, 51–62 (2014)
15. Mao, Q.-S.: A note on the Chebyshev-type inequality of Sugeno integrals. Appl. Math. Comput. **212**, 275–279 (2009)
16. Mesiar, R., Ouyang, Y.: General Chebyshev type inequalities for Sugeno integrals. Fuzzy Set Syst. **160**, 58–64 (2009)
17. Narukawa, Y., Murofushi, T.: Choquet integral and Sugeno integral as aggregation functions. In: Torra, V. (ed.) Information Fusion in Data Mining, Studies in Fuzziness and Soft Computing, vol. 123, pp. 27–39. Springer, Heidelberg (2003). https://doi.org/10.1007/978-3-540-36519-8_3
18. Ouyang, Y., Fang, J., Wang, L.: Fuzzy Chebyshev type inequality. Int. J. Approx. Reason. **48**, 829–835 (2008)
19. Ouyang, Y., Mesiar, R.: On the Chebyshev type inequality for seminormed fuzzy integral. Appl. Math. Lett. **22**, 1810–1815 (2009)
20. Ouyang, Y., Mesiar, R., Agahi, H.: An inequality related to Minkowski type for Sugeno integrals. Inform. Sci. **180**, 2793–2801 (2010)
21. Suárez-García, F., Álvarez-Gil, P.: Two families of fuzzy integrals. Fuzzy Set Syst. **18**, 67–81 (1986)
22. Wang, Z., Klir, G.: Generalized Measure Theory. Springer, New York (2009). https://doi.org/10.1007/978-0-387-76852-6
23. Wu, L., Sun, J., Ye, X., Zhu, L.: Hölder type inequality for Sugeno integral. Fuzzy Set Syst. **161**, 2337–2347 (2010)

Convergence in Measure Theorems of the Choquet Integral Revisited

Jun Kawabe[✉][iD]

Shinshu University, Nagano 3808553, Japan
jkawabe@shinshu-u.ac.jp

Abstract. The validity of the monotone convergence theorem, the Fatou and the reverse Fatou lemmas, and the dominated convergence theorem of the Choquet integral of measurable functions converging in measure are fully characterized by the conditional versions of the monotone autocontinuity and the autocontinuity. In those theorems the nonadditive measure may be infinite and the functions may be unbounded. The dual measure forms and the extension to symmetric and asymmetric Choquet integrals are also discussed.

Keywords: Nonadditive measure · Choquet integral · Conditional autocontinuity · Convergence in measure theorem

1 Introduction

The purpose of the paper is to give full descriptions of the monotone convergence theorem, the Fatou and the reverse Fatou lemmas, and the dominated convergence theorem of the Choquet integral of measurable functions converging in measure. Such theorems are collectively called the *convergence in measure theorems* and already discussed with a certain degree of generality when the measures are finite and/or the functions are bounded [2,4,5,7,8,11,15,16].

In this paper, those convergence in measure theorems are reconsidered in the general case where the nonadditive measure may be infinite and the functions may be unbounded. To this end, it is necessary to introduce the conditional versions of the already known characteristics of nonadditive measures such as the monotone autocontinuity and the autocontinuity. In fact, it will be shown that the validity of the monotone nonincreasing convergence in measure theorem is fully characterized by the conditional monotone autocontinuity from above, and the validity of the dominated convergence in measure theorem is fully characterized by the conditional autocontinuity.

The paper is organized as follows. Section 2 contains notation and the definition of a nonadditive measure and the Choquet integral. In Sect. 3 the conditional versions of the characteristics of nonadditive measures are introduced and their mutual relations are clarified. In next three sections the monotone convergence

This work was supported by JSPS KAKENHI Grant Number 17K05293.

V. Torra et al. (Eds.): MDAI 2019, LNAI 11676, pp. 17–28, 2019.
https://doi.org/10.1007/978-3-030-26773-5_2

in measure theorem, the Fatou and the reverse Fatou in measure lemmas, and the dominated convergence in measure theorem are fully characterized in terms of the conditional versions of the characteristics of nonadditive measures. The dual measure forms are considered in Sect. 7, and the extension to symmetric and asymmetric Choquet integrals are discussed in Sect. 8. The last section gives concluding remarks.

In this paper the detailed proofs are given only for typical results since the reader can prove others easily or in a similar way.

2   Preliminaries

Throughout the paper, $(X, \mathcal{A})$ is a measurable space. Let $\mathbb{R} = (-\infty, \infty)$ denote the set of the real numbers and $\mathbb{N}$ the set of the natural numbers. Let $\overline{\mathbb{R}} := [-\infty, \infty]$ be the set of the extended real numbers with usual total order and algebraic structure. Let us assume that $(\pm\infty) \cdot 0 = 0 \cdot (\pm\infty) = 0$ since this proves to be convenient in measure and integration theory. Let $\mathcal{F}(X)$ denote the set of all $\mathcal{A}$-measurable functions $f \colon X \to \overline{\mathbb{R}}$ and let $\mathcal{F}^+(X) := \{f \in \mathcal{F}(X) \colon f \geq 0\}$. For a subset A of X, the symbol χ_A denotes the characteristic function of A, that is, $\chi_A(x) = 1$ if $x \in A$ and $\chi_A(x) = 0$ otherwise, and the symbol A^c denotes the complement of A, that is, $A^c := X \setminus A$.

A nonadditive measure is a set function $\mu \colon \mathcal{A} \to [0, \infty]$ such that $\mu(\emptyset) = 0$ and $\mu(A) \leq \mu(B)$ whenever $A, B \in \mathcal{A}$ and $A \subset B$. It is called finite if $\mu(X) < \infty$ and infinite if $\mu(X) = \infty$. This type of set function is also called a monotone measure [16], a capacity [1], or a fuzzy measure [10,14] in the literature. Let $\mathcal{M}(X)$ denote the set of all nonadditive measures $\mu \colon \mathcal{A} \to [0, \infty]$ and let $\mathcal{M}_b(X) := \{\mu \in \mathcal{M}(X) \colon \mu(X) < \infty\}$. For any $\mu \in \mathcal{M}_b(X)$, its dual $\bar{\mu}$ is defined by $\bar{\mu}(A) := \mu(X) - \mu(A^c)$ for each $A \in \mathcal{A}$. Then $\bar{\mu}$ is also a nonadditive measure and $\bar{\bar{\mu}} = \mu$. If μ is finitely additive, then $\mu = \bar{\mu}$. A set $N \in \mathcal{A}$ is called μ-null if $\mu(N) = 0$ as usual. See [3,9,16] for further information on nonadditive measures.

The Choquet integral [1,13] is a typical nonlinear integral and widely used in nonadditive measure theory. For each $(\mu, f) \in \mathcal{M}(X) \times \mathcal{F}^+(X)$, the *Choquet integral* is defined by

$$\mathrm{Ch}(\mu, f) := \int_0^\infty \mu(\{f > t\}) dt,$$

where the right hand side is the Lebesgue integral or the improper Riemann integral. The nonincreasing distribution function $\mu(\{f > t\})$ may be replaced with $\mu(\{f \geq t\})$ without any change. A function $f \in \mathcal{F}(X)$ is called μ-integrable if $\mathrm{Ch}(\mu, |f|) < \infty$. The Choquet integral is equal to the Šipoš integral [12] for any nonadditive measure μ and also equal to the abstract Lebesgue integral if μ is σ-additive [6, Propositions 8.1 and 8.2].

For a sequence $\{f_n\}_{n \in \mathbb{N}} \subset \mathcal{F}(X)$ and $f \in \mathcal{F}(X)$ we say that f_n converges in μ-measure to f and write $f_n \xrightarrow{\mu} f$ if for every $x \in X$ and $n \in \mathbb{N}$ both $f_n(x)$ and $f(x)$ do not simultaneously take ∞ or $-\infty$ and $\lim_{n \to \infty} \mu(\{|f_n - f| > \varepsilon\}) = 0$ for every $\varepsilon > 0$. If $f_n \xrightarrow{\mu} f$, then $f_n^+ \xrightarrow{\mu} f^+$, $f_n^- \xrightarrow{\mu} f^-$, $|f_n| \xrightarrow{\mu} |f|$, and $|f_n - f| \xrightarrow{\mu} 0$.

3   The Conditional Autocontinuity of Nonadditive Measures

The autocontinuity of nonadditive measures and its monotone versions are known to play a crucial role when formulating the convergence in measure theorems of the Choquet integral; see, for instance, [5,7,8,11,15].

Definition 1. *Let $\mu \in \mathcal{M}(X)$.*

1. *μ is called autocontinuous from above if $\mu(A \cup B_n) \to \mu(A)$ whenever $A \in \mathcal{A}$, $\{B_n\}_{n \in \mathbb{N}} \subset \mathcal{A}$, and $\mu(B_n) \to 0$.*
2. *μ is called autocontinuous from below if $\mu(A \setminus B_n) \to \mu(A)$ whenever $A \in \mathcal{A}$, $\{B_n\}_{n \in \mathbb{N}} \subset \mathcal{A}$, and $\mu(B_n) \to 0$.*
3. *μ is called autocontinuous if it is autocontinuous from above and below.*
4. *μ is called monotone autocontinuous from above if $\mu(A \cup B_n) \to \mu(A)$ whenever $A \in \mathcal{A}$, $\{B_n\}_{n \in \mathbb{N}} \subset \mathcal{A}$ is nonincreasing, and $\mu(B_n) \to 0$.*
5. *μ is called monotone autocontinuous from below if $\mu(A \setminus B_n) \to \mu(A)$ whenever $A \in \mathcal{A}$, $\{B_n\}_{n \in \mathbb{N}} \subset \mathcal{A}$ is nonincreasing, and $\mu(B_n) \to 0$.*
6. *μ is called monotone autocontinuous if it is monotone autocontinuous from above and below.*

In order to give full description of the convergence in measure theorems in terms of the characteristics of nonadditive measures it is necessary to introduce the following conditional versions of the autocontinuity.

Definition 2. *Let $\mu \in \mathcal{M}(X)$.*

1. *μ is called conditionally autocontinuous from above if $\mu(A \cup B_n) \to \mu(A)$ whenever $A, C \in \mathcal{A}$, $\{B_n\}_{n \in \mathbb{N}} \subset \mathcal{A}$, $\mu(C) < \infty$, $\mu(B_n) \to 0$, and $A \cup B_n \subset C$ for every $n \in \mathbb{N}$.*
2. *μ is called conditionally autocontinuous from below if $\mu(A \setminus B_n) \to \mu(A)$ whenever $A, C \in \mathcal{A}$, $\{B_n\}_{n \in \mathbb{N}} \subset \mathcal{A}$, $\mu(C) < \infty$, $\mu(B_n) \to 0$, and $A \setminus B_n \subset C$ for every $n \in \mathbb{N}$.*
3. *μ is called conditionally autocontinuous if it is conditionally autocontinuous from above and below.*
4. *μ is called conditionally monotone autocontinuous from above if $\mu(A \cup B_n) \to \mu(A)$ whenever $A \in \mathcal{A}$, $\{B_n\}_{n \in \mathbb{N}} \subset \mathcal{A}$ is nonincreasing, $\mu(A \cup B_1) < \infty$, and $\mu(B_n) \to 0$.*
5. *μ is called conditionally monotone autocontinuous from below if $\mu(A \setminus B_n) \to \mu(A)$ whenever $A \in \mathcal{A}$, $\{B_n\}_{n \in \mathbb{N}} \subset \mathcal{A}$ is nonincreasing, $\mu(A) < \infty$, and $\mu(B_n) \to 0$.*
6. *μ is called conditionally monotone autocontinuous if it is conditionally monotone autocontinuous from above and below.*

The conditional versions of the autocontinuity are strictly weaker than the corresponding usual ones, and they coincide if μ is finite; see Example 1. Moreover, whether "conditional" or not, the monotone autocontinuity is strictly weaker than the autocontinuity; see Example 2.

Example 1. Let $X := \mathbb{R}$ and $\mathcal{A}$ be the σ-field of all Lebesgue measurable subsets of X. Let λ be the Lebesgue measure on $\mathbb{R}$.

(1) Let $\mu\colon \mathcal{A} \to [0, \infty]$ be the nonadditive measure defined by

$$\mu(A) := \begin{cases} \infty & \text{if } \lambda(A) \geq 1, \\ \lambda(A) & \text{if } \lambda(A) < 1, \end{cases}$$

then μ is conditionally autocontinuous from below, but is not monotone autocontinuous from below.

(2) Let $\mu\colon \mathcal{A} \to [0, \infty]$ be the nonadditive measure defined by

$$\mu(A) := \begin{cases} \infty & \text{if } \lambda(A) > 1, \\ \lambda(A) & \text{if } \lambda(A) \leq 1, \end{cases}$$

then μ is conditionally autocontinuous from above, but is not monotone autocontinuous from above.

Example 2. Let $X := [1, \infty)$ and $\mathcal{A}$ be the family of all subsets of X. Let $\mu\colon \mathcal{A} \to [0, \infty]$ be the nonadditive measure defined by

$$\mu(A) := \begin{cases} \infty & \text{if } A = X, \\ 3 & \text{if } A^c \text{ is a singleton}, \\ 2 & \text{if } A \text{ and } A^c \text{ both contain at least two points}, \\ 1/t & \text{if } A = \{t\} \text{ for some } t \in X, \\ 0 & \text{if } A = \emptyset. \end{cases}$$

Then μ is monotone autocontinuous since every nonincreasing $\{B_n\}_{n \in \mathbb{N}} \subset \mathcal{A}$ with $\mu(B_n) \to 0$ has the property that $B_n = \emptyset$ for almost all $n \in \mathbb{N}$. However, it is neither conditionally autocontinuous from above nor below. For instance, let $A := \{2\}$, $B_n := \{n + 2\}$ ($n = 1, 2, \dots$), and $C := (1, \infty)$. Then $\mu(A) = 1/2$, $\mu(C) = 3$, $\mu(B_n) = 1/(n + 2) \to 0$, and $\mu(A \cup B_n) = 2$ for every $n \in \mathbb{N}$. This means that μ is not conditionally autocontinuous from above. If we let $A := (1, \infty)$, $B_n := \{n + 1\}$ ($n = 1, 2, \dots$), and $C := (1, \infty)$, then $\mu(A) = \mu(C) = 3$, $\mu(B_n) = 1/(n + 1) \to 0$, and $\mu(A \setminus B_n) = 2$ for every $n \in \mathbb{N}$, which imply that μ is not conditionally autocontinuous from below.

Let us list some properties of various types of the autocontinuity to clarify their mutual relations. Recall the following characteristics of nonadditive measures; see, for instance, [3,9,16]: A nonadditive measure $\mu\colon \mathcal{A} \to [0, \infty]$ is called continuous from above if $\mu(A_n) \to \mu(A)$ whenever $A, A_n \in \mathcal{A}$ and $A_n \downarrow A$, continuous from below if $\mu(A_n) \to \mu(A)$ whenever $A, A_n \in \mathcal{A}$ and $A_n \uparrow A$, conditionally continuous from above if $\mu(A_n) \to \mu(A)$ whenever $A, A_n \in \mathcal{A}$, $A_n \downarrow A$, and $\mu(A_1) < \infty$, conditionally continuous from below if $\mu(A_n) \to \mu(A)$ whenever $A, A_n \in \mathcal{A}$, $A_n \uparrow A$, and $\mu(A) < \infty$, order continuous if $\mu(A_n) \to 0$ whenever $A_n \in \mathcal{A}$ and $A_n \downarrow \emptyset$, and conditionally order continuous if $\mu(A_n) \to 0$

whenever $A_n \in \mathcal{A}$, $A_n \downarrow \emptyset$, and $\mu(A_1) < \infty$. Moreover, μ is called null-additive if $\mu(A \cup B) = \mu(A)$ whenever $A, B \in \mathcal{A}$ and $\mu(B) = 0$, and subadditive if $\mu(A \cup B) \le \mu(A) + \mu(B)$ for every $A, B \in \mathcal{A}$.

- If μ is continuous from above (or below) and null-additive, then it is monotone autocontinuous from above (or below, respectively).
- If μ is conditionally continuous from above (or below) and null-additive, then it is conditionally monotone autocontinuous from above (or below, respectively).
- If μ is monotone autocontinuous from above (or below) and order continuous, then it is continuous from above (or below, respectively).
- If μ is conditionally monotone autocontinuous from above (or below) and conditionally order continuous, then it is conditionally continuous from above (or below, respectively).
- If μ is monotone autocontinuous from above or below, then it is null-additive.
- If μ is conditionally autocontinuous from below, then it is null-additive. However, this is not the case for the conditional autocontinuity from above and the conditional monotone autocontinuity from below; see Example 3

Example 3. Let X be an infinite set and $\mathcal{A}$ the family of all subsets of X. Let $\mu \colon \mathcal{A} \to [0, \infty]$ be the nonadditive measure defined by

$$\mu(A) := \begin{cases} \infty & \text{if } A = X, \\ 1 & \text{if } A \ne X \text{ and } A \text{ is infinite}, \\ 0 & \text{if } A = \emptyset \text{ or } A \text{ is finite.} \end{cases}$$

Then μ is conditionally autocontinuous from above and conditionally monotone autocontinuous from below, but is not null-additive. To see this, take any $x_0 \in X$ and let $A := \{x_0\}^c$ and $B := \{x_0\}$. Then $\mu(A) = 1$, $\mu(B) = 0$, but $\mu(A \cup B) = \infty$.

4 The Monotone Convergence in Measure Theorem

Let us explain our strategy of proving the convergence in measure theorems of the Choquet integral: First establish a proper distributional convergence theorem, which is a convergence theorem of distribution functions, and gives a bridge between the Choquet integral and the Lebesgue integral. Then the target theorem is derived from the convergence theorems of the Lebesgue integral.

Theorem 1. *Let $\mu \in \mathcal{M}(X)$. The following conditions are equivalent.*

1. *μ is conditionally monotone autocontinuous from above and null-additive.*
2. *For any $\{f_n\}_{n \in \mathbb{N}} \subset \mathcal{F}(X)$ and any $f \in \mathcal{F}(X)$, if they satisfy*
    (a) $f \le f_{n+1} \le f_n$ μ-a.e. for every $n \in \mathbb{N}$,
    (b) $f_n \xrightarrow{\mu} f$,
    then it follows that $\mu(\{f_n > t\}) \downarrow \mu(\{f > t\})$ for every continuity point $t \in \mathbb{R}$ of $\mu(\{f > t\})$ with $\mu(\{f_1 > t\}) < \infty$.

3. *The monotone nonincreasing convergence in measure theorem holds for μ, that is, for any $\{f_n\}_{n \in \mathbb{N}} \subset \mathcal{F}^+(X)$ and any $f \in \mathcal{F}^+(X)$, if they satisfy conditions (a) and (b) of assertion 2 and $\mathrm{Ch}(\mu, f_1) < \infty$, then it follows that $\mathrm{Ch}(\mu, f_n) \downarrow \mathrm{Ch}(\mu, f)$. Moreover, every f_n and f are μ-integrable.*

Proof. $1 \Rightarrow 2$: For each $t \in \mathbb{R}$, let $\varphi(t) := \mu(\{f > t\})$ and $\varphi_n(t) := \mu(\{f_n > t\})$ $(n = 1, 2, \dots)$. Let $t_0 \in \mathbb{R}$ be a continuity point of φ with $\mu(\{f_1 > t_0\}) < \infty$. Then the null-additivity of μ and condition (a) show that $\varphi(t_0) \leq \varphi_{n+1}(t_0) \leq \varphi_n(t_0)$ for every $n \in \mathbb{N}$, and hence, $\varphi(t_0) \leq \inf_{n \in \mathbb{N}} \varphi_n(t_0)$. Thus, it suffices to prove

$$\inf_{n \in \mathbb{N}} \varphi_n(t_0) \leq \varphi(t_0). \tag{1}$$

To see this, fix $\varepsilon > 0$ and let $A := \{f > t_0 - \varepsilon\}$, $B_n := \{|f_n - f| > \varepsilon\}$ $(n = 1, 2, \dots)$, and $C := \{f_1 > t_0\}$. Then $\mu(B_n) \to 0$ by condition (b), and $\mu(C) < \infty$. By condition (a) and the null-additivity of μ, one can find a nondecreasing sequence $\{N_n\}_{n \in \mathbb{N}}$ of μ-null sets such that $f(x) \leq f_{n+1}(x) \leq f_n(x)$ for every $x \notin N_n$. Then it is easy to verify that $\{B_n \setminus N_n\}_{n \in \mathbb{N}}$ is nonincreasing and $\{f_n > t_0\} \setminus N_n \subset (A \cup (B_n \setminus N_n)) \cap C$, so that

$$\mu(\{f_n > t_0\}) \leq \mu((A \cap C) \cup \{(B_n \setminus N_n) \cap C\}) \tag{2}$$

for every $n \in \mathbb{N}$. Since μ is conditionally monotone autocontinuous from above,

$$\mu(A \cap C) = \inf_{n \in \mathbb{N}} \mu((A \cap C) \cup \{(B_n \setminus N_n) \cap C\}). \tag{3}$$

Thus, (1) follows from (2) and (3).

$2 \Rightarrow 3$: Since $\mathrm{Ch}(\mu, f_1) < \infty$, there is a Lebesgue null set L such that $\mu(\{f_1 > t\}) < \infty$ for every $t \notin L$. Let D be the set of all discontinuity points of $\mu(\{f > t\})$. Then by assertion 2, $\mu(\{f_n > t\}) \downarrow \mu(\{f > t\})$ for every $t \notin L \cup D$. Consequently,

$$\mathrm{Ch}(\mu, f_n) = \int_0^\infty \mu(\{f_n > t\})dt \downarrow \int_0^\infty \mu(\{f > t\})dt = \mathrm{Ch}(\mu, f)$$

by the Lebesgue monotone convergence theorem.

$3 \Rightarrow 1$: Let $A, B_n \in \mathcal{A}$ and assume that $\{B_n\}_{n \in \mathbb{N}}$ is nonincreasing, $\mu(A \cup B_1) < \infty$, and $\mu(B_n) \to 0$. Let $f_n := \chi_{A \cup B_n}$ and $f := \chi_A$. Then they satisfy conditions (a) and (b). Moreover, $\mathrm{Ch}(\mu, f_1) = \mu(A \cup B_1) < \infty$. Therefore,

$$\mu(A \cup B_n) = \mathrm{Ch}(\mu, f_n) \to \mathrm{Ch}(\mu, f) = \mu(A),$$

which shows that μ is conditionally autocontinuous from above.

Let $A, B \in \mathcal{A}$ and $\mu(B) = 0$. If $\mu(A) = \infty$, then $\mu(A \cup B) = \mu(A) = \infty$, and hence, assume that $\mu(A) < \infty$. Let $f_n := \chi_A$ and $f := \chi_{A \cup B}$. Then f_n and f satisfy conditions (a), (b), and $\mathrm{Ch}(\mu, f_1) = \mu(A) < \infty$. Therefore,

$$\mu(A \cup B) = \mathrm{Ch}(\mu, f) = \lim_{n \to \infty} \mathrm{Ch}(\mu, f_n) = \mu(A),$$

and hence, μ is null-additive. $\qquad \square$

The monotone nondecreasing convergence in measure theorem is also derived in a similar way by the same strategy. Recall that every nonadditive measure that is monotone autocontinuous from below is null-additive.

Theorem 2. *Let* $\mu \in \mathcal{M}(X)$. *The following conditions are equivalent.*

1. μ *is monotone autocontinuous from below.*
2. *For any* $\{f_n\}_{n \in \mathbb{N}} \subset \mathcal{F}(X)$ *and any* $f \in \mathcal{F}(X)$, *if they satisfy*
   (a) $f_n \leq f_{n+1} \leq f$ μ-*a.e. for every* $n \in \mathbb{N}$,
   (b) $f_n \xrightarrow{\mu} f$,
   then it follows that $\mu(\{f_n > t\}) \uparrow \mu(\{f > t\})$ *for every continuity point* $t \in \mathbb{R}$ *of* $\mu(\{f > t\})$.
3. *The monotone nondecreasing convergence in measure theorem holds for* μ, *that is, for any* $\{f_n\}_{n \in \mathbb{N}} \subset \mathcal{F}^+(X)$ *and any* $f \in \mathcal{F}^+(X)$, *if they satisfy conditions (a) and (b) of assertion 2, then it follows that* $\mathrm{Ch}(\mu, f_n) \uparrow \mathrm{Ch}(\mu, f)$.

Remark 1. The classical Riesz theorem for a σ-additive measure μ states that every sequence $\{f_n\}_{n \in \mathbb{N}} \subset \mathcal{F}(X)$ converging in μ-measure to $f \in \mathcal{F}(X)$ has a subsequence $\{f_{n_k}\}_{k \in \mathbb{N}}$ converging μ-almost everywhere to f, so that the domination $f_n \leq f$ μ-a.e. follows from the monotonicity $f_n \leq f_{n+1}$ μ-a.e. for every $n \in \mathbb{N}$ and $f_n \xrightarrow{\mu} f$. However, this is not valid in general for a nonadditive measure unless μ has property (S); see, for instance, [4, Theorem 5.17]. Therefore, in the above theorems it is necessary to assume in advance that each f_n is dominated by the limit function f almost everywhere.

5   The Fatou and the Reverse Fatou in Measure Lemmas

The Fatou lemma can be proved by the same strategy as is stated in Sect. 4.

Theorem 3. *Let* $\mu \in \mathcal{M}(X)$. *The following conditions are equivalent.*

1. μ *is autocontinuous from below.*
2. *For any* $\{f_n\}_{n \in \mathbb{N}} \subset \mathcal{F}(X)$ *and any* $f \in \mathcal{F}(X)$, *if* $f_n \xrightarrow{\mu} f$ *then it follows that* $\mu(\{f > t\}) \leq \liminf_{n \to \infty} \mu(\{f_n > t\})$ *for every continuity point* $t \in \mathbb{R}$ *of* $\mu(\{f > t\})$.
3. *The Fatou in measure lemma holds for* μ, *that is, for any* $\{f_n\}_{n \in \mathbb{N}} \subset \mathcal{F}^+(X)$ *and any* $f \in \mathcal{F}^+(X)$, *if* $f_n \xrightarrow{\mu} f$ *then it follows that* $\mathrm{Ch}(\mu, f) \leq \liminf_{n \to \infty} \mathrm{Ch}(\mu, f_n)$.

To characterize the validity of the dominated convergence in measure theorem in terms of the characteristics of nonadditive measures, the following dominated versions of the Fatou and the reverse Fatou in measure lemmas are needed.

Theorem 4. *Let* $\mu \in \mathcal{M}(X)$. *The following conditions are equivalent.*

1. μ *is conditionally autocontinuous from below.*
2. *For any* $\{f_n\}_{n \in \mathbb{N}} \subset \mathcal{F}(X)$ *and any* $f \in \mathcal{F}(X)$, *if they satisfy*

(a) there is a $g \in \mathcal{F}(X)$ such that $f_n \leq g$ μ-a.e. for every $n \in \mathbb{N}$,

(b) $f_n \xrightarrow{\mu} f$,

then it follows that $\mu(\{f > t\}) \leq \liminf_{n \to \infty} \mu(\{f_n > t\})$ for every continuity point $t \in \mathbb{R}$ of $\mu(\{f > t\})$ with $\mu(\{g > t\}) < \infty$.

3. The dominated Fatou in measure lemma holds for μ, that is, for any $\{f_n\}_{n \in \mathbb{N}} \subset \mathcal{F}^+(X)$ and any $f \in \mathcal{F}^+(X)$, if they satisfy

(a) there is a $g \in \mathcal{F}^+(X)$ such that $f_n \leq g$ μ-a.e. for every $n \in \mathbb{N}$,

(b) $f_n \xrightarrow{\mu} f$,

(c) $\mathrm{Ch}(\mu, g) < \infty$,

then it follows that $\mathrm{Ch}(\mu, f) \leq \liminf_{n \to \infty} \mathrm{Ch}(\mu, f_n)$. Moreover, every f_n and f are μ-integrable.

Proof. $1 \Rightarrow 2$: For each $t \in \mathbb{R}$, let $\varphi(t) := \mu(\{f > t\})$ and $\varphi_n(t) := \mu(\{f_n > t\})$ $(n = 1, 2, \dots)$. Let $t_0 \in \mathbb{R}$ be a continuity point of φ with $\mu(\{g > t_0\}) < \infty$. Fix $\varepsilon > 0$ and let $A := \{f > t_0 + \varepsilon\}$, $B_n := \{|f_n - f| > \varepsilon\}$ $(n = 1, 2, \dots)$, and $C := \{g > t_0\}$. Then $\mu(B_n) \to 0$ by condition (b), and $\mu(C) < \infty$. By condition (a), for each $n \in \mathbb{N}$ there is a μ-null set N_n such that $f_n(x) \leq g(x)$ for every $x \notin N_n$. Then it is easy to verify that $A \setminus (B_n \cup N_n) \subset \{f_n > t_0\} \setminus N_n \subset C$ for every $n \in \mathbb{N}$. Since μ is conditionally autocontinuous from below, it follows that $\liminf_{n \to \infty} \varphi_n(t_0) \geq \varphi(t_0 + \varepsilon)$, which gives the conclusion.

$2 \Rightarrow 3$: Since $\mathrm{Ch}(\mu, g) < \infty$, there is a Lebesgue null set L such that $\mu(\{g > t\}) < \infty$ for every $t \notin L$. Let D be the set of all discontinuity points of $\mu(\{f > t\})$. Then by assertion 2, $\mu(\{f > t\}) \leq \liminf_{n \to \infty} \mu(\{f_n > t\})$ for every $t \notin L \cup D$. Consequently, it follows that

$$\mathrm{Ch}(\mu, f) \leq \int_0^\infty \liminf_{n \to \infty} \mu(\{f_n > t\}) dt$$

$$= \sup_{n \in \mathbb{N}} \int_0^\infty \inf_{k \geq n} \mu(\{f_k > t\}) dt \leq \liminf_{n \to \infty} \mathrm{Ch}(\mu, f_n)$$

by the Lebesgue monotone convergence theorem.

$3 \Rightarrow 1$: Let $A, C \in \mathcal{A}$, $\{B_n\}_{n \in \mathbb{N}} \subset \mathcal{A}$, $\mu(C) < \infty$, $\mu(B_n) \to 0$, and $A \setminus B_n \subset C$ for every $n \in \mathbb{N}$. Let $f_n := \chi_{A \setminus B_n}$, $f := \chi_A$, and $g := \chi_C$. Then they satisfy conditions (a), (b), and (c) of assertion 3. Therefore, it follows that

$$\mu(A) = \mathrm{Ch}(\mu, f) \leq \liminf_{n \to \infty} \mathrm{Ch}(\mu, f_n) = \liminf_{n \to \infty} \mu(A \setminus B_n),$$

which implies that μ is conditionally autocontinuous from below. $\qquad \square$

Theorem 5. *Let* $\mu \in \mathcal{M}(X)$. *The following conditions are equivalent.*

1. μ is conditionally autocontinuous from above and null-additive.

2. For any $\{f_n\}_{n \in \mathbb{N}} \subset \mathcal{F}(X)$ and any $f \in \mathcal{F}(X)$, if they satisfy

(a) there is a $g \in \mathcal{F}(X)$ such that $f_n \leq g$ μ-a.e. for every $n \in \mathbb{N}$,

(b) $f_n \xrightarrow{\mu} f$,

then it follows that $\limsup_{n \to \infty} \mu(\{f_n > t\}) \leq \mu(\{f > t\})$ for every continuity point $t \in \mathbb{R}$ of $\mu(\{f > t\})$ with $\mu(\{g > t\}) < \infty$.

3. *The reverse Fatou in measure lemma holds for μ, that is, for any $\{f_n\}_{n\in\mathbb{N}} \subset$*
   $\mathcal{F}^+(X)$ and any $f \in \mathcal{F}^+(X)$, if they satisfy
   (a) *there is a $g \in \mathcal{F}^+(X)$ such that $f_n \leq g$ μ-a.e. for every $n \in \mathbb{N}$,*
   (b) *$f_n \xrightarrow{\mu} f$,*
   (c) *$\mathrm{Ch}(\mu, g) < \infty$,*
   then it follows that $\limsup_{n\to\infty} \mathrm{Ch}(\mu, f_n) \leq \mathrm{Ch}(\mu, f)$. Moreover, every f_n is
   μ-integrable.

6   The Dominated Convergence in Measure Theorem

The dominated convergence in measure theorem now follows from the dominated
Fatou and the reverse Fatou in measure lemmas.

Theorem 6. *Let $\mu \in \mathcal{M}(X)$. The following conditions are equivalent.*

1. *μ is conditionally autocontinuous.*
2. *The dominated convergence in measure theorem holds for μ, that is, for any*
   $\{f_n\}_{n\in\mathbb{N}} \subset \mathcal{F}^+(X)$ and any $f \in \mathcal{F}^+(X)$, if they satisfy
   (a) *there is a $g \in \mathcal{F}^+(X)$ such that $f_n \leq g$ μ-a.e. for every $n \in \mathbb{N}$,*
   (b) *$f_n \xrightarrow{\mu} f$,*
   (c) *$\mathrm{Ch}(\mu, g) < \infty$,*
   then it follows that $\mathrm{Ch}(\mu, f_n) \to \mathrm{Ch}(\mu, f)$. Moreover, every f_n and f are
   μ-integrable.

Proof. $1 \Rightarrow 2$: By Theorems 4 and 5, every f_n and f are μ-integrable, and

$$\mathrm{Ch}(\mu, f) \leq \liminf_{n\to\infty} \mathrm{Ch}(\mu, f_n) \leq \limsup_{n\to\infty} \mathrm{Ch}(\mu, f_n) \leq \mathrm{Ch}(\mu, f),$$

which gives the conclusion.

$2 \Rightarrow 1$: It follows from Theorems 4 and 5.   □

7   The Dual Measure Forms

The dual measure forms of the convergence in measure theorems are inevitable in
our strategy to obtain convergence theorems of the symmetric and asymmetric
Choquet integrals. Whether "conditional" or not, the dual measure $\bar{\mu}$ is not
necessarily autocontinuous even if the original measure μ is autocontinuous. This
means that the dual measure forms are not direct consequences of the results
in previous sections. For instance, the dual measure forms of the Fatou and the
reverse Fatou in measure lemmas are stated and proved as follows.

Theorem 7. *Let $\mu \in \mathcal{M}_b(X)$. The following conditions are equivalent.*

1. *μ is autocontinuous from above.*
2. *For any $\{f_n\}_{n\in\mathbb{N}} \subset \mathcal{F}(X)$ and any $f \in \mathcal{F}(X)$, if $f_n \xrightarrow{\mu} f$ then it follows*
   that $\limsup_{n\to\infty} \mu(\{f_n < t\}) \leq \mu(\{f < t\})$ for every continuity point $t \in \mathbb{R}$
   of $\mu(\{f < t\})$.

3. *The dual Fatou in measure lemma holds for μ, that is, for any $\{f_n\}_{n\in\mathbb{N}} \subset \mathcal{F}^+(X)$ and any $f \in \mathcal{F}^+(X)$, if $f_n \xrightarrow{\mu} f$ then it follows that $\mathrm{Ch}(\bar{\mu}, f) \leq \liminf_{n\to\infty} \mathrm{Ch}(\bar{\mu}, f_n)$.*

Theorem 8. *Let $\mu \in \mathcal{M}_b(X)$. The following conditions are equivalent.*

1. *μ is autocontinuous from below.*
2. *For any $\{f_n\}_{n\in\mathbb{N}} \subset \mathcal{F}(X)$ and any $f \in \mathcal{F}(X)$, if $f_n \xrightarrow{\mu} f$ then it follows that $\mu(\{f < t\}) \leq \liminf_{n\to\infty} \mu(\{f_n < t\})$ for every continuity point $t \in \mathbb{R}$ of $\mu(\{f < t\})$.*
3. *The dual reverse Fatou in measure lemma holds for μ, that is, for any $\{f_n\}_{n\in\mathbb{N}} \subset \mathcal{F}^+(X)$ and any $f \in \mathcal{F}^+(X)$, if they satisfy*
   (a) *there is a $g \in \mathcal{F}^+(X)$ such that $f_n \leq g$ μ-a.e. for every $n \in \mathbb{N}$,*
   (b) *$f_n \xrightarrow{\mu} f$,*
   (c) *$\mathrm{Ch}(\bar{\mu}, g) < \infty$,*
   then it follows that $\limsup_{n\to\infty} \mathrm{Ch}(\bar{\mu}, f_n) \leq \mathrm{Ch}(\bar{\mu}, f)$. Moreover, every f_n is $\bar{\mu}$-integrable.

Proof. $1 \Rightarrow 2$: Observe that assertion 2 of this theorem is equivalent to that of Theorem 3.

$2 \Rightarrow 3$: By assertion 2, for almost all $t \in \mathbb{R}$ it follows that $\mu(\{f < t\}) \leq \liminf_{n\to\infty} \mu(\{f_n < t\})$, and hence, $\limsup_{n\to\infty} \bar{\mu}(\{f_n \geq t\}) \leq \bar{\mu}(\{f \geq t\})$. Moreover, condition (a) and the null-additivity of μ imply that $\bar{\mu}(\{f_n \geq t\}) \leq \bar{\mu}(\{g \geq t\})$ for every $t \in \mathbb{R}$ and $n \in \mathbb{N}$, so that condition (c) gives

$$\int_0^\infty \sup_{n\in\mathbb{N}} \bar{\mu}(\{f_n \geq t\})dt \leq \int_0^\infty \bar{\mu}(\{g \geq t\})dt = \mathrm{Ch}(\bar{\mu}, g) < \infty.$$

Then the conclusion follows from the Lebesgue monotone convergence theorem.

$3 \Rightarrow 1$: Let $A, B_n \in \mathcal{A}$ and $\mu(B_n) \to 0$. Let $f_n := \chi_{A^c \cup B_n}$, $f := \chi_{A^c}$, and $g := 1$. Then they satisfy conditions (a), (b), and (c). Therefore,

$$\bar{\mu}(A^c) = \mathrm{Ch}(\bar{\mu}, f) \geq \limsup_{n\to\infty} \mathrm{Ch}(\bar{\mu}, f_n) = \limsup_{n\to\infty} \bar{\mu}(A^c \cup B_n),$$

and hence, $\bar{\mu}(A^c \cup B_n) \to \bar{\mu}(A^c)$. Consequently,

$$\mu(A \setminus B_n) = \mu(X) - \bar{\mu}(A^c \cup B_n) \to \mu(X) - \bar{\mu}(A^c) = \mu(A),$$

which implies that μ is autocontinuous from below. $\qquad\square$

In the same way as Theorem 6, the dual measure form of the dominated convergence in measure theorem follows from Theorem 8 and the dominated version of Theorem 7.

Theorem 9. *Let $\mu \in \mathcal{M}_b(X)$. The following conditions are equivalent.*

1. *μ is autocontinuous.*
2. *For any $\{f_n\}_{n\in\mathbb{N}} \subset \mathcal{F}^+(X)$ and any $f \in \mathcal{F}^+(X)$, if they satisfy*
   (a) *there is a $g \in \mathcal{F}^+(X)$ such that $f_n \leq g$ μ-a.e. for every $n \in \mathbb{N}$,*
   (b) *$f_n \xrightarrow{\mu} f$,*
   (c) *$\mathrm{Ch}(\bar{\mu}, g) < \infty$,*
   then it follows that $\mathrm{Ch}(\bar{\mu}, f_n) \to \mathrm{Ch}(\bar{\mu}, f)$. Moreover, every f_n and f are $\bar{\mu}$-integrable.

8   The Extension to Symmetric and Asymmetric Integrals

The Choquet integral $\mathrm{Ch}: \mathcal{M}(X) \times \mathcal{F}^+(X) \to [0, \infty]$ can be extended in the following two ways:

$$\mathrm{Ch}^s(\mu, f) := \mathrm{Ch}(\mu, f^+) - \mathrm{Ch}(\mu, f^-), \quad (\mu, f) \in \mathcal{M}(X) \times \mathcal{F}(X),$$
$$\mathrm{Ch}^a(\mu, f) := \mathrm{Ch}(\mu, f^+) - \mathrm{Ch}(\bar{\mu}, f^-), \quad (\mu, f) \in \mathcal{M}_b(X) \times \mathcal{F}(X).$$

The functional Ch^s is called the *symmetric Choquet integral*, while Ch^a is called the *asymmetric Choquet integral*. They are not defined if the right hand side is of the form $\infty - \infty$. The symmetric Choquet integral Ch^s is symmetric in the sense that

$$\mathrm{Ch}^s(\mu, -f) = -\mathrm{Ch}^s(\mu, f)$$

and the asymmetric integral I^a is asymmetric in the sense that

$$\mathrm{Ch}^a(\mu, -f) = -\mathrm{Ch}^a(\bar{\mu}, f).$$

Given $\mu \in \mathcal{M}(X)$, a function $f \in \mathcal{F}(X)$ is called *symmetrically μ-integrable* if $\mathrm{Ch}^s(\mu, f) < \infty$, and *asymmetrically μ-integrable* if $\mathrm{Ch}^a(\mu, f) < \infty$. In this section the dominated convergence in measure theorem is extended to symmetric and asymmetric integrals.

Theorem 10. *Let $\mu \in \mathcal{M}(X)$. The following conditions are equivalent.*

1. *μ is conditionally autocontinuous.*
2. *For any $\{f_n\}_{n \in \mathbb{N}} \subset \mathcal{F}(X)$ and any $f \in \mathcal{F}(X)$, if they satisfy*
   (a) *there are $p, q \in \mathcal{F}(X)$ such that $q \leq f_n \leq p$ μ-a.e. for every $n \in \mathbb{N}$,*
   (b) *$f_n \xrightarrow{\mu} f$,*
   (c) *$\mathrm{Ch}(\mu, p^+) < \infty$ and $\mathrm{Ch}(\mu, q^-) < \infty$,*
   then it follows that $\mathrm{Ch}^s(\mu, f_n) \to \mathrm{Ch}^s(\mu, f)$. Moreover, every f_n and f are symmetrically μ-integrable.

Assume that μ is finite. Then assertion 1 is equivalent to the following assertion.

3. *For any $\{f_n\}_{n \in \mathbb{N}} \subset \mathcal{F}(X)$ and any $f \in \mathcal{F}(X)$, if they satisfy conditions (a) and (b) of assertion 2, and $\mathrm{Ch}(\mu, p^+) < \infty$ and $\mathrm{Ch}(\bar{\mu}, q^-) < \infty$ then it follows that $\mathrm{Ch}^a(\mu, f_n) \to \mathrm{Ch}^a(\mu, f)$. Moreover, every f_n and f are asymmetrically μ-integrable.*

Proof. We give only the proof of $1 \Rightarrow 3$. Observe that condition (a) implies that $f_n^+ \leq p^+$ and $f_n^- \leq q^-$ μ-a.e. for every $n \in \mathbb{N}$, and condition (b) implies that $f_n^+ \xrightarrow{\mu} f^+$ and $f_n^- \xrightarrow{\mu} f^-$. Therefore it follows from Theorems 6 and 9 that

$$\mathrm{Ch}^a(\mu, f_n) = \mathrm{Ch}(\mu, f_n^+) - \mathrm{Ch}(\bar{\mu}, f_n^-)$$
$$\to \mathrm{Ch}(\mu, f^+) - \mathrm{Ch}(\bar{\mu}, f^-) = \mathrm{Ch}^a(\mu, f)$$

and f_n and f are asymmetrically μ-integrable. $\square$

9 Concluding Remarks

In this paper, in the case where the nonadditive measure may be infinite and the functions may be unbounded, the validity of the monotone convergence in measure theorem, the Fatou and the reverse Fatou in measure lemmas, and the dominated convergence in measure theorem of the Choquet integral are fully characterized by the characteristics of nonadditive measures. Our strategy is based on suitable distributional convergence theorems, which are convergence theorems of distribution functions, and the convergence theorems of the Lebesgue integral. This strategy is also applicable for other convergence theorems such as the bounded convergence theorem and the Vitali convergence theorem [2,5,7,8], and for other modes of convergence of functions such as pointwise convergence, almost everywhere convergence, and almost uniform convergence [3,4,6,9,11,12, 15,16].

References

1. Choquet, G.: Theory of capacities. Ann. Inst. Fourier (Grenoble) **5**, 131–295 (1953–1954)
2. Couso, I., Montes, S., Gil, P.: Stochastic convergence, uniform integrability and convergence in mean on fuzzy measure spaces. Fuzzy Sets Syst. **129**, 95–104 (2002)
3. Denneberg, D.: Non-additive Measure and Integral, 2nd edn. Kluwer Academic Publishers, Dordrecht (1997)
4. Li, J., Mesiar, R., Pap, E., Klement, E.P.: Convergence theorems for monotone measures. Fuzzy Sets Syst. **281**, 103–127 (2015)
5. Kawabe, J.: The bounded convergence in measure theorem for nonlinear integral functionals. Fuzzy Sets Syst. **271**, 31–42 (2015)
6. Kawabe, J.: A unified approach to the monotone convergence theorem for nonlinear integrals. Fuzzy Sets Syst. **304**, 1–19 (2016)
7. Kawabe, J.: The Vitali type theorem for the Choquet integral. Linear Nonlinear Anal. **3**, 349–365 (2017)
8. Murofushi, T., Sugeno, M., Suzaki, M.: Autocontinuity, convergence in measure, and convergence in distribution. Fuzzy Sets Syst. **92**, 197–203 (1997)
9. Pap, E.: Null-additive Set Functions. Kluwer Academic Publishers, Bratislava (1995)
10. Ralescu, D., Adams, G.: The fuzzy integral. J. Math. Anal. Appl. **75**, 562–570 (1980)
11. Rébillé, Y.: Autocontinuity and convergence theorems for the Choquet integral. Fuzzy Sets Syst. **194**, 52–65 (2012)
12. Šipoš, J.: Integral with respect to a pre-measure. Math. Slovaca **29**, 141–155 (1979)
13. Schmeidler, D.: Integral representation without additivity. Proc. Am. Math. Soc. **97**, 255–261 (1986)
14. Sugeno, M.: Theory of fuzzy integrals and its applications. Ph.D. dissertation, Tokyo Institute of Technology, Tokyo (1974)
15. Wang, Z.: Convergence theorems for sequences of Choquet integrals. Int. J. Gen. Syst. **26**, 133–143 (1997)
16. Wang, Z., Klir, G.J.: Generalized Measure Theory. Springer, New York (2009)

Risk-Sensitive Markov Decision Under Risk Constraints with Coherent Risk Measures

Yuji Yoshida[✉]

University of Kitakyushu,
4-2-1 Kitagata, Kokuraminami, Kitakyushu 802-8577, Japan
yoshida@kitakyu-u.ac.jp

Abstract. A Markov decision process with constraints of coherent risk measures is discussed. Risk-sensitive expected rewards under utility functions are approximated by weighted average value-at-risks, and risk constraints are described by coherent risk measures. In this paper, coherent risk measures are represented as weighted average value-at-risks with the best risk spectrum derived from decision maker's risk averse utility, and the risk spectrum can inherit the risk averse property of the decision maker's utility as weighting. To find risk levels for feasible ranges, firstly a risk-minimizing problem is discussed by mathematical programming. Next dynamic risk-sensitive reward maximization under risk constraints is investigated. Dynamic programming can not be applied to this dynamic optimization model, and we try other approaches. A few numerical examples are given to understand the obtained results.

1 Introduction

This paper discusses risk-sensitive decision making which will be useful for artificial intelligence's quick and responsible reasoning. Risk management in decision making is studied by several approaches. One is risk-sensitive expected rewards, which were introduced by Howard and Matheson [4]. Risk-sensitive expectation is given by

$$f^{-1}(E(f(\cdot))), \tag{1}$$

where f and f^{-1} are decision maker's utility function and its inverse function and $E(\cdot)$ is an expectation. This approach is to estimate risky aspects of states through utility functions, and it is studied by several authors (Bäuerle and Rieder [3]). However this criterion (1) with non-linear utility functions f has complexity regarding computation in general.

Another approach is risk criteria given by risk measures such as value-at-risks, conditional value-at-risks and so on. Risk measure is one of the most important concepts in economic theory, financial analysis, asset management, engineering and so on, and it has been improved from practical aspects as well

© Springer Nature Switzerland AG 2019
V. Torra et al. (Eds.): MDAI 2019, LNAI 11676, pp. 29–40, 2019.
https://doi.org/10.1007/978-3-030-26773-5_3

as theoretical researches. The variance was used as a risk measure in decision processes. Nowadays drastic declines of asset prices are studied, and *value-at-risk (VaR)* is used widely to estimate the risk of asset price decline in practical management. VaR is defined by percentiles at a specified probability, however it does not have coherency. *Coherent risk measures* have been studied to improve the criterion of risks with worst scenarios (Artzner et al. [2]). Several improved risk measures based on VaR are proposed: for example, conditional value-at-risks, expected shortfall, entropic value-at-risk (Rockafellar and Uryasev [6], Tasche [7]). Recently Kusuoka [5] gave a spectral representation for coherent risk measures. Further Yoshida [10] has introduced a *spectral weighted average value-at-risk* as the best coherent risk measure derived from decision maker's utility functions. Using this derived coherent risk measure, the risk measure can inherit the risk averse property of the decision maker's utility function as risk spectrum weighting. Yoshida [12] has also applied it to portfolio selection in finance. In this paper we adopt the spectral weighted average value-at-risks when we estimate a dynamic risk-sensitive rewards under risk constraints given by coherent risk measures. This paper is a risk-sensitive extension of objective functions in Yoshida [9], and it also extends the results in Yoshida [11] to dynamic Markov decision models.

2   Coherent Risk Measure Derived from Risk Averse Utility

In this section we introduce basic concepts from Yoshida [10]. Let $\mathbb{R} = (-\infty, \infty)$ and let P be a non-atomic probability on a sample space Ω. Let $\mathcal{X}$ be the family of all integrable real-valued random variables X on Ω with a continuous distribution function $x \mapsto F_X(x) = P(X < x)$ for which there exists a non-empty open interval I such that $F_X(\cdot) : I \to (0, 1)$ is strictly increasing and onto. *Value-at-risk (VaR)* at a positive probability p is given by the percentile of the distribution F_X, i.e.

$$\mathrm{VaR}_p(X) = \sup\{x \in I \mid F_X(x) \le p\} = F_X^{-1}(p) \tag{2}$$

for $p \in (0, 1)$ and $\mathrm{VaR}_1(X) = \sup I$, where F_X^{-1} is the inverse function of F_X. Then *average value-at-risk (AVaR)* at a positive probability p is given by

$$\mathrm{AVaR}_p(X) = \frac{1}{p} \int_0^p \mathrm{VaR}_q(X) \, dq. \tag{3}$$

Definition 1 is introduced to characterize risk measures (Artzner et al. [2]).

Definition 1. A map $\rho : \mathcal{X} \mapsto \mathbb{R}$ is called a *coherent risk measure* if it satisfies the following (i)–(iv):

(i) $\rho(X) \ge \rho(Y)$ for $X, Y \in \mathcal{X}$ satisfying $X \le Y$ (*monotonicity*).
(ii) $\rho(cX) = c\rho(X)$ for $X \in \mathcal{X}$ and $c \in \mathbb{R}$ satisfying $c \ge 0$ (*positive homogeneity*).

(iii) $\rho(X + c) = \rho(X) - c$ for $X \in \mathcal{X}$ and $c \in \mathbb{R}$ (*translation invariance*).
(iv) $\rho(X + Y) \leq \rho(X) + \rho(Y)$ for $X, Y \in \mathcal{X}$ (*sub-additivity*).

It is known in Artzner et al. [2] that $-\mathrm{AVaR}_p(\cdot)$ *is a coherent risk measure* however $-\mathrm{VaR}_p(\cdot)$ *is not coherent* because sub-additivity (iv) does not hold. Conditional value-at-risks and expected shortfall are also famous coherent risk measures. Now, for a probability $p \in (0, 1]$ and a non-increasing right-continuous function λ on $[0, 1]$ satisfying $\int_0^1 \lambda(q)\, dq = 1$, we define *a weighted average value-at-risk with weighting* λ on $(0, p)$ by

$$\mathrm{AVaR}_p^\lambda(X) = \int_0^p \mathrm{VaR}_q(X)\, \lambda(q)\, dq \bigg/ \int_0^p \lambda(q)\, dq. \tag{4}$$

Then λ is called a *risk spectrum*. Hence coherent risk measures are represented by weighted average value-at-risks with risk spectra (Kusuoka [5], Yoshida [10]).

Lemma 1. *Let* $\rho : \mathcal{X} \mapsto \mathbb{R}$ *be a law invariant, comonotonically additive, continuous coherent risk measure. Then there exists a risk spectrum* λ *such that*

$$\rho(X) = -\mathrm{AVaR}_1^\lambda(X) \tag{5}$$

for $X \in \mathcal{X}$. *Further,* $-\mathrm{AVaR}_p^\lambda$ *is a coherent risk measure on* $\mathcal{X}$ *for* $p \in (0, 1)$.

In this paper we use a law invariant, comonotonically additive, continuous coherent risk measure ρ for risk constraints, and we also deal with a case when value-at-risks are represented as

$$\mathrm{VaR}_p(X) = E(X) + \kappa(p) \cdot \sigma(X) \tag{6}$$

with the mean $E(X)$ and the standard deviation $\sigma(X)$ of random variables $X \in \mathcal{X}$, where $\kappa : (0, 1) \mapsto (-\infty, \infty)$ is an increasing function. We assume there exists a probability distribution ψ on $\mathbb{R} \times [0, \infty)$ of means $E(X)$ and standard deviations $\sigma(X)$ of random variables $X \in \mathcal{X}$. From (4) and (6) we have

$$\mathrm{AVaR}_p^\lambda(X) = E(X) + \kappa^\lambda(p) \cdot \sigma(X), \tag{7}$$

where $\kappa^\lambda(p) = \int_0^p \kappa(q)\, \lambda(q)\, dq / \int_0^p \lambda(q)\, dq$. Let $f : I \mapsto \mathbb{R}$ be a C^2-class risk averse utility function satisfying $f' > 0$ and $f'' \leq 0$ on I, where I is an open interval. For a probability $p \in (0, 1]$ and a random variable $X(\in \mathcal{X})$, a non-linear risk-sensitive form $f^{-1}(\frac{1}{p} \int_0^p f(\mathrm{VaR}_q(X))\, dq)$ is an average value-at-risk of X on the downside $(0, p)$ under the utility f. The following lemma is from Yoshida [10].

Lemma 2. *A risk spectrum* λ *which minimizes the distance between the non-linear risk-sensitive form and weighted average value-at-risk (4):*

$$\sum_{X \in \mathcal{X}} \left(f^{-1} \left(\frac{1}{p} \int_0^p f(\mathrm{VaR}_q(X))\, dq \right) - \mathrm{AVaR}_p^\lambda(X) \right)^2 \tag{8}$$

for $p \in (0, 1]$ *is given by*

$$\lambda(p) = e^{-\int_p^1 C(q)\, dq} C(p) \tag{9}$$

with a component function C *in [10] if* λ *is non-increasing.*

If utility function f is specified, the component function C is given concretely in Example 2. In Lemma 2 the coherent risk measure $-\text{AVaR}_p^\lambda$ has a kind of semi-linearity such as Definition 1(ii)(iii) and it is useful for computation, and the risk spectrum λ can also inherit the risk averse property of the non-linear utility function f as weighting on $(0, p)$. Regarding risk-sensitive rewards, in the next section we use the risk spectrum λ given in Lemma 2 because $-\text{AVaR}_p^\lambda$ is the best coherent risk measure derived from risk averse utility f.

3 Markov Decision Processes with Risk Constraints

We deal with the following Markov decision process. Let n be a positive integer. Let a *state space* be $\mathbb{R}$ and let an *action space* be $\mathcal{A} = \{(x^1, x^2, \cdots, x^n) \in [0,1]^n \mid \sum_{i=1}^n x^i = 1 \text{ and } x^i \geq 0 (i = 1, 2, \cdots, n)\}$. A positive integer T denotes a *terminal time*. Let $\{X_t^i\}_{t=0}^T (\subset \mathcal{X})$ be a *reward process* for *asset* $i(= 1, 2, \cdots, n)$ such that X_t^i is independent of the past information $\mathcal{M}_{t-1}$, which is the σ-field generated by $\{X_s^j \mid s = 0, 1, 2, \cdots, t-1; j = 1, 2, \cdots, n\}$. Then an *immediate reward* is $R_t^i = X_t^i - X_{t-1}^i$. Hence we put their expectations and covariances respectively by $\mu_t^i = E(R_t^i)$ and $\sigma_t^{ij} = E((R_t^i - \mu_t^i)(R_t^j - \mu_t^j))$ for $i, j = 1, 2, \cdots, n$. We give *Markov policies* by $\pi = \{\pi_t\}_{t=1}^T$ where mappings $\pi_t = (\pi_t^1, \pi_t^2, \cdots, \pi_t^n) : \Omega \mapsto \mathcal{A}$ for $t = 1, 2, \cdots, T$, and then π_t is called a *strategy*. They are chosen depending only on the current state X_{t-1}^π. Put a collection of all Markov policies by Π. For a Markov policy $\pi = \{(\pi_t^1, \pi_t^2, \cdots, \pi_t^n)\}_{t=1}^T \in \Pi$, a weighted sum of n immediate rewards and the reward at next time t are given by

$$R_t^\pi = \sum_{i=1}^n \pi_t^i R_t^i \quad \text{and} \quad X_t^\pi = X_{t-1}^\pi + R_t^\pi. \tag{10}$$

The expectation and the standard deviation of immediate reward R_t^π are respectively

$$E(R_t^\pi) = \sum_{i=1}^n \pi_t^i \mu_t^i \quad \text{and} \quad \sigma(R_t^\pi) = \sqrt{\sum_{i=1}^n \sum_{j=1}^n \pi_t^i \pi_t^j \sigma_t^{ij}}. \tag{11}$$

Let ρ be a coherent risk measure in Lemma 1 and let f be a C^2-class risk averse utility functions in the previous section. Let δ be a positive constant. Then we investigate the following problem.

Problem 1. Maximize the risk-sensitive expected terminal reward

$$f^{-1}(E(f(X_T^\pi))) \tag{12}$$

with respect to Markov policies $\pi \in \Pi$ under risk constraints

$$\rho(R_t^\pi) \leq \delta \tag{13}$$

for all $t = 1, 2, \cdots, T$.

In (12), $f^{-1}(E(f(\cdot))) = f^{-1}(\int_0^1 \mathrm{VaR}_q(f(\cdot))\,dq) = f^{-1}(\int_0^1 f(\mathrm{VaR}_q(\cdot))\,dq)$ is approximated, using Lemma 2, by $\mathrm{AVaR}_1^\lambda(\cdot)$ with a risk spectrum λ. Regarding (13) by Lemma 1 we give a coherent risk measure ρ by $\rho = -\mathrm{AVaR}_p^\nu$ with a risk spectrum ν. Hence we estimate the downside risks on $(0, p)$. Thus we discuss the following optimization instead of Problem 1.

Problem 2. Maximize the weighted average value-at-risk

$$\mathrm{AVaR}_1^\lambda(X_T^\pi) \tag{14}$$

with respect to Markov policies $\pi \in \Pi$ under risk constraints

$$-\mathrm{AVaR}_p^\nu(R_t^\pi) \le \delta \tag{15}$$

for all $t = 1, 2, \cdots, T$.

4   Feasibility of Risk Constraints and Risk-Sensitive Expected Rewards

Let $\Pi_t(\delta)$ be the collection of all Markov policies $\pi \in \Pi$ satisfying the risk constraints (15), and let $\Pi_t = \bigcup_{\delta > 0} \Pi_t(\delta)$. In this section we firstly discuss the feasibility for risk constraints (15), i.e. $\Pi_t(\delta) \ne \emptyset$. Fix a probability $p \in (0, 1)$ and time $t(= 1, 2, \cdots, T)$. We investigate the following maximization problem for $\mathrm{AVaR}_p^\nu(R_t^\pi)$ to find the lower bound of $-\mathrm{AVaR}_p^\nu(R_t^\pi)$.

Problem 3. Maximize weighted average value-at-risk

$$\mathrm{AVaR}_p^\nu(R_t^\pi) = \sum_{i=1}^n \pi_t^i \mu_t^i + \kappa^\nu(p)\sqrt{\sum_{i=1}^n \sum_{j=1}^n \pi_t^i \pi_t^j \sigma_t^{ij}} \tag{16}$$

with respect to strategies $\pi_t = (\pi_t^1, \pi_t^2, \cdots, \pi_t^n) \in \Pi_t$.

Let $\gamma \in \mathbb{R}$. Under a constraint

$$E(R_t^\pi) = \sum_{i=1}^n \pi^i \mu^i = \gamma, \tag{17}$$

Problem 3 is solved by quadratic programming (Yoshida [8]), and then the corresponding value (16) is reduced to

$$\gamma + \kappa^\nu(p)\sqrt{\frac{A_t \gamma^2 - 2B_t \gamma + C_t}{\Delta_t}}, \tag{18}$$

where

$$\mu_t = \begin{bmatrix} \mu_t^1 \\ \mu_t^2 \\ \vdots \\ \mu_t^n \end{bmatrix}, \quad \Sigma_t = \begin{bmatrix} \sigma_t^{11} & \sigma_t^{12} & \cdots & \sigma_t^{1n} \\ \sigma_t^{21} & \sigma_t^{22} & \cdots & \sigma_t^{2n} \\ \vdots & \vdots & \ddots & \vdots \\ \sigma_t^{n1} & \sigma_t^{n2} & \cdots & \sigma_t^{nn} \end{bmatrix}, \quad \mathbf{1} = \begin{bmatrix} 1 \\ 1 \\ \vdots \\ 1 \end{bmatrix},$$

$A_t = \mathbf{1}^{\mathrm{T}} \Sigma_t^{-1} \mathbf{1}, B_t = \mathbf{1}^{\mathrm{T}} \Sigma_t^{-1} \mu_t, C_t = \mu_t^{\mathrm{T}} \Sigma_t^{-1} \mu_t, \Delta_t = A_t C_t - B_t^2$ and T denotes the transpose of a vector. If $A_t > 0$, $\Delta_t > 0$ and $\kappa^\nu(p) < -\sqrt{\Delta_t/A_t}$ are satisfied, we can easily check the real-valued function (18) of γ is concave and it has the

maximum $\dfrac{B_t - \sqrt{A_t \kappa^\nu(p)^2 - \Delta_t}}{A_t}$ at $\gamma = \dfrac{B_t}{A_t} + \dfrac{\Delta_t}{A_t \sqrt{A_t \kappa^\nu(p)^2 - \Delta_t}}$. Since $\sup_{\pi_t \in \Pi_t}(18) = \sup_\gamma \{\sup_{\pi_t \in \Pi_t: \sum_{i=1}^n \pi_t^i \mu_t^i = \gamma}(18)\}$, we obtain the following analytical solutions for Problem 3 (Yoshida [8]).

Lemma 3. *Let $A_t > 0$, $\Delta_t > 0$, $\kappa^\nu(p) \leq \kappa^\lambda(1) \leq 0$ and $\kappa^\nu(p) < -\sqrt{\Delta_t/A_t}$. The maximum weighted average value-st-risk in Problem 3 is*

$$\frac{B_t - \sqrt{A_t \kappa^\nu(p)^2 - \Delta_t}}{A_t} \tag{19}$$

at the expected immediate reward

$$\gamma = \frac{B_t}{A_t} + \frac{\Delta_t}{A_t \sqrt{A_t \kappa^\nu(p)^2 - \Delta_t}}. \tag{20}$$

We define the lower bound of risk values $-\mathrm{AVaR}_p^\nu(R_t^\pi)$ by a constant $\underline{\delta}_t(p) = \inf_{\pi_t \in \Pi_t}(-\mathrm{AVaR}_p^\nu(R_t^\pi)) = -\sup_{\pi_t \in \Pi_t} \mathrm{AVaR}_p^\nu(R_t^\pi)$. From Lemma 3, it follows

$$\underline{\delta}_t(p) = -\frac{B_t}{A_t} + \frac{\sqrt{A_t \kappa^\nu(p)^2 - \Delta_t}}{A_t}. \tag{21}$$

Therefore its feasible range is $\{\delta \mid \Pi_t(\delta) \neq \emptyset\} = [\underline{\delta}_t(p), \infty)$. Let a risk level $\delta \in [\underline{\delta}_t(p), \infty)$. Next we discuss the following optimization at time t.

Problem 4. Maximize the weighted average value-at-risk

$$\mathrm{AVaR}_1^\lambda(R_t^\pi) = E(R_t^\pi) + \kappa^\lambda(1) \cdot \sigma(R_t^\pi) \tag{22}$$

with respect to Markov policies $\pi \in \Pi$ under risk constraints

$$\mathrm{AVaR}_p^\nu(R_t^\pi) = E(R_t^\pi) + \kappa^\nu(p) \cdot \sigma(R_t^\pi)) \geq -\delta. \tag{23}$$

Since $\kappa^\lambda(1) \leq 0$ and $\kappa^\nu(p) < 0$, from the viewpoint of (18), Problem 4 is reduced to the following problem.

Problem 5. Maximize the expected risk-sensitive expected reward

$$\gamma_t + \kappa^\lambda(1) \sqrt{\frac{A_t \gamma_t^2 - 2B_t \gamma_t + C_t}{\Delta_t}} \tag{24}$$

with respect to $\gamma_t \in \mathbb{R}$ under risk constraint

$$\gamma_t + \kappa^\nu(p) \sqrt{\frac{A_t \gamma_t^2 - 2B_t \gamma_t + C_t}{\Delta_t}} \geq -\delta. \tag{25}$$

Hence (25) is equivalent to $\gamma_t \in [\gamma_t^-, \gamma_t^+]$, where

$$\gamma_t^\pm = \frac{B_t \kappa^\nu(p)^2 + \Delta_t \delta \mp \sqrt{\Delta_t} \kappa^\nu(p) \sqrt{A_t \delta^2 + 2B_t \delta + C_t - \kappa^\nu(p)^2}}{A_t \kappa^\nu(p)^2 - \Delta_t}. \quad (26)$$

By solving Problem 5, we obtain the following lemma (Refer to Yoshida [11]).

Theorem 1. *Let A_t and Δ_t be positive and let $\kappa^\lambda(1)$ and $\kappa^\nu(p)$ satisfy $\kappa^\nu(p) \leq \kappa^\lambda(1) \leq 0$ and $\kappa^\nu(p) < -\sqrt{\Delta_t/A_t}$. Then the following (i) and (ii) hold:*

(i) The maximum the weighted average value-at-risk, which implies risk-sensitive expected reward, in Problem 4 is

$$\varphi_t^\circ = \begin{cases} \dfrac{B_t}{A_t} - \dfrac{\sqrt{A_t \kappa^\lambda(1)^2 - \Delta_t}}{A_t} & \text{at a reward } \gamma_t^\circ = \dfrac{B_t}{A_t} + \dfrac{\Delta_t}{A_t \sqrt{A_t \kappa^\lambda(1)^2 - \Delta_t}} \\ \qquad \text{if } \delta_t^+ \leq \delta \text{ and } \kappa^\lambda(1) < -\sqrt{\Delta_t/A_t}, \\ \gamma_t^+ - \dfrac{\kappa^\lambda(1)}{\kappa^\nu(p)}(\delta + \gamma_t^+) & \text{at a reward } \gamma_t^\circ = \gamma_t^+ \quad \text{otherwise,} \end{cases} \quad (27)$$

where $\delta_t^+ = -\dfrac{B_t}{A_t} + \dfrac{A_t \kappa^\lambda(1) \kappa^\nu(p) - \Delta_t}{A_t \sqrt{A_t \kappa^\lambda(1)^2 - \Delta_t}}.$

(ii) Further an optimal strategy is given by

$$\pi_t^\circ = \xi_t^\circ \Sigma_t^{-1} 1 + \eta_t^\circ \Sigma_t^{-1} \mu_t, \quad (28)$$

where $\xi_t^\circ = \dfrac{C_t - B_t \gamma_t^\circ}{\Delta_t}$ and $\eta_t^\circ = \dfrac{A_t \gamma_t^\circ - B_t}{\Delta_t}$, if $\pi_t \geq 0$.

5   Dynamic Risk-Sensitive Rewards Under Risk Constraints

Let a time space $\mathbb{T} = \{1, 2, \cdots, T\}$. In this section we assume $A_t > 0$, $\Delta_t > 0$, $\kappa^\nu(p) \leq \kappa^\lambda(1) \leq 0$ and $\kappa^\nu(p) < -\sqrt{\Delta_t/A_t}$ for all $t \in \mathbb{T}$. Let the initial state be a real number $X_0^\pi = x_0$. Then $E(X_0) = \gamma_0 = x_0$ and $\sigma(X_0)^2 = 0$. For a Markov policy $\pi = \{\pi_t\}_{t=1}^T \in \Pi$, the expectation and the standard deviation of terminal rewards $X_T^\pi = x_0 + \sum_{t=1}^T R_t^\pi$ are

$$E(X_T^\pi) = x_0 + \sum_{t=1}^T \left(\sum_{i=1}^n \pi_t^i \mu_t^i \right) \quad \text{and} \quad \sigma(X_T^\pi) = \sqrt{\sum_{t=1}^T \left(\sum_{i=1}^n \sum_{j=1}^n \pi_t^i \pi_t^j \sigma_t^{ij} \right)}. (29)$$

In similar approach to Problem 5, Problem 2 is reduced to Problem 6 using (29).

Problem 6. Maximize the *risk-sensitive expected reward*

$$\Phi(\gamma_1, \gamma_2, \cdots \gamma_T) = x_0 + \sum_{t=1}^T \gamma_t + \kappa^\lambda(1) \sqrt{\sum_{t=1}^T \frac{A_t \gamma_t^2 - 2B_t \gamma_t + C_t}{\Delta_t}} \quad (30)$$

with respect to $(\gamma_1, \gamma_2, \cdots \gamma_T) \in \mathbb{R}^T$ under risk constraint

$$\gamma_t \in [\gamma_t^-, \gamma_t^+] \tag{31}$$

for all $t \in \mathbb{T} = \{1, 2, \cdots, T\}$.

It is *difficult to solve Problem* 6 *by dynamic programming* because of the form (30). If T is not large, we can calculate solutions of Problem 6 by *multi-parameter optimization*. In the rest of this section, we investigate Problem 6 taking into account of large T. We can easily check the following lemma.

Lemma 4. *Let* $t \in \mathbb{T}$ *and constants* $E, D \in \mathbb{R}$ *satisfying* $D \geq 0$. *Define a function*

$$\psi(\gamma) = E + \gamma + \kappa^\lambda(1)\sqrt{\frac{A_t \gamma^2 - 2B_t \gamma + C_t}{\Delta_t} + D} \tag{32}$$

for $\gamma \in \mathbb{R}$. *Then the following* (i) *and* (ii) *hold:*

(i) *If* $\kappa^\lambda(1)^2 \leq \Delta_t / A_t$, *then function* ψ *is non-decreasing on* $\mathbb{R}$.
(ii) *If* $\kappa^\lambda(1)^2 > \Delta_t / A_t$, *then function* ψ *is concave and it has a maximum at* $\gamma = \gamma^D$, *where*

$$\gamma^D = \frac{B_t}{A_t} + \frac{\Delta_t \sqrt{1 + A_t D}}{A_t \sqrt{A_t \kappa^\lambda(1)^2 - \Delta_t}}. \tag{33}$$

Further we can obtain the following result from the first-order necessary condition for optimal solutions in Problem (P6).

Theorem 2. *Assume* $A_t > 0$, $\Delta_t > 0$, $\kappa^\nu(p) \leq \kappa^\lambda(1) \leq 0$ *and* $\kappa^\nu(p) < -\sqrt{\Delta_t / A_t}$ *for all* $t \in \mathbb{T}$. *Let* Φ *has a maximum at a point* $(\gamma_1^*, \gamma_2^*, \cdots, \gamma_T^*) \in \prod_{t=1}^T [\gamma_t^-, \gamma_t^+]$ *in Problem (P6). Then the following* (i) *and* (ii) *hold:*

(i) *There exists a subset* $\mathbb{T}^*$ *of* $\mathbb{T} = \{1, 2, \cdots, T\}$ *for which* (34) *and* (35) *hold:*

$$\kappa^\lambda(1)^2 > \sum_{t \notin \mathbb{T}^*} \frac{\Delta_t}{A_t} \tag{34}$$

and the point $(\gamma_1^*, \gamma_2^*, \cdots, \gamma_T^*)$ *satisfies*

$$\gamma_t^* = \begin{cases} \gamma_t^+ & for\ t \in \mathbb{T}^* \\ \dfrac{\Delta_t \theta^* + B_t}{A_t} \ (< \gamma_t^+) & for\ t \notin \mathbb{T}^*, \end{cases} \tag{35}$$

where

$$D^* = \sum_{t \in \mathbb{T}^*} \frac{A_t (\gamma_t^+)^2 - 2B_t \gamma_t^+ + C_t}{\Delta_t} \quad and \quad \theta^* = \frac{\sqrt{\sum_{t \notin \mathbb{T}^*} \frac{1}{A_t} + D^*}}{\sqrt{\kappa^\lambda(1)^2 - \sum_{t \notin \mathbb{T}^*} \frac{\Delta_t}{A_t}}}. \tag{36}$$

(ii) Further an optimal strategy *is given by*

$$\pi_t^* = \xi_t^* \Sigma_t^{-1} \mathbf{1} + \eta_t^* \Sigma_t^{-1} \mu_t, \tag{37}$$

where $\xi_t^* = \frac{C_t - B_t \gamma_t^*}{\Delta_t}$ *and* $\eta_t^* = \frac{A_t \gamma_t^* - B_t}{\Delta_t}$, *if* $\pi_t \geq \mathbf{0}$.

Remark. From (35) in Theorem 2, the point $(\gamma_1^*, \gamma_2^*, \cdots, \gamma_T^*)$ is given by $\gamma_t^* = \gamma_t^+$ or $\gamma_t^* = \frac{\Delta_t \theta + B_t}{A_t}$. Therefore we need to investigate solutions (35) with 2^T combination, however 2^T becomes numerous if T is larger. Hence the following theorem, which is from Lemma 4 and a sufficient condition $\frac{\partial \Phi}{\partial \gamma_t}\Big|_{\gamma_t = \gamma_t^+} \geq 0$ for all t, brings us easy computation of optimal solutions in Problem (P6) when T is large.

Theorem 3. *Assume* $A_t > 0$, $\Delta_t > 0$, $\kappa^\nu(p) \leq \kappa^\lambda(1) \leq 0$ *and* $\kappa^\nu(p) < -\sqrt{\Delta_t/A_t}$ *for all* $t \in \mathbb{T}$. *If an inequality condition*

$$\kappa^\lambda(1)^2 \max_{t \in \mathbb{T}} \left(\frac{A_t \gamma_t^+ - B_t}{\Delta_t} \right)^2 \leq \sum_{t \in \mathbb{T}} \frac{A_t (\gamma_t^+)^2 - 2B_t \gamma_t^+ + C_t}{\Delta_t} \tag{38}$$

holds, then the optimal solution in Problem (P6) is $\gamma_t^* = \gamma_t^+$ *for all* $t \in \mathbb{T}$.

Remark. The left term in (38) is bounded upper, however the right term ($\geq \sum_{t \in \mathbb{T}} \frac{1}{A_t} > 0$) becomes larger and it goes to infinity as $T \to \infty$. Therefore in *actual cases we can check* (38)*is satisfied for large* T.

6   Numerical Examples

Example 1. Let a domain $I = \mathbb{R}$ and let f be a *risk neutral utility function* $f(x) = ax + b$ for $x \in \mathbb{R}$ with constants $a(> 0)$ and $b(\in \mathbb{R})$. Then its risk spectrum in Lemma 2 is given by $\lambda(p) = 1$. The corresponding weighted average value-at-risk (5) is reduced to the *average value-at-risk* (3) and the *risk-sensitive expectation is reduced to the usual expectation*: Let $\hat{\kappa}(p) = \int_0^p \kappa(q) \, dq / \int_0^p dq = \frac{1}{p} \int_0^p \kappa(q) \, dq$. Then (7) follows $\text{AVaR}_p^\lambda(\cdot) = \text{AVaR}_p(\cdot) = E(\cdot) + \hat{\kappa}(p) \cdot \sigma(\cdot)$ with $\hat{\kappa}(p) = \kappa^\lambda(p)$. Hence we note $\hat{\kappa}(1) = 0$ in Fig. 2 and $f^{-1}(E(f(\cdot))) = \text{AVaR}_1^\lambda(\cdot) = \text{AVaR}_1(\cdot) = E(\cdot)$.

Example 2. Let a domain $I = \mathbb{R}$ and let a *risk averse exponential utility function*

$$f(x) = \frac{1 - e^{-\tau x}}{\tau} \tag{39}$$

for $x \in \mathbb{R}$ with a positive constant τ. Then $-\frac{f''}{f'} = \tau$ is the *degree of the absolute risk averse* of decision maker's utility (39) (Arrow [1]). Figure 1 illustrates utility functions. Let $\mathcal{X}$ be a family of random variables X which have normal distribution functions, and let $G : \mathbb{R} \to (0, 1)$ be the *cumulative distribution function of the standard normal distribution*, i.e. $G(x) = \frac{1}{\sqrt{2\pi}} \int_{-\infty}^{x} e^{-\frac{z^2}{2}} dz$ for $x \in \mathbb{R}$. Define an increasing function $\kappa : (0,1) \mapsto \mathbb{R}$ by its inverse function $\kappa(p) = G^{-1}(p)$ for probabilities $p \in (0,1)$. Then we have value-at-risk $\mathrm{VaR}_p(X) = \mu + \kappa(p) \cdot \sigma$ for $X \in \mathcal{X}$ with mean μ and standard deviation σ. Suppose there exists a distribution function $\psi : \mathbb{R} \times (0,\infty) \mapsto [0,\infty)$ such that $\psi(\mu,\sigma) = \phi(\mu) \cdot \frac{2^{1-n/2}}{\Gamma(n/2)} \sigma^{n-1} e^{-\frac{\sigma^2}{2}}$ for $(\mu,\sigma) \in \mathbb{R} \times [0,\infty)$, where $\phi(\mu)$ is some probability distribution, $\Gamma(\cdot)$ is a gamma function and $\frac{2^{1-n/2}}{\Gamma(n/2)} \sigma^{n-1} e^{-\frac{\sigma^2}{2}}$ is a chi distribution with degree of freedom n. Let $n = 4$ be the number of assets. Hence we put the expectations μ_t^i and the covariances σ_t^{ij} of immediate rewards $R_t^i (\in \mathcal{X})$ as Table 1. We take a utility $f(x) = \frac{1-e^{-0.05x}}{0.05}$ with $\tau = 0.05$ in (39), and by Lemma 2 there exists a risk spectrum λ satisfying $f^{-1}(E(f(\cdot))) \approx \mathrm{AVaR}_1^\lambda(\cdot)$. Then, by Yoshida [10], the best risk spectrum in Lemma 2 is given by

$$\lambda(p) = e^{-\int_p^1 C(q)\,dq} C(p) \tag{40}$$

for $p \in (0,1]$, where the component function C is given by

$$C(p) = \frac{1}{p} \cdot \frac{\int_0^\infty \left(1 - \frac{1}{\frac{1}{p}\int_0^p e^{\tau\sigma(\kappa(p)-\kappa(q))}\,dq}\right) \sigma^n e^{-\frac{\sigma^2}{2}}\,d\sigma}{\int_0^\infty \log\left(\frac{1}{p}\int_0^p e^{\tau\sigma(\kappa(p)-\kappa(q))}\,dq\right) \sigma^n e^{-\frac{\sigma^2}{2}}\,d\sigma} \tag{41}$$

with $\tau = 0.05$. From (40) and (41), we have $\kappa^\lambda(1) = \int_0^1 \kappa(q)\,\lambda(q)\,dq / \int_0^1 \lambda(q)\,dq = -0.03$. On the other hand for risk measures ρ we use another utility $g(x) = 1 - e^{-x}$ with $\tau = 1$ in (39). Then we give a coherent risk measure ρ by $\rho = -\mathrm{AVaR}_p^\nu$ with a risk spectrum ν given by the downside estimate $g^{-1}(\frac{1}{p}\int_0^p g(\mathrm{VaR}_q(\cdot))\,dq) \approx \mathrm{AVaR}_p^\nu(\cdot)$. We discuss a case of risk probability $p = 0.05$ in the normal distribution, and then similarly we have $\kappa^\nu(0.05) = \int_0^{0.05} \kappa(q)\,\nu(q)\,dq / \int_0^{0.05} \nu(q)\,dq = -2.29701$. Hence we have $A_t = 13.534 > 0$ and $\Delta_t = 0.00230221 > 0$, and we can easily check $\kappa^\nu(0.05) < \kappa^\lambda(1) < -\sqrt{\Delta_t/A_t} = -0.0130425$. From (25) we also have $\underline{\delta}_t(p) = 0.533748$. Therefore now we take a *risk level* $\delta = 1$ *in the feasible range* $[0.533748, \infty)$.

Table 1. The expectations μ_t^i and the covariances σ_t^{ij} of immediate rewards.

μ_t^i		σ_t^{ij}	$j=1$	$j=2$	$j=3$	$j=4$
$i=1$	0.096	$i=1$	0.41	-0.08	-0.06	0.05
$i=2$	0.085	$i=2$	-0.08	0.39	-0.07	0.06
$i=3$	0.093	$i=3$	-0.06	-0.07	0.38	-0.05
$i=4$	0.087	$i=4$	0.05	0.06	-0.05	0.37

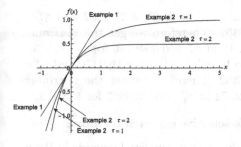

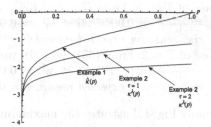

Fig. 1. Utility functions $f(x)$.

Fig. 2. Functions $\hat{\kappa}(p)$ and $\kappa^\lambda(p)$

From Theorem 1, the *maximum risk-sensitive reward in Problem* 4 *is* $\varphi_t^\circ = 0.0832788$ *at the reward* $\gamma_t^\circ = 0.092334$ in $[\gamma_t^-, \gamma_t^+] = [0.0855802, 0.0957352]$ with an *optimal strategy* $\pi_t^\circ = (0.391686, 0.179112, 0.36118, 0.0680223)$. Here Fig. 3 illustrates the maximum risk-sensitive reward φ_t° and the expected reward γ_t° for δ, which are connected at $\delta_t^+ = 0.600997$. We see φ_t° is smaller than γ_t° because γ_t° *implies actual expected rewards and the maximum risk-sensitive reward* φ_t° *contains decision maker's risk aversion under his utility.*

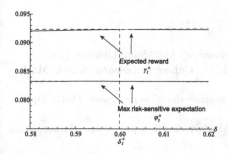

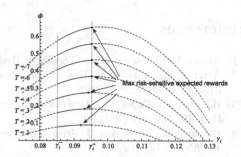

Fig. 3. Maximum risk-sensitive expected reward φ_t° and the expected reward γ_t°.

Fig. 4. Maximum Φ at γ_t^* ($T = 1, 2, \cdots, 7$)

We investigate a dynamic case with a terminal time T in Sect. 5. We need to find expected rewards γ_t^* in constraint $[\gamma_t^-, \gamma_t^+] = [0.0855802, 0.0957352]$. Hence we have $\kappa^\lambda(1)^2 = 0.0009 > \Delta_t/A_t = 0.00017$ for all $t = 1, 2, \cdots, T$ and $\kappa^\lambda(1)^2 > \sum_{t=1}^T \Delta_t/A_t$ for $T = 1, 2, \cdots, 5$. Now we check solutions for each T.

Case of $T \geq 4$: Inequality condition (38) holds since

$$\kappa^\lambda(1)^2 \max_{t \in \mathbb{T}} \left(\frac{A_t \gamma_t^+ - B_t}{\Delta_t} \right)^2 = 0.813024 < 0.910218 \leq T \times 0.227555$$

$$= \sum_{t \in \mathbb{T}} \frac{A_t(\gamma_t^+)^2 - 2B_t \gamma_t^+ + C_t}{\Delta_t}.$$

By Theorem 3 we get the *expected reward* $\gamma_t^* = \gamma_t^+ = 0.0957352$ for all $t = 1, 2 \cdots, T$. Then the *maximum risk-sensitive expected reward is* $\Phi = 0.354319, 0.446676, 0.539357, 0.632284$ respectively for $T = 4, 5, 6, 7$.

Case of $T = 3$: From (35) and (36), we have $\mathbb{T}^* = \emptyset$, $\theta^* = 23.8502$ and $\gamma_t^* = \frac{\Delta_t \theta^* + B_t}{A_t} = 0.0946796 < \gamma_t^+ = 0.0957352$. Therefore we get the *maximum risk-sensitive expected reward* $\Phi = 0.17215$ at $\gamma_t^* = 0.0946796$ for $t = 1, 2, 3$.

Case of $T = 2$: From (35) and (36), we also have $\mathbb{T}^* = \emptyset$, $\theta^* = 16.2476$ and $\gamma_t^* = \frac{\Delta_t \theta^* + B_t}{A_t} = 0.0933863 < \gamma_t^+ = 0.0957352$. Therefore we get the *maximum risk-sensitive expected reward* $\Phi = 0.262574$ at $\gamma_t^* = 0.0933863$ for $t = 1, 2$.

Finally Fig. 4 illustrates the maximum risk-sensitive expected rewards Φ.

Concluding remark. Maximization of risk-sensitive expected rewards in Problem 1 is a traditional and famous problem in Markov decision processes. Using Lemmas 1 and 2, we can incorporate the decision maker's risk averse attitude into coherent risk measures as weighting for average value-at-risks, and we apply it to Problem 1. Once spectra λ in (40) and $\kappa^\lambda(p)$ in Fig. 2 are prepared, we can obtain solutions in various optimization quickly. The proposed method brings quick and responsible decision making for artificial intelligence reasoning.

Acknowledgments. This research is supported from JSPS KAKENHI Grant Number JP 16K05282.

References

1. Arrow, K.J.: Essays in the Theory of Risk-Bearing. Markham, Chicago (1971)
2. Artzner, P., Delbaen, F., Eber, J.-M., Heath, D.: Coherent measures of risk. Math. Finance **9**, 203–228 (1999)
3. Bäuerle, N., Rieder, U.: More risk-sensitive Markov decision processes. Math. Oper. Res. **39**, 105–120 (2014)
4. Howard, R., Matheson, J.: Risk-sensitive Markov decision processes. Manag. Sci. **18**, 356–369 (1972)
5. Kusuoka, S.: On law-invariant coherent risk measures. Adv. Math. Econ. **3**, 83–95 (2001)
6. Rockafellar, R.T., Uryasev, S.: Optimization of conditional value-at-risk. J. Risk **2**, 21–41 (2000)
7. Tasche, D.: Expected shortfall and beyond. J. Bank. Finance **26**, 1519–1533 (2002)
8. Yoshida, Y.: A dynamic risk allocation of value-at-risks with portfolios. J. Adv. Comput. Intell. Intell. Inform. **16**, 800–806 (2012)
9. Yoshida, Y.: Maximization of returns under an average value-at-risk constraint in fuzzy asset management. Procedia Comput. Sci. **112**, 11–20 (2017)
10. Yoshida, Y.: Coherent risk measures derived from utility functions. In: Torra, V., Narukawa, Y., Aguiló, I., González-Hidalgo, M. (eds.) MDAI 2018. LNCS (LNAI), vol. 11144, pp. 15–26. Springer, Cham (2018). https://doi.org/10.1007/978-3-030-00202-2_2
11. Yoshida, Y.: Risk-sensitive decision making under risk constraints with coherent risk measures. In: Czarnowski, I. et al. (ed.) Smart Innovation, Systems and Technologies 143. Springer, Cham, June 2019, to appear
12. Yoshida, Y.: Portfolio optimization in fuzzy asset management with coherent risk measures derived from risk averse utility. Neural Computing and Applications. https://doi.org/10.1007/s00521-018-3683-y

Set-Based Extended Functions

Radko Mesiar[1], Anna Kolesárová[2], Adam Šeliga[1](✉), Javier Montero[3],
and Daniel Gómez[4]

[1] Faculty of Civil Engineering, Slovak University of Technology,
Radlinského 11, 810 05 Bratislava, Slovakia
{radko.mesiar,adam.seliga}@stuba.sk

[2] Faculty of Chemical and Food Technology, Slovak University of Technology,
Radlinského 9, 812 37 Bratislava, Slovakia
anna.kolesarova@stuba.sk

[3] Instituto de Matematica Interdisciplinar,
Departamento de Estadística e Investigación Operativa,
Fac. de Ciencias Matemáticas, Universidad Complutense de Madrid,
Plaza de las Ciencias 3, 28040 Madrid, Spain
monty@mat.ucm.es

[4] Departamento de Estadística y Ciencia de los Datos, Fac. de Estudios Estadísticos,
Universidad Complutense de Madrid, Av. Puerta de Hierro s/n, 28040 Madrid, Spain
dagomez@estad.ucm.es

Abstract. In this paper, inspired by the Zadeh approach to the fuzzy
connectives in fuzzy set theory and by some applications, we introduce
and study set-based extended functions on different universes. After presenting some results for set-based extended functions on a general universe, we focus our investigation on set-based extended functions on some
particular universes, including lattices and (bounded) chains. A special
attention is devoted to characterization of set-based extended aggregation functions on the unit interval $[0, 1]$.

Keywords: Aggregation function · Extended aggregation function ·
Extended function · Set-based extended aggregation function ·
Set-based extended function

1 Introduction

Lotfi Zadeh proposed in his seminal paper [13] to use the minimum and maximum
operators for modeling fuzzy intersection and fuzzy union, respectively. This
paper focuses on such kinds of fusion procedures that share with Zadeh's proposal
a particular property, namely, that these fuzzy connectives can be seen as func-
tions which, for any $n, m \in \mathbb{N}$ and any input vectors $\mathbf{x} = (x_1, \ldots, x_n) \in [0,1]^n$
and $\mathbf{z} = (z_1, \ldots, z_m) \in [0,1]^m$ such that the sets $\{x_1, \ldots, x_n\}$ and $\{z_1, \ldots, z_m\}$
coincide, provide for input vectors $\mathbf{x}$ and $\mathbf{z}$ the same output values, i.e.,

$$Min(\mathbf{x}) = Min(\mathbf{z}) \text{ and } Max(\mathbf{x}) = Max(\mathbf{z}).$$

© Springer Nature Switzerland AG 2019
V. Torra et al. (Eds.): MDAI 2019, LNAI 11676, pp. 41–51, 2019.
https://doi.org/10.1007/978-3-030-26773-5_4

In statistics, for a sample $(x_1, \ldots, x_n)$ several kinds of mean values have been introduced. For example, the arithmetic mean $AM(\mathbf{x}) = \frac{1}{n} \sum_{i=1}^{n} x_i$ is the minimizer of the sum of squares $\sum_{i=1}^{n} (x_i - a)^2$ (Least Squares Method). Minimizing the maximal deviation, i.e., looking for the minimizer of $\max\{|x_i - a| \mid i = 1, \ldots, n\}$ leads to the resulting mean M given by

$$M(\mathbf{x}) = \frac{\min\{x_1, \ldots, x_n\} + \max\{x_1, \ldots, x_n\}}{2}.$$

Observe that repeating or rearrangement of observations does not have any influence on the output of M, i.e., for example, taking a sample

$$\mathbf{z} = (x_1, x_1, x_1, x_2, x_2, x_3, \ldots, x_n),$$

we obtain $M(\mathbf{z}) = M(\mathbf{x})$.

Inspired by the mentioned observations, and taking into account that in most fusion problems the number of values to be fused cannot be fixed a priori, in this paper we will work with extended functions $F \colon \bigcup_{n \in \mathbb{N}} X^n \to X$, $X \neq \emptyset$, satisfying, in addition, the above discussed property. They will be called set-based extended functions on X (for the definition see below). Evidently, each such set-based extended function depends on the set $\{y_1, \ldots, y_k\}$ of values related to the input vector $(x_1, \ldots, x_n)$, where $\{x_1, \ldots, x_n\} = \{y_1, \ldots, y_k\}$ and $\mathrm{card}(\{y_1, \ldots, y_k\}) = k$. Hence, neither the repetition of arguments to be fused nor their rearrangement have any influence on the output result.

We will proceed as follows. First, we propose the concept of set-based extended functions defined for arbitrary but finitely many inputs from some non-empty universe X, with outputs also from X. In the beginning, we examine properties of set-based extended functions acting on a general universe X. The obtained results are contained in Sect. 2. The next section is devoted to the investigation of set-based extended functions on a (bounded) lattice X. In Sect. 4, X is considered to be a (bounded) chain. This section also contains a characterization of set-based extended aggregation functions on $X = [0, 1]$. Finally, some concluding remarks are added.

2   Set-Based Extended Functions on a General Universe

Suppose that we classify some products and their samples as *good* or *bad* only, i.e., we deal with the universe $X = \{g, b\}$. A function $F \colon \bigcup_{n \in \mathbb{N}} X^n \to X$ assigns to a sample $\mathbf{x} = (x_1, \ldots, x_n) \in X^n$ either the value *good*—if all the inputs $x_1, \ldots, x_n$ are *good*, or the value *bad*—in all other cases. The output value $F(\mathbf{x})$ depends on the set $\{x_1, \ldots, x_n\}$ only, namely,

$$F(x_1, \ldots, x_n) = \begin{cases} b & \text{if } b \in \{x_1, \ldots, x_n\}, \\ g & \text{otherwise.} \end{cases}$$

Moreover, if we add any other inputs $y_1, \ldots, y_k$, but such that each of them has already appeared in the original sample, i.e., $y_1, \ldots, y_k \in \{x_1, \ldots, x_n\}$, then

$$F(x_1, \ldots, x_n, y_1, \ldots, y_k) = F(x_1, \ldots, x_n).$$

In what follows, we formalize the above described situation, and define the notion of set-based extended function on a general universe X. We start by recalling the notion of extended function on X.

Definition 2.1. *Let $X \neq \emptyset$. Any function $F \colon \bigcup\limits_{n \in \mathbb{N}} X^n \to X$ will be called an extended function on X.*

Extended functions have open arity, i.e., they can work for any finite number of arguments.

Definition 2.2. *Let $X \neq \emptyset$. A function $F \colon \bigcup\limits_{n \in \mathbb{N}} X^n \to X$ is called a set-based extended function on X if $F(\mathbf{y}) = F(\mathbf{x})$ for any $n, k \in \mathbb{N}$ and all $\mathbf{x} = (x_1, \ldots, x_n) \in X^n$, $\mathbf{y} = (y_1, \ldots, y_k) \in X^k$, such that $\{x_1, \ldots, x_n\} = \{y_1, \ldots, y_k\}$.*

Example 2.1. Consider a set X with cardinality $\mathrm{card}(X) > 2$. Let E be a proper subset of X, and $a, b \in X$, $a \neq b$. Define $F_{E,a,b} \colon \bigcup\limits_{n \in \mathbb{N}} X^n \to X$ by

$$F_{E,a,b}(x_1, \ldots, x_n) = \begin{cases} a & \text{if } E \cap \{x_1, \ldots, x_n\} \neq \emptyset, \\ b & \text{otherwise.} \end{cases}$$

Then $F_{E,a,b}$ is a set-based extended function on X. Note that $F_{E,a,b}$ is associative if and only if $a \in E$, where the associativity of a function $F \colon \bigcup\limits_{n \in \mathbb{N}} X^n \to X$ means that

$$F(\mathbf{x}, \mathbf{y}) = F(F(\mathbf{x}), F(\mathbf{y}))$$

for all $\mathbf{x}, \mathbf{y} \in \bigcup\limits_{n \in \mathbb{N}} X^n$.

Example 2.1 is an example of a particular case of the construction of set-based extended functions described in the following proposition.

Proposition 2.1. *Let $X \neq \emptyset$. Let $\mathcal{P} = \{E_1, \ldots, E_k\}$ be a partition of X and $a_1, \ldots, a_k \in X$. Define $F \colon \bigcup\limits_{n \in \mathbb{N}} X^n \to X$ by*

$$F(\mathbf{x}) = a_i, \quad \text{where } i = \min\{j \in \{1, \ldots, k\} \mid \{x_1, \ldots, x_n\} \cap E_j \neq \emptyset\}. \quad (1)$$

Then F is a set-based extended function on X.

Example 2.2. Let $p \in \mathbb{N}$ and $X = \{1, \ldots, p\}$. Then

- if we consider the partition $\mathcal{P} = \{E_i\}_{i=1}^{p}$, where $E_i = \{i\}$, and $a_i = i$, then (1) defines the function $Min \colon \bigcup\limits_{n \in \mathbb{N}} X^n \to X$ given by $Min(x_1, \ldots, x_n) = \min\{x_1, \ldots, x_n\}$;

– if $\mathcal{P} = \{E_i\}_{i=1}^p$, where $E_i = \{p-i+1\}$ and $a_i = p-i+1$, then (1) yields the function Max, $Max(x_1, \ldots, x_n) = \max\{x_1, \ldots, x_n\}$.

Lemma 2.1. *Let $X \neq \emptyset$ and $\mathcal{H}(X) = \{\emptyset \neq E \subseteq X \mid E$ is finite$\}$. Then each set-based extended function F on X corresponds in a one-to-one correspondence to a set function $G\colon \mathcal{H}(X) \to X$ given, for each $E = \{x_1, \ldots, x_n\}$ in $\mathcal{H}(X)$, by*

$$G(E) = F(x_1, \ldots, x_n).$$

Clearly, $\mathcal{H}(X)$ is the power set of X except the empty set whenever X is finite.

Note that properties of the set function $G\colon \mathcal{H}(X) \to X$ can be transformed into new kinds of properties of the related set-based extended function F on X, as is shown in the following example.

Example 2.3. Consider $X = \mathbb{N}$ and define $G\colon \mathcal{H}(\mathbb{N}) \to \mathbb{N}$ by $G(E) = \sum\limits_{i \in E} i$.

Obviously, G is monotone non-decreasing, because for all E_1, E_2 in $\mathcal{H}(\mathbb{N})$, $G(E_1) \leq G(E_2)$ whenever $E_1 \subseteq E_2$. G is also additive, i.e.,

$$G(E_1 \cup E_2) = G(E_1) + G(E_2) \quad \text{whenever} \quad E_1 \cap E_2 = \emptyset.$$

The set-based extended function $F\colon \bigcup\limits_{n \in \mathbb{N}} \mathbb{N}^n \to \mathbb{N}$ corresponding to G, is given by

$$F(x_1, \ldots, x_n) = \sum_{i \in \mathbb{N}} i \cdot \min \left\{ 1, \sum_{j=1}^n \mathbf{1}_{\{i\}}(x_j) \right\},$$

and is neither monotone non-decreasing nor additive in the standard case, because, given any $n \in \mathbb{N}$, the relation $\mathbf{x} \leq \mathbf{y}$ does not imply $F(\mathbf{x}) \leq F(\mathbf{y})$ for all $\mathbf{x}, \mathbf{y} \in \mathbb{N}^n$, and similarly, the additivity property $F(\mathbf{x}+\mathbf{y}) = F(\mathbf{x}) + F(\mathbf{y})$ does not hold for all $\mathbf{x}, \mathbf{y} \in \mathbb{N}^n$.

However, F is monotone non-decreasing with respect to the partial order $\preceq$ on $\bigcup\limits_{n \in \mathbb{N}} \mathbb{N}^n$, defined as follows: for any $n, k \in \mathbb{N}$ and all $\mathbf{x} \in \mathbb{N}^n$, $\mathbf{y} \in \mathbb{N}^k$,

$$\mathbf{x} \preceq \mathbf{y} \text{ whenever } n \leq k \text{ and } x_i = y_i \text{ for all } i \leq n.$$

Indeed, then for all $\mathbf{x}, \mathbf{y} \in \bigcup\limits_{n \in \mathbb{N}} \mathbb{N}^n$, if $\mathbf{x} \preceq \mathbf{y}$ then $F(\mathbf{x}) \leq F(\mathbf{y})$.

Similarly, F is concatenation additive, i.e., if $\{x_1, \ldots, x_n\} \cap \{y_1, \ldots, y_k\} = \emptyset$, then $F(\mathbf{x}, \mathbf{y}) = F(\mathbf{x}) + F(\mathbf{y})$.

We still give another example illustrating Lemma 2.1.

Example 2.4. Consider $X = \{0, 1\}$. Then a function $F\colon \bigcup\limits_{n \in \mathbb{N}} \{0,1\}^n \to \{0, 1\}$ is an extended Boolean function. The cardinality of X is $\text{card}(X) = 2$, $\mathcal{H}(X) = \{\{0\}, \{1\}, \{0,1\}\}$, i.e., $\text{card}(\mathcal{H}(X)) = 3$, thus there are exactly $2^3 = 8$ set functions $G_i\colon \mathcal{H}(X) \to \{0,1\}$, $i = 1, \ldots, 8$. Consequently, there are 8 set-based extended Boolean functions F_i, where F_i corresponds to G_i by Lemma 2.1. The results are summarized in Table 1.

Table 1. Set-based extended Boolean functions

$G_i \backslash E$	{0}	{1}	{0,1}	$F_i(\mathbf{x})$
G_1	0	0	0	0
G_2	0	0	1	$\bigvee_{j,k} \lvert x_j - x_k \rvert$
G_3	0	1	0	$\bigwedge_j x_j$
G_4	0	1	1	$\bigvee_j x_j$
G_5	1	0	0	$1 - F_4(\mathbf{x})$
G_6	1	0	1	$1 - F_3(\mathbf{x})$
G_7	1	1	0	$1 - F_2(\mathbf{x})$
G_8	1	1	1	$1 - F_1(\mathbf{x})$

Proposition 2.2. *Fix* $X = \{1, 2, \ldots, k\}$. *Consider a permutation* $\sigma \colon X \to X$ *and a total order* $\preceq_\sigma$ *on* X *determined by* σ, *given by*

$$x \preceq_\sigma y \text{ if and only if } \sigma^{-1}(x) \le \sigma^{-1}(y).$$

Let $G_\sigma \colon \mathcal{H}(X) \to X$, $G_\sigma(E) = \min_{\preceq_\sigma}\{x \mid x \in E\}$. *Then the set-based extended function* $F_\sigma \colon \bigcup_{n \in \mathbb{N}} X^n \to X$, $F_\sigma(\mathbf{x}) = G_\sigma(\{x_1, \ldots, x_n\})$, *is symmetric, associative, and with neutral element* $e = \sigma(n)$, *but in general,* F_σ *need not be monotone.*

Recall that $e \in X$ is a neutral element of an extended function F on X, if for all $n \in \mathbb{N}$, and all $\mathbf{x} \in X^n$, with $e = x_i$ for some $i \in \{1, \ldots, n\}$, we have

$$F(x_1, \ldots, x_{i-1}, e, x_{i+1}, \ldots, x_n) = F(x_1, \ldots, x_{i-1}, x_{i+1}, \ldots, x_n).$$

Obviously, in Proposition 2.2, there are $k!$ set-based extended functions F_σ.

Remark 2.1. In Proposition 2.2, if for each $x, y \in X$,

$$x < y < e \implies \sigma^{-1}(x) < \sigma^{-1}(y) \text{ and } x > y > e \implies \sigma^{-1}(x) < \sigma^{-1}(y),$$

then F_σ is an idempotent uninorm (and only in that case). There are 2^{k-1} idempotent uninorms on X.

Note that the previous result for idempotent uninorms was also proved by Zemánková in [12].

We now summarize some properties related to general set-based functions.

Proposition 2.3. *Let* $X \neq \emptyset$. *Set-based extended functions on* X *have the following properties.*

(i) *Each set-based extended function on* X *is symmetric.*

(ii) For any function $V: X^k \to X$ and any set-based extended functions $F_1, \ldots, F_k$ on X, also the composite $F = V(F_1, \ldots, F_k): \bigcup_{n \in \mathbb{N}} X^n \to X$ is a set-based extended function on X.

(iii) For any function $V: X \to X$ and a any set-based extended function F on X, also the composites $V(F), F(V): \bigcup_{n \in \mathbb{N}} X^n \to X$, given by

$$V(F)(\mathbf{x}) = V(F(\mathbf{x})) \quad \text{and} \quad F(V)(\mathbf{x}) = F(V(x_1), \ldots, V(x_n)),$$

respectively, are set-based extended functions.

Proposition 2.4. Let $X_i \neq \emptyset$, $i = 1, \ldots, k$, and let X be the Cartesian product of X_i, $X = X_1 \times \cdots \times X_k$. For any set-based extended functions F_i on X_i, $i = 1, \ldots, k$, the function $F: \bigcup_{n \in \mathbb{N}} X^n \to X$, defined by

$$F((x_1^{(1)}, \ldots, x_k^{(1)}), \ldots, (x_1^{(n)}, \ldots, x_k^{(n)})) = (F_1(x_1^{(1)}, \ldots, x_1^{(n)}), \ldots, F_k(x_k^{(1)}, \ldots, x_k^{(n)})),$$

is a set-based extended function on X.

The following theorem shows that some algebraic properties of a function $F: \bigcup_{n \in \mathbb{N}} X^n \to X$ already ensure that F is a set-based extended function on X.

Theorem 2.1. Let $X \neq \emptyset$. Let $F: \bigcup_{n \in \mathbb{N}} X^n \to X$ be symmetric, idempotent and associative. Then F is a set-based extended function on X.

Proof: Let F satisfy the given assumptions. For any $n \in \mathbb{N}$ and each $\mathbf{x} = (x_1, \ldots, x_n) \in X^n$ with $\text{card}(\{x_1, \ldots, x_n\}) = k$, let $\{x_1, \ldots, x_n\} = \{y_1, \ldots, y_k\}$. Then there is a partition $\{I_1, \ldots, I_k\}$ of $\{x_1, \ldots, x_n\}$ given by

$$I_i = \{j \in \{1, \ldots, n\} \mid x_j = y_i\}.$$

Then, writing $I_i = \{j_{i1}, \ldots, j_{im_i}\}$, where $m_i = \text{card}(I_i)$, we have

$$
\begin{aligned}
F(\mathbf{x}) &= F(x_{j_{11}}, \ldots, x_{j_{1m_1}}, x_{j_{21}}, \ldots, x_{j_{2m_2}}, \ldots, x_{j_{k1}}, \ldots, x_{j_{km_k}}) \\
&= F(F(x_{j_{11}}, \ldots, x_{j_{1m_1}}), F(x_{j_{21}}, \ldots, x_{j_{2m_2}}), \ldots, F(x_{j_{k1}}, \ldots, x_{j_{km_k}})) \\
&= F(y_1, \ldots, y_k),
\end{aligned}
$$

where the first equality follows from the symmetry of F, the second one from its associativity, and the third one follows from the idempotency of F. Obviously, for all $\mathbf{x}, \mathbf{z} \in \bigcup_{n \in \mathbb{N}} X^n$, such that $\{x_1, \ldots, x_n\} = \{y_1, \ldots, y_k\} = \{z_1, \ldots, z_m\}$, we have $F(\mathbf{x}) = F(\mathbf{z})$, and hence F is a set-based extended function on X. $\square$

Note that neither idempotency nor associativity are necessary properties for being F a set-based extended function, see Example 2.1 and Proposition 2.1.

3   Set-Based Extended Functions on Lattices

In this section we consider X to be a carrier of a lattice $(X, \leq)$. For any fixed $a \in X$, we define a function $F_a \colon \bigcup_{n \in \mathbb{N}} X^n \to X$ by

$$F_a(\mathbf{x}) = \begin{cases} \bigvee_i x_i & \text{if } \bigvee_i x_i < a, \\ \bigwedge_i x_i & \text{if } \bigwedge_i x_i > a, \\ a & \text{otherwise.} \end{cases}$$

Obviously, F_a is symmetric and idempotent, and its associativity can also be verified. By Theorem 2.1, F_a is a set-based extended function on X. Moreover, F_a is monotone non-decreasing, and thus it is an extended aggregation function on X, see [7] (because of the idempotency of F_a we need not consider X to be a bounded lattice). Observe that if X is bounded, with top and bottom elements $\mathbf{1}_X$ and $\mathbf{0}_X$, respectively, then $F_{\mathbf{1}_X} = \vee$ is the standard join on X, and $F_{\mathbf{0}_X} = \wedge$ is the standard meet on X. By Theorem 2.1, any idempotent uninorm F on a bounded (distributive) lattice X [8], is a set-based extended function on X. Similarly, idempotent nullnorms on bounded lattices, see [9], are set-based extended functions.

Proposition 3.1. *Let $(X, \leq)$ be an ordinal sum of lattices $(X_i, \leq_i)_{i \in I}$, and let for any $i \in I$, $F_i \colon \bigcup_{n \in \mathbb{N}} X_i^n \to X_i$ be a set-based extended function on X_i. Define $F \colon \bigcup_{n \in \mathbb{N}} X^n \to X$ by*

$$F(x_1, \ldots, x_n) = F_i(y_1, \ldots, y_k),$$

where

$$i = \min\{j \in I \mid \{x_1, \ldots, x_n\} \cap X_j \neq \emptyset\},$$
$$k = \operatorname{card}(\{j \in \{1, \ldots, n\} \mid x_j \in X_i\}),$$
$$\{y_1, \ldots, y_k\} = \{x_j \mid x_j \in X_i\}.$$

Then F is a set-based extended function on X. Moreover, F is monotone non-decreasing if and only if all F_i, $i \in I$, are of that property, and it is idempotent if and only if all F_i, $i \in I$, are idempotent.

More information on ordinal sum of lattices can be found, e.g., in [3].

4   Set-Based Extended Aggregation Functions on Chains

In this section we consider X to be a (bounded) chain. A total order on X has an important impact on characterization of monotone set-based extended functions on X.

Proposition 4.1. *Let X be a chain. Then $F\colon \bigcup_{n\in\mathbb{N}} X^n \to X$ is a monotone non-decreasing (non-increasing) set-based extended function if and only if for each $\mathbf{x} \in \bigcup_{n\in\mathbb{N}} X^n$ we have*

$$F(\mathbf{x}) = D(Min(\mathbf{x}), Max(\mathbf{x})), \tag{2}$$

for some monotone non-decreasing (non-increasing) function $D\colon X^2 \to X$.

Proof: It is not difficult to see that representation of F in the form (2) is sufficient for being F a monotone non-decreasing (non-increasing) set-based extended function on X. We only prove a necessary condition.

Let F be a monotone non-decreasing set-based extended function on a chain X. As F is symmetric, with no loss of generality, we can only consider elements $\mathbf{x} \in \bigcup_{n\in\mathbb{N}} X^n$ such that $x_1 \leq \cdots \leq x_n$. Then $x_1 = Min(\mathbf{x})$, $x_n = Max(\mathbf{x})$ and we can write

$$F(x_1, x_n) = F(x_1, \ldots, x_1, x_n) \leq F(x_1, x_2, \ldots, x_{n-1}, x_n) \leq F(x_1, x_n, \ldots, x_n)$$
$$= F(x_1, x_n), \tag{3}$$

which yields $F(\mathbf{x}) = F(Min(\mathbf{x}), Max(\mathbf{x}))$. Putting $D = F|_{X^2}$, we obtain the required representation in the form (2). The monotonicity of D follows from the monotonicity of F. To get the result for a monotone non-increasing F, it is enough to reverse the inequalities in (3).   □

Now we provide a characterization of set-based extended aggregation functions acting on a bounded chain X, in particular on $X = [0, 1]$. In what follows, we only recall the notion of extended aggregation function on $[0, 1]$, for more details on (extended) aggregation functions and their properties we recommend, e.g., [4,7,10], see also [1,2].

Definition 4.1. *A function $A\colon \bigcup_{n\in\mathbb{N}} [0, 1]^n \to [0, 1]$ is an extended aggregation function on $[0, 1]$ if A is monotone non-decreasing and satisfies the boundary conditions, i.e.,*

(i) for all elements $\mathbf{0} = (0, \ldots, 0), \mathbf{1} = (1, \ldots, 1) \in \bigcup_{n\in\mathbb{N}} [0, 1]^n$, $A(\mathbf{0}) = 0$ and $A(\mathbf{1}) = 1$;

(ii) for all $\mathbf{x}, \mathbf{y} \in \bigcup_{n\in\mathbb{N}} [0, 1]^n$ we have $A(\mathbf{x}) \leq A(\mathbf{y})$ whenever $\mathbf{x} \leq \mathbf{y}$.

Note that for $\mathbf{x}, \mathbf{y} \in \bigcup_{n\in\mathbb{N}} [0, 1]^n$ we have $\mathbf{x} \leq \mathbf{y}$ if and only if $\mathbf{x}$ and $\mathbf{y}$ are n-tuples of the same arity n satisfying $x_i \leq y_i$ for each $i = 1, \ldots, n$.

We will also work with n-ary aggregation functions on $[0, 1]$, i.e., functions

$$A_{(n)}\colon [0, 1]^n \to [0, 1]$$

which satisfy boundary conditions (i) and monotonicity conditions (ii) from Definition 4.1 for a considered fixed $n \in \mathbb{N}$. Clearly, given an extended aggregation function A on $[0, 1]$, the function $A_{(n)} = A|_{[0,1]^n}$ is an n-ary aggregation function.

Definition 4.2. *A function* $A: \bigcup_{n \in \mathbb{N}} [0,1]^n \to [0,1]$ *is a set-based extended aggregation function if A is an extended aggregation function on $[0,1]$ satisfying the set-based property, i.e., for all $n, k \in \mathbb{N}$, and all $\mathbf{x} = (x_1, \ldots, x_n) \in [0,1]^n$ and $\mathbf{y} = (y_1, \ldots, y_k) \in [0,1]^k$, $A(\mathbf{y}) = A(\mathbf{x})$ whenever $\{x_1, \ldots, x_n\} = \{y_1, \ldots, y_k\}$.*

It can be shown that set-based extended aggregation functions on $[0,1]$ can be completely characterized as follows.

Theorem 4.1. *Let $A: \bigcup_{n \in \mathbb{N}} [0,1]^n \to [0,1]$ be an extended aggregation function on $[0,1]$. A is a set-based extended aggregation function on $[0,1]$ if and only if for all $\mathbf{x} \in \bigcup_{n \in \mathbb{N}} [0,1]^n$ we have*

$$A(\mathbf{x}) = A(Min(\mathbf{x}), Max(\mathbf{x})). \tag{4}$$

For more results on set-based extended aggregation functions on $[0,1]$, see [11].

By the previous theorem, set-based extended aggregation functions on $[0,1]$ are generated by binary aggregation functions; there is a one-to-one correspondence between the set of all set-based extended aggregation functions and the set of all symmetric binary aggregation functions. Observe that in the case of an associative symmetric binary aggregation function $A: [0,1]^2 \to [0,1]$ there are two possible ways how to extend it into an extended aggregation function. On the one hand, based on formula (2), one can define the function $A_\square: \bigcup_{n \in \mathbb{N}} [0,1]^n \to [0,1]$ by

$$A_\square(\mathbf{x}) = A(Min(\mathbf{x}), Max(\mathbf{x})),$$

and on the other hand, using the associativity of A, one can define the function $A_\triangle: \bigcup_{n \in \mathbb{N}} [0,1]^n \to [0,1]$ by

$$A_\triangle(x_1) = x_1, \quad A_\triangle(x_1, x_2) = A(x_1, x_2),$$

and for all $n \geq 3$,

$$A_\triangle(x_1, \ldots, x_n) = A(A_\triangle(x_1, \ldots, x_{n-1}), x_n).$$

Due to Proposition 2.1, $A_\square = A_\triangle$ if and only if a binary aggregation function A is idempotent, i.e., $A(x,x) = x$ for all $x \in [0,1]$. Note that this is, e.g., the case of idempotent uninorms [6,12], and also the case of idempotent nullnorms [5] (compare F_a introduced in Sect. 3). As a negative example, consider the standard product $A(x_1, x_2) = x_1 x_2$. Then $A_\triangle(x_1, \ldots, x_n) = \prod_{i=1}^{n} x_i$ is the standard product, which, if $n \neq 2$, differs from $A_\square(\mathbf{x}) = (Min(\mathbf{x})) \cdot Max(\mathbf{x})$.

5 Concluding Remarks

In this paper, we have introduced and discussed set-based extended functions, which can be seen as a generalization of extended functions $F\colon \bigcup_{n\in\mathbb{N}} X^n \to X$, which are symmetric, idempotent and associative. In the case when X is a lattice, the introduced set-based extended functions can be viewed as a particular generalization of joins, meets, idempotent uninorms and idempotent nullnorms. In the case of bounded chains, we have shown the existence of a one-to-one correspondence between set-based aggregation functions A and symmetric binary aggregation functions D given by

$$A(\mathbf{x}) = D(Min(\mathbf{x}), Max(\mathbf{x})).$$

Based on the presented approach, in our future research we intend to solve how to relate aggregation of input values $x_1, \ldots, x_n$ to aggregation of inputs $x_1, \ldots, x_n, x_{n+1}, \ldots, x_{n+k}$, where $x_{n+1}, \ldots, x_{n+k}$ are some additionally obtained observations.

Acknowledgement. R. Mesiar and A. Šeliga kindly acknowledge the support of the grant VEGA 1/0006/19, and A. Kolesárová is grateful for the support of the grant VEGA 1/0614/18. All these three authors also acknowledge the support of the project of Science and Technology Assistance Agency under the contract No. APVV–18–0052. D. Gómez and J. Montero kindly acknowledge the support of the projects TIN205-66471-P (Government of Spain), S2013/ICE-2845 (State of Madrid) and Complutense University research group GR3/14-910149. Moreover, the authors thank M. Botur for inspirative personal discussion.

References

1. Beliakov, G., Pradera, A., Calvo, T.: Aggregation Functions: A Guide for Practitioners. Springer, Heidelberg (2007). https://doi.org/10.1007/978-3-540-73721-6
2. Beliakov, G., Bustince, H., Calvo, T.: A Practical Guide to Averaging Functions. Springer, Heidelberg (2016). https://doi.org/10.1007/978-3-319-24753-3
3. Birkhoff, G.: Lattice Theory, 3rd edn. American Mathematical Society, Providence (1973). Sec. Printing
4. Calvo, T., Kolesárová, A., Komorníková, M., Mesiar, R.: Aggregation operators: properties, classes and construction methods. In: Calvo, T., Mayor, G., Mesiar, R. (eds.) Aggregation Operators, pp. 3–107. Physica, Heidelberg (2002). https://doi.org/10.1007/978-3-7908-1787-4_1
5. Calvo, T., De Baets, B., Fodor, J.: The functional equations of Frank and Alsina for uninorms and nullnorms. Fuzzy Sets Syst. **120**, 385–394 (2001)
6. De Baets, B.: Idempotent uninorms. Eur. J. Oper. Res. **180**, 631–642 (1999)
7. Grabisch, M., Marichal, J.-L., Mesiar, R., Pap, E.: Aggregation Functions. Cambridge University Press, Cambridge (2009)
8. Karaçal, F., Mesiar, R.: Uninorms on bounded lattices. Fuzzy Sets Syst. **261**, 33–43 (2015)

9. Karaçal, F., Akif Ince, M., Mesiar, R.: Nullnorms on bounded lattices. Inf. Sci. **325**, 227–236 (2015)
10. Mesiar, R., Kolesárová, A., Komorníková, M., Calvo, T.: Aggregation functions on [0, 1]. In: Kacprzyk, J., Pedrycz, W. (eds.) Handbook of Computational Intelligence, pp. 61–73. Springer, Heidelberg (2015). https://doi.org/10.1007/978-3-662-43505-2_4
11. Mesiar, R., Kolesárová, A., Gómez, D., Montero, J.: Set-based extended aggregation functions. Int. J. Intell. Syst. (2019, accepted)
12. Mesiarová-Zemánková, A.: A note on decomposition of idempotent uninorms into an ordinal sum of singleton semigroups. Fuzzy Sets Syst. **299**, 140–145 (2016)
13. Zadeh, L.A.: Fuzzy sets. Inform. Control **8**, 338–353 (1965)

Fuzzy Confirmation Measures (a)symmetry Properties

Emilio Celotto[1], Andrea Ellero[1] , and Paola Ferretti[2(✉)]

[1] Department of Management, Ca' Foscari University of Venice, Venice, Italy
emi.web@tin.it, ellero@unive.it
[2] Department of Economics, Ca' Foscari University of Venice, Venice, Italy
ferretti@unive.it

Abstract. While Bayesian Confirmation Measures assess the degree to which an antecedent E supports a conclusion H in a rule $E \Rightarrow H$ by means of probabilities, Fuzzy Confirmation Measures evaluate the quality of fuzzy association rules between the fuzzy antecedent A and fuzzy consequence B. Fuzzy Confirmation Measures defined in terms of confidence can be compared in different ways, among them symmetry properties evaluations play an important role. We first focus on symmetry properties for Fuzzy Confirmation Measures and then on the evaluation of possible levels of asymmetry. We suggest a way to measure the level of asymmetry and we also provide some examples to illustrate its possible use.

Keywords: Fuzzy Association Rules · Fuzzy Confirmation Measures · Symmetries · Asymmetry degree

1 Introduction

The fact that a large number of records in a database that possess a (set of) attribute(s) A, possess also a (set of) attribute(s) B, highlighting a regularity in the dataset, can be expressed in terms of an association rule $A \Rightarrow B$ that turns out to be relevant when exceptions to it are *rare* (see [1, 22]).

The relevance of association rules hidden in a dataset can be evaluated with the so called interestingness measures; among them two remarkable examples are *support* and *confidence*.

To assess the relevance of association rules, a class of interestingness measures is the so called family of Bayesian Confirmation Measures (BCMs); since they were defined in different contexts, many attempts can be found in literature aimed at understanding which one performs better, in which context and with respect to which criteria. Some relevant criteria suggested to classify a BCM as preferable (or not preferable) are symmetry properties as proposed and discussed, e.g., in [10, 12, 19, 21].

© Springer Nature Switzerland AG 2019
V. Torra et al. (Eds.): MDAI 2019, LNAI 11676, pp. 52–63, 2019.
https://doi.org/10.1007/978-3-030-26773-5_5

The notion of Bayesian Confirmation Measure inspired the definition of Fuzzy Confirmation Measures (FCMs) for fuzzy association rules due to Glass [15]: while in the case of crisp association rules a record can possess or not an attribute, fuzzy association rules allow a record to possess an attribute with a certain *degree*. Rules become in this case much more flexible in describing information hidden in the data and new interestingness measures can be defined in order to assess their relevance. This implies, in particular, a new definition of support and of confidence of the association rule. Besides Fuzzy Confirmation Measures, Glass also introduces symmetry properties for Fuzzy Confirmation Measures, suggesting which of them should be considered as desirable and which should not.

In this paper we will further discuss symmetry properties for Fuzzy Confirmation Measures, expanding the set of symmetries to be considered and, for measures that are not symmetric, providing a way to measure the degree of their asymmetry.

We will recall in Sect. 2 the definitions of Bayesian and of Fuzzy Confirmation Measures, furthermore some new FCMs will be proposed. Symmetry properties will be presented and studied in Sect. 3, while in Sect. 4 we will suggest a way to evaluate the degree of asymmetry of FCMs, providing some examples of its computation. Concluding remarks are presented in Sect. 5.

2 Bayesian and Fuzzy Confirmation Measures

As mentioned, Bayesian Confirmation Measures (BCMs) are interestingness measures aimed at evaluating the degree to which an antecedent E supports or contradicts the conclusion H, using prior probability $Pr(H)$, posterior probability $Pr(H|E)$ and $Pr(E)$, the probability of antecedent E. The emerging of antecedent E may change the knowledge about the occurrence of H, since conclusion H may be confirmed when $Pr(H|E) > Pr(H)$, or disconfirmed when $Pr(H|E) < Pr(H)$. This way, it is natural to define a measure of confirmation as follows (see, e.g., [11,14]).

Definition 1. *A function c of antecedent E and conclusion H, is a Bayesian Confirmation Measure (BCM) when*

$$c(E,H) > 0 \text{ if } Pr(H|E) > Pr(H) \qquad \textit{(confirmation case)}$$
$$c(E,H) = 0 \text{ if } Pr(H|E) = Pr(H) \qquad \textit{(neutrality case)}$$
$$c(E,H) < 0 \text{ if } Pr(H|E) < Pr(H) \qquad \textit{(disconfirmation case)}$$

Very well-known BCMs are, e.g.,

$$d(E,H) = Pr(H|E) - Pr(H) \qquad K(E,H) = \frac{Pr(E|H) - Pr(E|\neg H)}{Pr(E|H) + Pr(E|\neg H)} \qquad (1)$$

defined by Carnap in [3] and by Kemeny and Oppenheim in [23], respectively.

If a BCM evaluates the degree to which an antecedent E supports or contradicts the hypothesis H, its fuzzy counterpart, i.e. a Fuzzy Confirmation Measure (FCM), is required to weigh the strength of a fuzzy association rule $A \Rightarrow B$, namely a rule for which a record admitting a fuzzy antecedent A also has fuzzy consequence B (see Glass in [15]). Before recalling the definition of a FCM it is necessary to fix the notion of confidence of a fuzzy association rule.

Given a set of records $X = \{x_1, x_2, \ldots, x_n\}$ and a set $\mathcal{A}$ of attributes, each attribute A_i $(i = 1, \ldots, m)$ in $\mathcal{A}$ can be represented as a fuzzy set defined by a membership function $A_i : X \to [0, 1]$, where $A_i(x_j)$ denotes the degree to which the ith attribute A_i applies to record x_j. Similarly, the standard negator $n(A_i) :$ $X \to [0, 1]$ defined by $n(A_i(x_j)) = 1 - A_i(x_j)$ identifies the fuzzy complement set $n(A_i)$ of the fuzzy set A_i.

This way, given the set $\mathcal{F}$ of all the fuzzy sets in X, a fuzzy association rule $A \Rightarrow B$ between rule antecedent A and rule consequence B, where $A, B \in \mathcal{F}$, can be evaluated by measures like *support* (*supp*) and *confidence* (*conf*):

$$supp(A \Rightarrow B) = \frac{\sum_{x \in X} A(x) \otimes B(x)}{|X|} \qquad conf(A \Rightarrow B) = \frac{\sum_{x \in X} A(x) \otimes B(x)}{\sum_{x \in X} A(x)}$$

where $\otimes$ denotes a t-norm used to define the intersection of fuzzy sets (see [15])[1].

It is precisely the definition of confidence for the fuzzy association rule $A \Rightarrow B$ that is used in [15] to propose a definition of Fuzzy Confirmation Measure.

Since in the crisp case the confidence measure can be considered as an estimate of the conditional probability $Pr(B|A)$, based on the relative frequency of B given A, in [15] Glass considers $conf(A \Rightarrow B)$ as the fuzzy counterpart of $Pr(B|A)$ and the confidence of the default rule $T \Rightarrow A$ as the fuzzy counterpart of the estimate of $Pr(A)$[2].

In the Bayesian confirmation setting the fact that the antecedent E confirms the conclusion H can be equivalently expressed by one of the following inequalities (see e.g. [24])

a. $Pr(H|E) > Pr(H)$
b. $Pr(H|E) > Pr(H|\neg E)$
c. $Pr(E|H) > Pr(E)$
d. $Pr(E|H) > Pr(E|\neg H)$.

These conditions constitute different perspectives of confirmation (*Bayesian confirmation, strong Bayesian confirmation, likelihoodist confirmation* and *strong likelihoodist confirmation*, respectively, see [22]) with different philosophical motivations. Nevertheless, from the logical point of view, those perspectives are equivalent (provided they do not lead to undefined values). It is with reference to this aspect that the idea of considering confidence as the fuzzy counterpart of probability requires some caution. In fact, the choice of the t-norm used to define the confidence $conf(A \Rightarrow B)$ has a direct impact on the equivalence among the fuzzy counterparts of the four above defined perspectives:

a. $conf(A \Rightarrow B) > conf(T \Rightarrow B)$

---

[1] Throughout the paper, the formulas are assumed to be well defined, i.e. we assume as granted that denominators do not vanish.
[2] Here T is the totally true constant, that is $T(x) = 1 \, \forall x \in X$.

b. $conf(A \Rightarrow B) > conf(n(A) \Rightarrow B)$
c. $conf(B \Rightarrow A) > conf(T \Rightarrow A)$
d. $conf(B \Rightarrow A) > conf(n(B) \Rightarrow A)$

provided $conf(T \Rightarrow A)$ and $conf(T \Rightarrow B)$ are neither 0 nor 1.

In particular, the equivalence among inequalities $conf(A \Rightarrow B) > conf(T \Rightarrow B)$ and $conf(B \Rightarrow A) > conf(T \Rightarrow A)$ is verified by fuzzy confidence measures that satisfy *weak dependence* (see [15]). If in addition also $conf(A \Rightarrow B) > conf(T \Rightarrow B)$ is equivalent to $conf(A \Rightarrow B) > conf(n(A) \Rightarrow B)$, the fuzzy confidence measure is said to satisfy *strong dependence*. The use of the product t-norm ensures weak and strong dependence: it was therefore chosen by Glass to set the definition of a Fuzzy Confirmation Measure.

Definition 2. *A Fuzzy Confirmation Measure of the degree to which a fuzzy set A confirms a fuzzy set B is a function $c_f : \mathcal{F} \times \mathcal{F} \to \mathbb{R}$ that satisfies:*

(i) $c_f(A, B) > 0$ *if* $conf(A \Rightarrow B) > conf(n(A) \Rightarrow B)$ *(confirmation case)*
(ii) $c_f(A, B) = 0$ *if* $conf(A \Rightarrow B) = conf(n(A) \Rightarrow B)$ *(neutrality case)*
(iii) $c_f(A, B) < 0$ *if* $conf(A \Rightarrow B) < conf(n(A) \Rightarrow B)$ *(disconfirmation case)*

where conf is the product based fuzzy confidence measure

$$conf(A \Rightarrow B) = \frac{\sum_{x \in X} A(x) \cdot B(x)}{\sum_{x \in X} A(x)}.$$

In the following we assume that $conf(A \Rightarrow B)$ is defined by means of the product t-norm. Fuzzy Confirmation Measures can be obtained by adapting the analogous Bayesian confirmation measures to the fuzzy environment [15]:

$$d(A, B) = conf(A \Rightarrow B) - conf(T \Rightarrow B) \tag{2}$$

whose corresponding BCM was defined by Carnap [3],

$$G(A, B) = \log \left[\frac{conf(B \Rightarrow A)}{conf(n(B) \Rightarrow A)} \right] \tag{3}$$

which recalls the BCM proposed by Good [18],

$$K(A, B) = \frac{conf(B \Rightarrow A) - conf(n(B) \Rightarrow A)}{conf(B \Rightarrow A) + conf(n(B) \Rightarrow A)} \tag{4}$$

which corresponds to the BCM defined by Kemeny and Oppenheim [23], and

$$Z(A, B) = \begin{cases} Z_1(A, B) = \dfrac{conf(A \Rightarrow B) - conf(T \Rightarrow B)}{1 - conf(T \Rightarrow B)} & \text{in case of confirmation} \\[2ex] Z_2(A, B) = \dfrac{conf(A \Rightarrow B) - conf(T \Rightarrow B)}{conf(T \Rightarrow B)} & \text{otherwise.} \end{cases} \tag{5}$$

The last measure, Z, was defined in the BCM framework by Rescher [27] and further analysed in [11] and [20]. Remark that, besides d and Z, with the help of some algebraic manipulation also G and K can be expressed in terms of $conf(T \Rightarrow B)$ and $conf(A \Rightarrow B)$ only.

The FCMs that can be written as functions of $conf(A \Rightarrow B)$ and $conf(T \Rightarrow B)$ only, constitute a special class of FCMs, let'us call them IFC (Initial Final Confidence) Fuzzy Confirmation Measures.[3] In order to provide a set of FCMs to be used as *guinea pigs* in the next sessions, we add some FCMs that are not necessarily IFC Fuzzy Confirmation Measures.

A first example of a not-IFC Fuzzy Confirmation Measure is

$$b(A, B) = [conf(A \Rightarrow B) - conf(T \Rightarrow B)]conf(T \Rightarrow A), \qquad (6)$$

proposed in [15], in which a third variable, namely $conf(T \Rightarrow A)$, needs to be considered in the definition.

Other examples of not-IFC measures are

$$M(A, B) = conf(B \Rightarrow A) - conf(T \Rightarrow A) \qquad (7)$$

and

$$N(A, B) = conf(B \Rightarrow A) - conf(n(B) \Rightarrow A) \qquad (8)$$

which correspond in the BCM framework to the measures $M(E, H)$ and $N(E, H)$ proposed in [25] and [26]. The measures M and N cannot be rewritten using only $conf(A \Rightarrow B)$ and $conf(T \Rightarrow B)$, but can be formulated in terms of $conf(A \Rightarrow B)$, $conf(T \Rightarrow B)$ and $conf(T \Rightarrow A)$ as

$$M(A, B) = conf(T \Rightarrow A) \frac{conf(A \Rightarrow B) - conf(T \Rightarrow B)}{conf(T \Rightarrow B)}$$

and

$$N(A, B) = conf(T \Rightarrow A) \frac{conf(A \Rightarrow B) - conf(T \Rightarrow B)}{conf(T \Rightarrow B)(1 - conf(T \Rightarrow B))}.$$

Further fuzzy measures can be defined referring to other BCMs (for the corresponding measures in the BCM setting, see [9,13,22,28]):

$$F(A, B) = [conf(A \Rightarrow B) - conf(T \Rightarrow B)]/conf(T \Rightarrow B) \qquad (9)$$

$$C(A, B) = \frac{conf(A \Rightarrow B)conf(T \Rightarrow A) + conf(n(A) \Rightarrow n(B))conf(T \Rightarrow n(A)}{1 - conf(T \Rightarrow A)conf(T \Rightarrow B) - conf(T \Rightarrow n(A))conf(T \Rightarrow n(B))}$$
$$- \frac{conf(T \Rightarrow A)conf(T \Rightarrow B) + conf(T \Rightarrow n(A))conf(T \Rightarrow n(B))}{1 - conf(T \Rightarrow A)conf(T \Rightarrow B) - conf(T \Rightarrow n(A))conf(T \Rightarrow n(B))} \qquad (10)$$

$$R(A, B) = [conf(T \Rightarrow n(B)) - conf(A \Rightarrow n(B))]/conf(T \Rightarrow n(B)) \qquad (11)$$

$$Gr(A, B) = \sqrt{conf(B \Rightarrow A)} - \sqrt{conf(T \Rightarrow A)} \qquad (12)$$

---

[3] The IFC definition recalls the analogous class of BCMs that can be written as functions of $Pr(H|E)$ and $Pr(H)$ only, which are called IFPD (Initial Final Probability Dependence) confirmation measures [11].

Theorem 1. *The measures $C(A,B)$, $F(A,B)$, $Gr(A,B)$, $M(A,B)$, $N(A,B)$ and $R(A,B)$ are Fuzzy Confirmation Measures.*

Proof. We present the proof for the measure $M(A,B)$.
By definition, $M(A,B) > 0$ if and only if

$$conf(B \Rightarrow A) > conf(T \Rightarrow A)$$

that is, by weak dependence property which is satisfied by the product-based confidence measure $conf$, if and only if

$$conf(A \Rightarrow B) > conf(T \Rightarrow B).$$

Moreover, by strong dependence definition proposed in [15] and the assumption of product t-norm in the definition of $conf$, the previous inequality is satisfied if and only if

$$conf(A \Rightarrow B) > conf(n(A) \Rightarrow B).$$

Therefore the measure M satisfies the sign condition which is required in the case of confirmation. Analogous observations can be used to prove the sign conditions for the case of neutrality and of disconfirmation.

□

3 Symmetries

Symmetry properties of confirmation measures have been widely studied in the literature (see, e.g., [20] and [17]), discussing on the reasons why some of them should be required while some other ones should be considered as undesirable. The main symmetry properties have been introduced in [3] and then investigated in [19]. In [10] the list of symmetry properties has been extended to the set specified in the following definition[4].

Definition 3. *A confirmation measure $c(E,H)$ satisfies*

Evidence Symmetry (ES) if $c(E,H) = -c(\neg E, H)$;
Hypothesis Symmetry (HS) if $c(E,H) = -c(E, \neg H)$;
Evidence Hypothesis Symmetry (EHS) if $c(E,H) = c(\neg E, \neg H)$;
Inversion Symmetry (IS) if $c(E,H) = c(H,E)$;
Evidence Inversion Symmetry (EIS) if $c(E,H) = -c(H, \neg E)$;
Hypothesis Inversion Symmetry (HIS) if $c(E,H) = -c(\neg H, E)$;
Evidence Hypothesis Inversion Symmetry (EHIS) if $c(E,H) = c(\neg H, \neg E)$.

---

[4] *Inversion Symmetry (IS) is also called Commutativity Symmetry (see e.g. [12]).*

Let us observe that adding to the set of the symmetries listed in Definition 3 the *trivial symmetry* $c(E, H) = c(E, H)$, the set can be seen as a group isomorphic to the dihedral group D_8 (see [2,6,30]). As a consequence, symmetry properties are related; for example, if ES and IS are satisfied then also EIS holds.

As mentioned, several authors have analyzed the desirability and, conversely, the undesirability of symmetry properties of confirmation measures. Among them, Eells and Fitelson [12] and Glass [16] stated that an *acceptable* measure should not exhibit symmetries ES, EHS and IS while HS represents the only desirable property in the context of probabilistic confirmation measures and in the context of association rule strength. Note that $EHIS$, EIS and HIS symmetries have been considered and that those Authors implicitly assert that the measures that solely satisfy HS are acceptable. Carnap's d is an example of an acceptable confirmation measure in Eells and Fitelson's framework. Crupi et al. [10] widened the set of the symmetries and distinguished the cases of confirmation and disconfirmation in their analysis. They considered as desirable only HS, HIS and $EHIS$ in case of confirmation and only HS, EIS and IS in case of disconfirmation, while all other symmetries were be considered undesirable. Rescher's Z, defined in [27], is an example of BCM that meets all the symmetry requirements considered desirable in [10]. Observe that, again, only HS was considered a desirable symmetry in both confirmation/disconfirmation cases.

In the field of rule interestingness, Greco et al. [21] considered ES, EHS and HS as desirable properties, while they classified as undesirable all the other symmetry properties. This way, a good measure appears to be, for example, Nozick's confirmation measure $N(E, H) = Pr(E|H) - Pr(E|\neg H)$ [26].

Given that, depending on the context, different symmetry properties are considered as desirable, it is interesting to study which of them can coexist: this has been addressed in [6,8,30].

In the framework of fuzzy association rules, the same set of symmetry properties of Definition 3 can be expressed in terms of the fuzzy sets A, $n(A)$, B and $n(B)$.

Definition 4. *A Fuzzy Confirmation Measure $c_f(A, B)$ satisfies*

> **Evidence Symmetry** (ES) *if* $c_f(A, B) = -c_f(n(A), B)$;
> **Hypothesis Symmetry** (HS) *if* $c_f(A, B) = -c_f(A, n(B))$;
> **Evidence Hypothesis Symmetry** (EHS) *if* $c_f(A, B) = c_f(n(A), n(B))$;
> **Inversion Symmetry** (IS) *if* $c_f(A, B) = c_f(B, A)$;
> **Evidence Inversion Symmetry** (EIS) *if* $c_f(A, B) = -c_f(B, n(A))$;
> **Hypothesis Inversion Symmetry** (HIS) *if* $c_f(A, B) = -c_f(n(B), A)$;
> **Evidence Hypothesis Inversion Symmetry** $(EHIS)$ *if* $c_f(A, B) = c_f(n(B), n(A))$.

In Table 1 we list all the symmetry properties satisfied by the FCMs considered in Sect. 2, either proposed by Glass or introduced in this paper for the first time (those properties can be deduced using Theorem 3 in [15] and Proposition 3 in [6]).

Table 1. Symmetries of some FCMs

FCM	ES	HS	EHS	IS	EIS	HIS	EHIS
$b(A, B)$	Yes	Yes	Yes	Yes	Yes	Yes	Yes
$C(A, B)$	No	No	Yes	Yes	No	No	Yes
$d(A, B)$	No	Yes	No	No	No	No	No
$F(A, B)$	No	No	No	Yes	No	No	No
$G(A, B)$	No	Yes	No	No	No	No	No
$Gr(A, B)$	No	No	No	No	No	No	No
$K(A, B)$	No	Yes	No	No	No	No	No
$M(A, B)$	Yes	No	No	No	No	No	No
$N(A, B)$	Yes	Yes	Yes	No	No	No	No
$R(A, B)$	No	No	No	No	No	No	Yes
$Z(A, B)$	No	Yes	No	No	No	No	No

The symmetry properties in Definition 4 are provided in terms of the fuzzy sets A and B and their negations $n(A)$ and $n(B)$, respectively, but can also be expressed in terms of the confidence measures which are involved in the definition of a Fuzzy Confirmation Measure (see [4] and [5] for the analysis presented in the BCM setting). More precisely, if we define

$$\alpha = conf(T \Rightarrow A), \qquad \beta = conf(T \Rightarrow B), \qquad \gamma = conf(A \Rightarrow B)$$

where $conf$ is the product based fuzzy confidence measure, it is possible to express in terms of α, β, γ all the above defined symmetries.

For example, if we consider symmetry HS, i.e.,

$$c_f(A, B) = -c_f(A, n(B))$$

we can rewrite HS as

$$c_f(conf(T \Rightarrow A), conf(T \Rightarrow B), conf(A \Rightarrow B))$$
$$= -c_f(conf(T \Rightarrow A), conf(T \Rightarrow n(B)), conf(A \Rightarrow n(B)))$$

and, since $conf(T \Rightarrow n(B)) = 1 - conf(T \Rightarrow B)$, also as

$$c_f(conf(T \Rightarrow A), conf(T \Rightarrow B), conf(A \Rightarrow B))$$
$$= -c_f(conf(T \Rightarrow A), 1 - conf(T \Rightarrow B), 1 - conf(A \Rightarrow B))$$

that is, $c_f(\alpha, \beta, \gamma) = -c_f(\alpha, 1 - \beta, 1 - \gamma)$. Similar algebraic manipulations allow to rewrite each symmetry property of a FCM in terms of α, β and γ, as reported in Table 2.

4   Lack of Symmetry and Degree of Asymmetry

Confirmation measures may satisfy a symmetry property or even combinations of symmetry properties, but we focus now on symmetry properties that a confirmation measure does *not* meet: how far is the measure from satisfying them? For

Table 2. Symmetries in the (α, β, γ) setting

Symmetry	Definition	Expression in terms of α, β, γ
ES	$c_f(A,B) = -c_f(n(A),B)$	$c_f(\alpha,\beta,\gamma) = -c_f(1-\alpha,\beta,(\beta-\alpha\gamma)/(1-\alpha))$
HS	$c_f(A,B) = -c_f(A,n(B))$	$c_f(\alpha,\beta,\gamma) = -c_f(\alpha,1-\beta,1-\gamma)$
EHS	$c_f(A,B) = c_f(n(A),n(B))$	$c_f(\alpha,\beta,\gamma) = c_f(1-\alpha,1-\beta,1-(\beta-\alpha\gamma)/(1-\alpha))$
IS	$c_f(A,B) = c_f(B,A)$	$c_f(\alpha,\beta,\gamma) = c_f(\beta,\alpha,\alpha\gamma/\beta)$
EIS	$c_f(A,B) = -c_f(B,n(A))$	$c_f(\alpha,\beta,\gamma) = -c_f(\beta,1-\alpha,1-\alpha\gamma/\beta)$
HIS	$c_f(A,B) = -c_f(n(B),A)$	$c_f(\alpha,\beta,\gamma) = -c_f(1-\beta,\alpha,(1-\gamma)\alpha/(1-\beta))$
$EHIS$	$c_f(A,B) = c_f(n(B),n(A))$	$c_f(\alpha,\beta,\gamma) = c_f(1-\beta,1-\alpha,1-(1-\gamma)\alpha/(1-\beta))$

example, $C(A,B)$ and $F(A,B)$ both satisfy property IS, but neither of them satisfies HS: how far are the two measures from satisfying it? Is their *lack of symmetry* comparable? In other terms, we want to define a *degree of asymmetry* of a confirmation measure, with respect to a given symmetry.

Let us consider, as an example, HS symmetry, since a FCM c_f fulfills Hypothesis Symmetry if

$$c_f(\alpha,\beta,\gamma) = \tilde{c}_f(\alpha,\beta,\gamma) \quad \forall \alpha, \beta, \gamma$$

with $\tilde{c}_f(\alpha,\beta,\gamma) = -c_f(\alpha,1-\beta,1-\gamma)$, when the condition is not satisfied, we can evaluate the degree of asymmetry considering how much the values of $c_f(\alpha,\beta,\gamma)$ are far from $\tilde{c}_f(\alpha,\beta,\gamma)$, as α, β, γ vary.

More in general, a Fuzzy Confirmation Measure c_f satisfies symmetry S, with $S \in \{ES, HS, EHS, IS, EIS, HIS, EHS\}$, if

$$c_f(\alpha,\beta,\gamma) = \tilde{c}_f(\alpha,\beta,\gamma) \tag{13}$$

and $\tilde{c}_f(\alpha,\beta,\gamma)$ is a suitably defined function, which depends on both the chosen symmetry S and the Fuzzy Confirmation Measure c_f under consideration; the functions $\tilde{c}_f$ to be considered in each symmetry are reported in the right hand side term of the last column of Table 2. Accordingly, when condition (13) is not satisfied, c_f will be called S-asymmetric (or, asymmetric with respect to S).

To evaluate the degree of asymmetry of a FCM we first introduce an order of asymmetry (see [7]), which will be used in the definition of asymmetry measure (Definition 6) to avoid the possibility of facing two FCMs which, with respect to the same symmetry S, are inversely ordered by different asymmetry measures [29].

Definition 5. *A Fuzzy Confirmation Measure c_{f_1} is called to be less S-asymmetric than a Fuzzy Confirmation Measure c_{f_2}, written $c_{f_1} \preceq_{Sa} c_{f_2}$, if*

$$\frac{|c_{f_1}(\alpha,\beta,\gamma) - \tilde{c}_{f_1}(\alpha,\beta,\gamma)|}{|c_{f_1}(\alpha,\beta,\gamma)|} \leq \frac{|c_{f_2}(\alpha,\beta,\gamma) - \tilde{c}_{f_2}(\alpha,\beta,\gamma)|}{|c_{f_2}(\alpha,\beta,\gamma)|}$$

for each (α,β,γ).

Observe that $\preceq_{Sa}$ is a preorder, that is the relation is reflexive and transitive. Note that it is not antisymmetric: in fact, if we consider, for example, the FCMs $C(A, B)$ and $F(A, B)$ satisfying Inversion Symmetry, this implies that $C(\alpha, \beta, \gamma) - \check{C}(\alpha, \beta, \gamma) = 0 = F(\alpha, \beta, \gamma) - \tilde{F}(\alpha, \beta, \gamma)$, that is $C\preceq_{ISa}F$ and $F\preceq_{ISa}C$ when IS is considered, but the two FCMs are different.

Finally, we can propose a general definition of asymmetry measure which is compatible with the preorder.

Definition 6. *A measure of S-asymmetry is a function μ_{Sa} that satisfies the following conditions:*

1. $\mu_{Sa}(c_f) = 0$ if and only if c_f is S-symmetric;
2. if $c_{f_1}\preceq_{Sa}c_{f_2}$ then $\mu_{Sa}(c_{f_1}) \leq \mu_{Sa}(c_{f_2})$.

Let us consider the L^p-norm of function $c_f(x, y, z) - \tilde{c}_f(x, y, z)$ as a measure of the distance among c_f and $\tilde{c}_f$. More precisely we propose the set of S-asymmetry measures defined as:

$$\mu_{Sa}{}^p(c_f) = \frac{\|c_f - \tilde{c}_f\|_p}{\|c_f\|_p} \tag{14}$$

where $\| \cdot \|_p$ denotes the L^p-norm, $p \in [1, \infty]$, and $\|c_f\|_p$ is used to allow comparisons of the asymmetry measures μ_{Sa} among different FCM functions.

The proposed measures $\mu_{Sa}{}^p$ are clearly compatible with the partial order $\preceq_{Sa}$ as required in Definition 6.

Observe that different choices of p allow to emphasize different kinds of S-asymmetries: for example, when $p = 1$ attention is put on the average degree of S-asymmetry, while $p = \infty$ drives the focus on the maximal degree of asymmetry that can be attained on single points.

As an example, considering the measures C and F, and the L^1-norm, we can compare them with respect to the Hypothesis Symmetry; we obtain[5]

$$\mu_{HSa}{}^1(C) = 0,33206841 \qquad \mu_{HSa}{}^1(F) = 1,07571540$$

by which the lack of HS is highlighted; moreover, it is possible to compare their degree of asymmetry and claim that $C\preceq_{HSa}F$.

Roughly speaking, since HS is considered as a desirable property (e.g., in [15]), C could be considered *preferable* to F if one of the two measures has to be chosen. If instead we consider ES, which is considered as undesirable by some Authors, F performs better than C given that

$$\mu_{ESa}{}^1(C) = 0,33206841 \qquad \mu_{ESa}{}^1(F) = 0,80840290$$

from which the suggestion is, instead, to choose F since $C\preceq_{ESa}F$.

---

[5] Asymmetry degree computations were performed with Wolfram's software *Mathematica* (version 11.0.1.0).

5   Conclusions

The large number of Bayesian Confirmation Measures available in the literature suggest possible definitions of associated FCMs. As with the BCMs, symmetry properties may help in identifying the FCMs which are more apt to be chosen in different contexts. In this paper we extended the set of symmetry properties for FCMs, which was proposed by Glass [15]. Moreover, we focused on suitable measures of the asymmetry of FCMs. We provided some preliminary numerical computations that highlight the possible different levels of asymmetry displayed by a FCM when symmetry properties are checked. It is therefore possible to observe how close a FCM is in meeting a symmetry property which is considered as desirable or how far it is in meeting a undesirable symmetry property.

References

1. Agrawal, R., Imieliński, T., Swami, A.: Mining association rules between sets of items in large databases. In: Buneman, P., Jajodia, S. (eds.) ACM SIGMOD International Conference on Management of Data, SIGMOD 1993. ACM, New York (1993). Record 22(2), 207–216
2. Beardon, A.F.: Algebra and Geometry. Cambridge University Press, Cambridge (2005)
3. Carnap, R.: Logical Foundations of Probability. University of Chicago Press, Chicago (1950)
4. Celotto, E.: Visualizing the behavior and some symmetry properties of Bayesian confirmation measures. Data Min. Knowl. Disc. **31**(3), 739–773 (2016)
5. Celotto, E., Ellero, A., Ferretti, P.: Monotonicity and symmetry of IFPD Bayesian confirmation measures. In: Torra, V., Narukawa, Y., Navarro-Arribas, G., Yañez, C. (eds.) MDAI 2016. LNCS (LNAI), vol. 9880, pp. 114–125. Springer, Cham (2016). https://doi.org/10.1007/978-3-319-45656-0_10
6. Celotto, E., Ellero, A., Ferretti, P.: Coexistence of symmetry properties for Bayesian confirmation measures. 17/WP/2018, Department of Economics, University Ca' Foscari Venezia, pp. 1–14 (2018). ISSN 1827-3580
7. Celotto, E., Ellero, A., Ferretti, P.: Asymmetry degree as a tool for comparing interestingness measures in decision making: the case of Bayesian confirmation measures. In: Esposito, A., Faundez-Zanuy, M., Morabito, F.C., Pasero, E. (eds.) WIRN 2017 2017. SIST, vol. 102, pp. 289–298. Springer, Cham (2019). https://doi.org/10.1007/978-3-319-95098-3_26
8. Celotto, E., Ellero, A., Ferretti, P.: Concurrent symmetries of Bayesian confirmation measures: a new perspective (2019, submitted)
9. Cohen, J.: A coefficient of agreement for nominal scales. Educ. Psychol. Measur. **20**(1), 37–46 (1960)
10. Crupi, V., Tentori, K., Gonzalez, M.: On Bayesian measures of evidential support: theoretical and empirical issues. Philos. Sci. **74**(2), 229–252 (2007)
11. Crupi, V., Festa, R., Buttasi, C.: Towards a grammar of Bayesian confirmation. In: Suárez, M., Dorato, M., Rédei, M. (eds.) Epistemology and Methodology of Science, pp. 73–93. Springer, Dordrecht (2010). https://doi.org/10.1007/978-90-481-3263-8_7
12. Eells, E., Fitelson, B.: Symmetries and asymmetries in evidential support. Philos. Stud. **107**(2), 129–142 (2002)

13. Finch, H.A.: Confirming power of observations metricized for decisions among hypotheses. Philos. Sci. **27**(4), 293–307 (1960)
14. Geng, L., Hamilton, H.J.: Interestingness measures for data mining: a survey. ACM Comput. Surv. **38**(3), 1–32 (2006)
15. Glass, D.H.: Fuzzy confirmation measures. Fuzzy Sets Syst. **159**(4), 475–490 (2008)
16. Glass, D.H.: Confirmation measures of association rule interestingness. Knowl.-Based Syst. **44**, 65–77 (2013)
17. Glass, D.H.: Entailment and symmetry in confirmation measures of interestingness. Inform. Sci. **279**, 552–559 (2014)
18. Good, I.J.: Probability and the Weighing of Antecedent. Hafners, New York (1950)
19. Greco, S., Pawlak, Z., Słowiński, R.: Can Bayesian confirmation measures be useful for rough set decision rules? Eng. Appl. Artif. Intell. **17**(4), 345–361 (2004)
20. Greco, S., Słowiński, R., Szczęch, I.: Properties of rule interestingness measures and alternative approaches to normalization of measures. Inform. Sci. **216**, 1–16 (2012)
21. Greco, S., Słowiński, R., Szczęch, I.: Finding meaningful Bayesian confirmation measures. Fundam. Inform. **127**(1–4), 161–176 (2013)
22. Greco, S., Słowiński, R., Szczęch, I.: Measures of rule interestingness in various perspectives of confirmation. Inform. Sci. **346–347**, 216–235 (2016)
23. Kemeny, J., Oppenheim, P.: Degrees of factual support. Philos. Sci. **19**, 307–324 (1952)
24. Maher, P.: Confirmation Theory. The Encyclopedia of philosophy, 2nd edn. Macmillan Reference USA, New York (2005)
25. Mortimer, H.: The Logic of Induction. Prentice Hall, Paramus (1988)
26. Nozick, R.: Philosophical Explanations. Clarendon, Oxford (1981)
27. Rescher, N.: A theory of antecedent. Philos. Sci. **25**, 83–94 (1958)
28. Rips, L.: Two kinds of reasoning. Psychol. Sci. **12**(2), 129–134 (2001)
29. Siburg, K.F., Stehling, K., Stoimenov, P.A., Weiß, G.N.F.: An order of asymmetry in copulas, and implications for risk management. Insurance Math. Econom. **68**, 241–247 (2016)
30. Susmaga, R., Szczęch, I.: Selected group-theoretic aspects of confirmation measure symmetries. Inform. Sci. **346–347**, 424–441 (2016)

Fake News Detection in Microblogging Through Quantifier-Guided Aggregation

Marco De Grandis, Gabriella Pasi⬤, and Marco Viviani⁽✉⁾⬤

Department of Informatics, Systems, and Communication (DISCo),
University of Milano-Bicocca, Viale Sarca, 336, 20126 Milan, Italy
m.degrandis@campus.unimib.it, {gabriella.pasi,marco.viviani}@unimib.it

Abstract. Nowadays, big volumes of User-Generated Content (UGC) spread across various kinds of social media. In microblogging, UCG can be generated in the form of 'newsworthy' posts, i.e., related to information that has a public utility for the people. In this context, being the UGC diffused without almost any traditional form of trusted external control, the possibility of incurring in possible fake news is far from remote. For this reason, several approaches for fake news detection in microblogging have been proposed up to now, mostly based on machine learning techniques. In this paper, an ongoing work based on the use of the Multi-Criteria Decision Making (MCDM) paradigm to detect fake news is proposed. The aim is to reduce data dependency in building the model, and to have flexible control over the choices behind the fake news detection process.

Keywords: Credibility · Fake news · Social media ·
User-Generated Content · Multi-Criteria Decision Making ·
Aggregation operators

1 Introduction

In the Web 2.0 era, the interaction between users is promoted by a number of social media that facilitate the establishment of multiple social relationships [3], and the diffusion of information in the form of *User-Generated Content* (UGC). In most cases, UGC is referred to personal information, comments, interests that users share within the virtual community they are involved in, for reasons that are connected to their emotional sphere. These contents, namely *conversation* posts, usually have an interest only for friends, or people sharing for example the same interests, of the person who generated the content. In other cases, however, contents that are of more general interest can be diffused, the so-called *newsworthy* or *news* posts. In the Social Web scenario, news posts are diffused in particular by microblogging platforms, such as Twitter,[1] and the Chinese Sina Weibo,[2] where millions of users act as real-time news diffusers [15].

[1] https://twitter.com.
[2] https://weibo.com.

© Springer Nature Switzerland AG 2019
V. Torra et al. (Eds.): MDAI 2019, LNAI 11676, pp. 64–76, 2019.
https://doi.org/10.1007/978-3-030-26773-5_6

It is clear how, in a context where traditional intermediaries have disappeared, through the phenomenon of 'disintermediation' that characterizes social media, the possibility of incurring in fake news is always higher. Due to the social consequences that relying on fake news can have (think, for example, of the possibility of directing political elections, or of spreading conspiracy theories at the level of public opinion) several approaches have been proposed in the last years to combat this phenomenon [7, 24].

Some methods are based on Machine Learning (ML) (mostly supervised) algorithms to classify news based on their credibility. Other methods are based on the propagation of initial credibility values, usually learned from a classifier. Although both types of approaches have proved to be effective, they present some open issues. On the one hand, the issues are related to the (possible) inscrutability of some of the ML algorithms used for fake news detection and to the difficulty of gathering suitable and unbiased labeled training data for supervised solutions [14]; on the other hand, to the need of correctly identifying the credibility values associated with news to be propagated.

In this paper, the focus is on a classification-based approach relying on the *Multi-Criteria Decision Making* (MCDM) paradigm, and on the proposal, in particular, of a numerical solution based on aggregation operators guided by linguistic quantifiers. The proposed approach tackles the open issues related to data-driven approaches for fake news detection, by giving to users flexible control over the classification process, and less dependency on the available data in building the model.

To illustrate the approach and for evaluation purposes, the proposed MCDM solution has been instantiated over the CREDBANK dataset,[3] constituted by microblogging posts gathered from Twitter; nonetheless, it can be generalized to other microblogging sites, and, potentially, to other kinds of social media [23].

2   Related Work

Several approaches have been proposed in the last years for detecting fake news in microblogging sites. Among the works in the literature, some of them have considered as an information unit (to be evaluated in terms of credibility) a *single post* (e.g., a tweet); other works have considered a *thread* (e.g., a set of tweets on the same topic) that represents a single *news event*. As previously introduced, these approaches belong to two main categories, namely (*i*) *classification-based*, and (*ii*) *propagation-based*. Approaches belonging to the first category are based on the use of machine learning algorithms (mostly supervised) that consider multiple features connected to both users and the content they generated. In the second category fall those approaches that, starting from given credibility values (usually learned from a classifier) are based on their propagation over the social network structure [13, 15, 27]. In this paper, a solution that is alternative to ML classification-based approaches is considered; for this reason, in the following, only the main approaches belonging to the first category will be detailed.

---

[3] https://github.com/compsocial/CREDBANK-data.

Castillo *et al.* [4,5] were among the first to tackle in a structured way the problem of information credibility on microblogging sites, Twitter in particular, by using classification-based approaches. The authors focus on automatic methods for assessing the credibility of a given 'time-sensitive' set of tweets, i.e., a *trending topic*. Specifically, they analyze tweets related to trending topics, and they classify these topics as credible or not credible based on multiple features extracted from tweets and their authors. In [5], in particular, the authors extend the model presented in [4], and evaluate it on the scenario of the use of Twitter during a crisis event. The approaches are based on the use of Bayesian methods, Logistic Regression, J48, Random Forests, and Meta Learning based on clustering, trained over labeled data obtained using crowdsourcing tools.

Other classification-based approaches (mostly supervised) are those described in [9,11,12,16], each of which proposes different features, machine learning algorithms and evaluation datasets, depending on the considered problem, i.e., the assessment of the credibility of target-topics in Twitter [16], the identification of credible tweets during high impact events [12], the detection of spammers [11] and troll profiles [9] in microblogging sites. With respect to the above-mentioned approaches, the recent work by Buntain and Golbeck described in [2] aims at considering a large set of credibility features, which are employed to automatically identify fake news in Twitter threads. Their model is trained over large-scale labeled datasets, CREDBANK in particular [20], which is a set of Twitter threads about news events labeled with crowdsourced credibility assessments.

Despite their effectiveness, the above-described approaches present some open issues related to the possible inscrutability of some of the machine learning algorithms employed and to their data dependency in training the models. To tackle these issues, in the next section a classification-based approach focusing on Multi-Criteria Decision Making is described. It represents a solution aimed at demonstrating the comparable effectiveness of the model-driven paradigm with respect to the data-driven one in the considered research context.

3   MCDM and Aggregation Operators

In a *Multi-Criteria Decision Making* (MCDM) problem, there are usually a set of candidate solutions, i.e., *alternatives* that are available to a *decision maker* (DM), multiple *criteria*, which are satisfied in a different way by each alternative, and different *importance weights* associated with each criterion. Solving an MCDM problem means to provide the decision maker with one or more optimal solutions (alternatives) respecting her/his preferences [10].

In the literature, many solutions to assist decision makers in choosing among a finite set of alternatives by employing *numerical techniques* have been proposed. A solution often employed consists in assigning distinct scores, namely *performance scores*, to each alternative with respect to each criterion. Each score represents a *degree of satisfaction* that expresses to what extent an alternative is satisfactory with respect to a criterion. In the considered context, where the alternatives are the *news* to be evaluated, and the criteria are the *features* characterizing the news, these performance scores can be interpreted as *degrees of credibility* of the news with respect to each feature. These multiple *credibility scores* can then be subsequently *aggregated* to obtain an *overall credibility score*. Formally, let us assume that: $A = \{a_1, a_2, \ldots, a_m\}$ is the set of *alternatives*, i.e., the *news*; $C = \{c_1, c_2, \ldots, c_n\}$ is the set of *criteria*, i.e., the *features* characterizing the news; s_i is the *satisfaction function* that, for a criterion c_i $(1 \leq i \leq n)$, returns the *performance score* $s_i(a_j) \in I, I = [0, 1]$, to which the alternative a_j $(1 \leq j \leq m)$ satisfies the criterion c_i (i.e., the *credibility score* in this context).

In MCDM, a common solution to obtain an overall performance score σ_j for each alternative a_j, i.e., an overall credibility score for each piece of news, is to employ an n-ary function $\mathcal{A}^{(n)}$, called *aggregation operator* (or *aggregation function*), which is a mapping $\mathcal{A}^{(n)} : [0, 1]^n \rightarrow [0, 1]$ acting on a definite number n of performance scores to be aggregated (for $n \in \mathbb{N}_0$). Formally: $\sigma_j = \mathcal{A}^{(n)}(s_1(a_j), s_2(a_j), \ldots, s_n(a_j))$. In the considered context, being the number of scores to be aggregated always known, the simple notation $\mathcal{A}$ will be used to indicate an aggregation operator.

3.1 OWA Operators and Linguistic Quantifiers

In the literature, several classes of aggregation operators, *averaging operators* in particular, have been employed to solve Multi-Criteria Decision Making problems [6,8,10,19]. In the context of fake news detection, a family of aggregation operators that could be of potential interest is that of *Ordered Weighted Averaging* (OWA) operators, extensively studied in the literature [26].

Definition 1. *An aggregation operator* $\mathcal{A}_{OWA} : [0, 1]^n \rightarrow [0, 1]$ *is called an* Ordered Weighted Averaging *(OWA) operator of dimension n if it has associated a weighting vector* $W = [w_1, w_2, \ldots, w_n]$ *such that* $w_k \in [0, 1]$ *and* $\sum_{k=1}^{n} w_k = 1$, *and where* $\mathcal{A}_{OWA}(x_1, x_2, \ldots, x_n) = \sum_{k=1}^{n} w_k b_k$, *in which* b_k *is the kth largest of the x_i values to be aggregated.*

Employing OWA operators gives the possibility to guide the aggregation by *linguistic quantifiers* (e.g., *all, some, many, ...*), which allow to choose the best alternative(s) based on the satisfaction of a 'certain amount' of the criteria by the alternative(s). In general, in the process of *quantifier-guided aggregation*, the decision maker provides a linguistic quantifier Q indicating the number (*absolute* quantifier) or the proportion (*relative* quantifier) of criteria s/he believes is sufficient to have a good solution. The procedure of generating the weighting vector W from a linguistic quantifier Q depends on its type. In this paper, *Regular Increasing Monotone* (RIM) relative quantifiers are considered, such as *at least $k\%$* and *most*. A linguistic quantifier is said to be a RIM quantifier if [25]: $Q(0) = 0$, $Q(1) = 1$, and $Q(r) \geq Q(s)$ if $r > s$ $(r, s \in [0, 1])$.

Equal Importance of Criteria. Starting form the definition of a RIM quantifier Q, the weights w_i of a weighting vector W of dimension n (n values to be aggregated) can be defined as follows:

$$w_i = Q\left(\frac{i}{n}\right) - Q\left(\frac{i-1}{n}\right), \quad \text{for } i = 1, \ldots, n \tag{1}$$

Equation (1) allows to define the weighting vector W by assuming that all the considered criteria are *equally important* for the DM. In real scenarios, it is often crucial to be able to discriminate the importance of the criteria that concur in a decision making process, as detailed in section below (e.g., in fake news detection, not all the features connected with a piece of news are equally significant in terms of credibility assessment).

Unequal Importance of Criteria. In [25], a way has been proposed for aggregating n scores with *distinct importance* associated with the criteria that generated them. Let us consider an alternative a (i.e., a piece of news in the considered context) to be evaluated with respect to n criteria; the performance scores by a of the n criteria are denoted by $x_1, x_2, \ldots, x_n$, each $x_i \in [0, 1]$, while the numeric values denoting the importance of the n criteria are denoted by $V_1, V_2, \ldots, V_n$. In the reordering process of the x_i values, it is important to maintain the correct association between the values and the importance of the criteria that originated them. For this reason, by u_j it is denoted the importance originally associated with the criterion that has the jth largest satisfaction degree. E.g., assuming that x_5 is the highest value among the x_i values, thus $b_1 = x_5$ and $u_1 = V_5$. At this point, to obtain the weight w_j of the weighting vector with weighted criteria, it is possible to employ, for each alternative a, the following equation:

$$w_j = Q\left(\frac{\sum_{k=1}^{j} u_k}{T}\right) - Q\left(\frac{\sum_{k=1}^{j-1} u_k}{T}\right) \tag{2}$$

where $T = \sum_{k=1}^{n} u_k$ is the sum of the importance values u_js. The weighting vector used in this aggregation will generally be different for each a, i.e., for each considered piece of news.

4   An OWA-Based Approach for Fake News Detection

In the context of fake news detection, a user is confronted with multiple newsworthy posts of which s/he does not know a priori the level of credibility. As illustrated in Sect. 2, it is possible to evaluate credibility either with respect to a single post, or with respect to a thread representing a *news event*. If every single post is considered as an alternative, credibility features (i.e., criteria) are those associated with the considered post and/or with the user who generated it. In the case of a news event, the features describe 'global properties' of the event, i.e., of the posts that compose the thread and their authors. In general, each feature can be more or less important in the assessment of the credibility of a piece of news (be it a single post or a news event). Moreover, as illustrated in Sect. 3.1, only a 'certain amount' of the features (criteria) could be satisfied by the alternative to provide optimal solutions to the decision maker.

By considering these premises, in this section an MCDM approach based on the use of OWA aggregation operators is proposed; different *aggregation schemes* guided by distinct *linguistic quantifiers* are presented, which allow to tune the number of (important) features to be considered, and to provide an overall credibility score associated with each considered piece of news.

4.1 Fake News Event Detection on Twitter

The proposed approach aims at representing an alternative to classification-based approaches focusing on (supervised) machine learning techniques. Most of these solutions have referred to Twitter and employed (labeled) datasets built on this social media platform to prove their effectiveness. Therefore, in this paper, the same context has been taken into account to instantiate and evaluate the proposed approach, by focusing in particular on the assessment of the credibility of *news events* extracted from the CREDBANK dataset.

The CREDBANK Dataset. It is composed of about 80 millions of tweets and associated metadata, grouped into 1,376 news events (circa 60,000 tweets per event). To each news event, it is associated a 30-element vector of *credibility labels* (called *accuracy labels* in [20]) provided by 30 distinct experts. Each credibility label is expressed on a 5-point Likert scale ranging from -2 (*certainly false*) to 2 (*certainly true*). The final decision can be made considering, for example, the majority of the credibility label values expressed by the 30 experts. To illustrate the proposed approach, a 'reduced' version of the CREDBANK dataset is employed in this paper, i.e., the one described and provided in [2], where the authors have considered the most retweeted tweets in order to discard, among the 1,376 events, those provoking less reaction. To have an overall score associated with each news event, they have computed the *mean accuracy rating* based on the 30 accuracy labels provided by experts. This led the authors to finally select 156 news events, of which 99 are labeled as true and 57 as fake.[4] It is worth to be underlined that this dataset contains only the news events representing the most significant news (in terms of reactions), and each news event is made up of *thousands* of individual tweets, for a total of more than 9 million tweets.

Features Identification and Representation. In the literature, several features have been used for evaluating the credibility of Twitter threads (i.e., representing news events); they belong to the following macro-categories: *Structural* features [S]: they are specific to the structure of each Twitter thread; *User-related* features [U]: they represent attributes related to the users, their profiles, their connections and interactions; *Content-related* features [C]: they are based on textual properties extracted from the content of the tweets; *Temporal* features [T]: they allow to take into consideration how the values of the other types of features change over time. In this paper, only the most informative feature set is considered, as illustrated in [2], which is composed of the following 15 features: 1. *Media count* [S]: the frequency of tweets that contain media contents (images, videos, etc.); 2. *Mention count* [S]: the frequency of tweets that contain mentions; 3. *URL count* [S]: the frequency of tweets that contain URLs; 4. *Retweet count* [S]: the number of retweets for the event; 5. *Hashtag count* [S]: the frequency of tweets that contain hashtags; 6. *Status count* [S]: the average number of tweets with respect to each user profile (in the thread); 7.

---

[4] https://github.com/cbuntain/CREDBANK-data.

Tweet count [S]: the frequency of tweets that contain only text (no media, mentions, hashtags or URLs); 8. *Verified* [U]: the number of verified profiles (in the thread); 9. *Density* [U]: the density of the network w.r.t. users (nodes) and their interactions (edges, i.e., mentions, replies, etc.); 10. *Followers* [U]: the average number of followers with respect to each user profile (in the thread); 11. *Friends* also known as *Followees* [U]: the average number of followees with respect to each user profile (in the thread); 12. *Polarity* [C]: the average positive or negative feelings expressed by the tweets (in a thread); 13. *Objectivity* [C]: the score of whether a thread is objective or not; 14. *Ages* [T]: the author account age relative to a tweet creation; 15. *Lifespan* [T]: the minutes between the first and the last tweet of the thread.

The above-mentioned features are of a different nature, refer to distinct concepts and, therefore, are expressed on different numerical scales. In the proposed MCDM approach, starting from the values associated with features, it is necessary to select a suitable satisfaction function that is able to transform them into suitable performance scores in the $[0, 1]$ interval to be aggregated, as illustrated in Sect. 3. To do this, the *min-max* normalization function has been employed:

$$s_i(a_j) = \frac{x_{i,j} - min(x_{i,h})}{max(x_{i,h}) - min(x_{i,h})} \tag{3}$$

where, for a news event a_j, $s_i(a_j)$ is the performance score normalized in the $[0, 1]$ interval with respect to the feature c_i, $x_{i,j}$ is the value of the feature c_i for a_j, $h = 1, \ldots, m$, and m is the total number of news events. The performance scores obtained this way are considered as the degrees of satisfaction of each news event with respect to each feature in terms of credibility. The value '1' is assumed as the evidence of a full satisfaction in terms of credibility, and the value '0' as a complete dissatisfaction.[5]

4.2   Quantifier-Guided Aggregation Schemes

The different aggregation schemes employed by the proposed approach are illustrated in this section. Initially, three simple aggregation schemes have been considered, whose acronyms are denoted below in capital letters. These aggregation schemes employs OWA operators guided by the following linguistic quantifiers: (*i*) the *all* (min) quantifier – OWA_ALL; (*ii*) the *at least one* (max) quantifier – OWA_ONE; (*iii*) the *mean* (arithmetic mean) quantifier – OWA_MEAN.

In addition to aggregation schemes (*i*)–(*iii*), two additional schemes have been developed, based on OWA operators guided by the following linguistic quantifiers: (*iv*) the *more than k%* quantifier – OWA_MORE; (*v*) the *most* quantifier – OWA_MOST. The formal definitions of the quantifiers employed in the proposed approach are provided in the following.

---

[5] Empirically, for all the features, higher values can be interpreted as 'more credible'. Some theoretical justifications about the type of features and the values associated with them in the assessment of the credibility of information are provided in [18, 23].

More than k%. According to [1], the *more than k% (more)* quantifier can be defined as

$$Q_{more}(r) = \begin{cases} 0 & \text{for } 0 < r \le k \\ \frac{r-k}{1-k} & \text{for } k < r \le 1 \end{cases} \tag{4}$$

In this paper, two configurations of this quantifier have been considered, i.e., for $k = 50$ and $k = 75$, representing the percentages of the satisfied criteria. The shape of Q_{more} for both configurations is illustrated in Fig. 1(a) and (b).

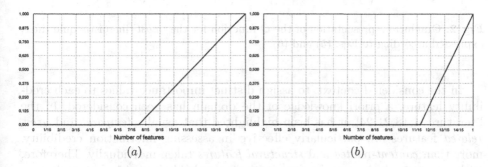

(a)   (b)

Fig. 1. Graphical representation of the Q function for the '*more than 50%*' (a), and the '*more than 75%*' (b) linguistic quantifiers.

Most. Two definitions of the *most* quantifier are considered in this paper. In a first definition [25], Q_{most} can be expressed as:

$$Q_{most}(r) = r^2 \tag{5}$$

According to [1], another definition of the *most* quantifier is the following:

$$Q_{most}(r) = \begin{cases} 0 & \text{for } 0 < r \le \alpha \\ \frac{r-\alpha}{\beta-\alpha} & \text{for } \epsilon < r < \beta \\ 1 & \text{for } r \ge \beta \end{cases} \tag{6}$$

The shape of Q_{most} under the two different definitions is illustrated in Fig. 2(a) and (b). In particular, Fig. 2(b) reports as an example the case of $\alpha = 0.3$ and $\beta = 0.8$.

When considering all criteria as *equally important*, the weighting vector W for aggregation schemes (*iv*) and (*v*) can be obtained according to Eq. (1), as illustrated in Sect. 3.1, 'Equal Importance of Criteria'. In this case, the above-defined linguistic quantifiers represent the proportion of criteria to be satisfied by the alternatives.

To consider the proportion of the *important* criteria to be satisfied, other aggregation schemes have been considered, where the weighting vector W is built by employing Eq. (2) illustrated in Sect. 3.1, 'Unequal Importance of Criteria', together with linguistic quantifiers defined by Eqs. (4)–(6). These additional schemes are denoted as: (*vi*) OWA_MORE_I; (*vii*) OWA_MOST_I.

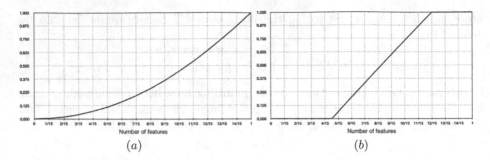

(a)   (b)

Fig. 2. Graphical representation of the Q function for the '*most*' linguistic quantifier, expressed according to Eqs. (5) and (6).

In the considered context, to assign distinct importance values to each credibility feature, a priori knowledge of the domain has been considered. In the literature [4,18,21,24], it has been highlighted that usually *temporal* and *user-related* features are particularly effective in assessing information credibility, more than *content-related* and *structural features* taken individually. Therefore, with respect to the proposed categorization provided in Sect. 4.1, discrete importance values in the set $\{1, 2, 3, 4\}$ have been assigned to each category of features: in particular, to temporal [T] features it has been assigned an importance value equal to 4; to user- [U] and content-related [C] features it has been assigned an importance value equal to 3 and 2 respectively; to structural [S] features it has been assigned an importance value equal to 1. It is important to notice that also continuous values, for example in the $[0, 1]$ interval, could have been employed.

5   Evaluation

To evaluate the proposed MCDM approach, a *binary classification* into fake news events and genuine ones has been performed by employing both the aggregation schemes (i)–(vii) proposed in this paper, and the data-driven approach proposed in [2], which employed the CREDBANK dataset and the same features used in this paper, and overcame several data-driven approaches in the literature. The evaluation metrics considered in this paper are *accuracy*, *precision*, *recall*, and *F1-score*, as defined in [17] for classification purposes.

5.1   Implementation Details

The experimental phases have been conducted by employing the *Python* programming language. In particular, to manage and to make numerical computations on data, such as the development of the proposed aggregation schemes, the *pandas* and *NumPy* libraries[6] has been used, while the *scikit-learn* library[7]

---

[6] https://pandas.pydata.org, http://www.numpy.org.
[7] http://scikit-learn.org/stable/index.html.

has been employed to evaluate the performance of the implemented classifiers with respect to the baseline through the computation of the evaluation metrics introduced before. It is worth to be underlined that the original code provided by Buntain and Golbeck[8] has been employed to perform the 5-fold cross-validation using the 100-tree Random Forest classifier at the basis of their approach, which is used in this paper as a baseline for the data-driven paradigm. With respect to the proposed MCDM approach, by applying aggregation schemes (i)–(vii) to the performance scores of criteria associated with news events, for each news event an overall credibility score in the $[0, 1]$ interval has been obtained. Then, news events have been classified as *genuine* or *fake* by selecting an optimal *threshold* over these overall scores. The threshold has been chosen in an experimental way, by selecting the one that maximizes classification effectiveness [22].

5.2   Summarization of Results and Discussion

In this section, the effectiveness of the binary classification (of news events into genuine or fake) provided by both the proposed aggregation schemes and the considered data-driven baseline is illustrated. Table 1 summarizes the results obtained with respect to the evaluation metrics previously introduced.

Table 1. Summarization of results of all the experiments.

	accuracy	*precision*	*recall*	*F1-score*
Baseline [2]	79%	80%	90%	85%
OWA_ALL	73%	**87%**	68%	76%
OWA_ONE	68%	69%	89%	78%
OWA_MEAN	70%	70%	93%	80%
OWA_MORE (50%)	76%	82%	81%	81%
OWA_MORE (75%)	79%	**87%**	79%	83%
OWA_MOST (exp)	65%	77%	65%	70%
OWA_MOST (0.5–0.6)	78%	79%	89%	84%
OWA_MORE_I (50%)	**83%**	**83%**	91%	**87%**
OWA_MORE_I (75%)	**83%**	**85%**	89%	**87%**
OWA_MOST_I (exp)	73%	78%	80%	79%
OWA_MOST_I (0.5–0.6)	82%	82%	91%	86%

As it emerges from the table, aggregation schemes guided by simple linguistic quantifiers (i.e., *all, at least one, mean*) do not provide significant results. For this reason, a detailed discussion will be provided only for non-trivial cases.

---

[8] https://github.com/cbuntain/CREDBANK-data/tree/master/src/main/python/Labeling.

With respect to aggregation schemes guided by the *more than k%* and *most* linguistic quantifiers, several configurations have been tested for each of them, i.e., the OWA_MORE (50%) and the OWA_MORE (75%), i.e., more than 50% and more than 75% of criteria satisfied, and the OWA_MOST (exp) and OWA_MOST (0.5–0.6), where in the first case the *most* quantifier is expressed according to Eq. (5), while in the second case it is expressed by means of Equation (6) where $\alpha = 0.5$ and $\beta = 0.6$ (these parameters provided the best results for the considered aggregation scheme). Aggregation schemes considering *different importance* associated with criteria have also been tested, i.e., OWA_MORE_I (50%), OWA_MORE_I (75%), OWA_MOST_I (exp), and OWA_MOST_I (0.5–0.6). A first consideration is that aggregation schemes based on OWA operators guided by the *more than k%* quantifier perform better with respect to those based on the *most* quantifier, in any case. Furthermore, at it was expected, those aggregation schemes considering different importance associated with criteria perform better than those considering all criteria as equal. With respect to this aspect, in particular, it is interesting to notice that the aggregation schemes for which importance values have been defined heuristically based on a prior knowledge, OWA_MORE_I (50%) and OWA_MORE_I (75%) in particular, perform better than the baseline that is completely data-driven. This suggests the potential feasibility and the effectiveness of the use of a completely model-driven approach to tackle the considered fake news detection problem.

6   Conclusions

In this paper, an approach for fake news detection in microblogging based on the Multi-Criteria Decision Making (MCDM) paradigm has been proposed. It is a model-driven classification-based approach, employing aggregation operators guided by linguistic quantifiers. In this model, news represent alternatives to be evaluated in terms of credibility; to each alternative are associated distinct credibility features that are satisfied by the alternative to a certain credibility extent, which can be expressed as a numerical credibility score. The overall credibility score of an alternative, i.e., a piece of news, is therefore obtained as the aggregation of the distinct credibility scores associated with the alternative. In this scenario, the decision maker can have a flexible control on the model; s/he can act: (i) on the choice of the aggregation operator (or the family of aggregation operators) to be used, (ii) on the choice of the number or the proportion of the features s/he estimates sufficient to have a good solution (by choosing absolute or relative linguistic quantifiers and their formal representations), and (iii) on the assignment of different importance values to different credibility features, for example exploiting prior knowledge s/he has of the considered domain. Furthermore, by proposing a model-driven solution, the proposed approach aims at tackling the difficulty to gather suitable and unbiased training data to be used by supervised data-driven approaches in assessing the credibility of UGC, as it has been discussed in the literature [14]. Since the construction of the model does not require learning data, the proposed approach also aims to be generalizable to other social media UGC, not only news. In the future, it will be interesting to learn the importance weights of the OWA operator, and to investigate other aggregation operators, such as WOWA operators and Fuzzy Integrals.

References

1. Ben-Arieh, D.: Sensitivity of multi-criteria decision making to linguistic quantifiers and aggregation means. Comput. Ind. Eng. **48**(2), 289–309 (2005)
2. Buntain, C., Golbeck, J.: Automatically identifying fake news in popular Twitter threads. In: IEEE Smart Cloud (SmartCloud 2017), pp. 208–215. IEEE (2017)
3. Carminati, B., Ferrari, E., Viviani, M.: A multi-dimensional and event-based model for trust computation in the social web. In: Aberer, K., Flache, A., Jager, W., Liu, L., Tang, J., Guéret, C. (eds.) SocInfo 2012. LNCS, vol. 7710, pp. 323–336. Springer, Heidelberg (2012). https://doi.org/10.1007/978-3-642-35386-4_24
4. Castillo, C., Mendoza, M., Poblete, B.: Information credibility on Twitter. In: Proceedings of the 20th International Conference on World Wide Web, pp. 675–684. ACM (2011)
5. Castillo, C., Mendoza, M., Poblete, B.: Predicting information credibility in time-sensitive social media. Internet Res. **23**(5), 560–588 (2012)
6. Ceravolo, P., Damiani, E., Viviani, M.: Adding a peer-to-peer trust layer to metadata generators. In: Meersman, R., Tari, Z., Herrero, P. (eds.) OTM 2005. LNCS, vol. 3762, pp. 809–815. Springer, Heidelberg (2005). https://doi.org/10.1007/11575863_102
7. Conroy, N.J., Rubin, V.L., Chen, Y.: Automatic deception detection: methods for finding fake news. Proc. AIST **52**(1), 1–4 (2015)
8. Damiani, E., Viviani, M.: Trading anonymity for influence in open communities voting schemata. In: SocInfo 2009, pp. 63–67. IEEE (2009)
9. Galán-García, P., et al.: Supervised machine learning for the detection of troll profiles in Twitter social network. Logic J. IGPL **24**(1), 42–53 (2016)
10. Greco, S., Ehrgott, M., Rui Figueira, J. (eds.): Multiple Criteria Decision Analysis: State of the Art Surveys. Springer, New York (2016). https://doi.org/10.1007/978-1-4939-3094-4
11. Gupta, A., Kaushal, R.: Improving spam detection in online social networks. In: Cognitive Computing and Information Processing (CCIP), pp. 1–6. IEEE (2015)
12. Gupta, A., Kumaraguru, P.: Credibility ranking of tweets during high impact events. In: Workshop on Privacy and Security in Online Social Media, p. 2. ACM (2012)
13. Gupta, M., Zhao, P., Han, J.: Evaluating event credibility on Twitter. In: SDM, pp. 153–164. SIAM (2012)
14. Heydari, A., Ali Tavakoli, M., Salim, N., Heydari, Z.: Detection of review spam: a survey. Expert Syst. Appl. **42**(7), 3634–3642 (2015)
15. Jin, Z., Cao, J., Jiang, Y.-G., Zhang, Y.: News credibility evaluation on Microblog with a hierarchical propagation model. In: ICDM 2014, pp. 230–239. IEEE (2014)
16. Kang, B., O'Donovan, J., Höllerer, T.: Modeling topic specific credibility on Twitter. In: Proceedings of IUI 2012, pp. 179–188. ACM (2012)
17. Kubat, M.: An Introduction to Machine Learning. Springer, Cham (2015). https://doi.org/10.1007/978-3-319-20010-1
18. Luca, M., Zervas, G.: Fake it till you make it: reputation, competition, and yelp review fraud. Manag. Sci. **62**(12), 3412–3427 (2016)

19. Marrara, S., Pasi, G., Viviani, M.: Aggregation operators in information retrieval. Fuzzy Sets Syst. **324**, 3–19 (2016)
20. Mitra, T., Gilbert, E.: CREDBANK: a large-scale social media corpus with associated credibility annotations. In: ICWSM, pp. 258–267 (2015)
21. Mukherjee, A., Venkataraman, V., Liu, B., Glance, N.: Fake review detection: classification and analysis of real and pseudo reviews, T.R. UIC-CS-03-2013 (2013)
22. Sebastiani, F.: Machine learning in automated text categorization. ACM Comput. Surv. (CSUR) **34**(1), 1–47 (2002)
23. Viviani, M., Pasi, G.: Quantifier guided aggregation for the veracity assessment of online reviews. Int. J. Intell. Syst. **32**(5), 481–501 (2016)
24. Viviani, M., Pasi, G.: Credibility in social media: opinions, news, and health information—a survey. Wiley Interdisc. Rev. Data Min. Knowl. Discovery **7**(5), e1209 (2017)
25. Yager, R.R.: Quantifier guided aggregation using OWA operators. Int. J. Intell. Syst. **11**(1), 49–73 (1996)
26. Yager, R.R., Kacprzyk, J.: The Ordered Weighted Averaging Operators: Theory and Applications. Springer, New York (2012). https://doi.org/10.1007/978-1-4615-6123-1
27. Zhao, L., Hua, T., Lu, C.-T., Chen, I.-R.: A topic-focused trust model for Twitter. Comput. Commun. **76**, 1–11 (2016)

Individual, Coalitional and Structural Influence in Group Decision-Making

Hang Luo[✉] [iD]

Peking University, Beijing 100871, China
hang.luo@pku.edu.cn

Abstract. We consider settings of group decision-making where agents' preferences/choices are influenced (and thus changed) by each other. As the influence of reality faced by an agent usually comes from more than one agent, previous work discussed at length multiple influences but in an individual way, which assumed that all influencing agents exert their own influences independently from each other and that the resulting preference/choice of the influenced agent could be a simple linear weighted aggregation of all influencing agents' preferences/choices. Some works discussed the influence of coalitions of multiple agents. As some agents hold the same beliefs, opinions or choices (such as in an "opinion alliance"), an extra influencing effect in addition to that of the separate individual influences should be considered. However, the *structural influence* has been ignored. The structure here mainly refers to the influencing relations among agents (which can be represented as links in social networks). Actually, previous work considers the structure (links) among agents just as the paths or channels of influence but ignores the fact that the structure itself can also exert an extra influencing effect. Moreover, it is not easy to address the influence of structures on an agent: as the influencing subject and the influenced object are disparate; the former are inter-relationships between agents, while the latter is the preference/choice of an individual agent. In this paper, we proposed a elementary framework to address the *three levels of influence* (*individual, coalitional and structural influence*) and their mixed effects.

Keywords: Group decision-making · Social network ·
Individual influence · Coalitional influence ·
Structural influence

1 Introduction

In the context of multi-agent (such as group decision-making) systems, the influence among agents' preferences or choices is quite common and has been studied by scholars from various disciplines [17], including artificial intelligence (particularly multi-agent system) [1,13,17,19–22], economics, decision theory and social network [2–12,14,15], and even politics [16,18]. In real-world situations, influences are diversified in polarity and strength [17], such as a positive influence

© Springer Nature Switzerland AG 2019
V. Torra et al. (Eds.): MDAI 2019, LNAI 11676, pp. 77–91, 2019.
https://doi.org/10.1007/978-3-030-26773-5_7

from a friend (ally) vs. a negative influence from an enemy (opponent) [16–18] and a strong influence from a close friend (family, relative) vs. a weak influence from a common friend (even a nodding acquaintance) [16,17]. Moreover, the influence faced by an agent is usually not from only a single agent (at a time) but simultaneously from more than one agent [16–18].

Previous work extensively discussed influences from multiple agents, but nearly all assumed that the influencing agents exert their own influences independently from each other. Some works discussed the influence of coalitions of agents [7–12] possessing the same belief, opinion or choice. Below, we sketch a more sophisticated model of influence (*three levels of influence*) where not only the influencing individual agents and their coalitions but also the structures among influencing agents are considered. The structures here in the context of a group decision with mutual influence indicate the influencing relations among agents, which can be represented as links or ties in a social network. Actually, the structures should not be perceived as just channels or paths of influence, but they themselves can exert some specific influencing effects, which will affect the influence result significantly. However, this *structural influence* has been ignored in previous work. Formally representing and computing this special influence is the essential work in this paper.

However, how to address the influencing effect from structures to an agent is not a straightforward question, as the influencing subject and influenced object are two disparate categories of variables. The former is the interpersonal relationship among agents (usually expressed by links in a networked graph), but the latter is the preference or choice of an individual (usually expressed by a normalized value or an alternative out of a set). It is relatively simple to address the influence of multiple agents' preferences/choices on another agent's preference/choice, but how to address the influencing effect from structures among agents on an individual agent's preference/choice to successfully close the gap and achieve the transformation between these two disparate things is a key question.

2   Influence Model in Group Decision-Making

In the settings of group decision-making, the influence model captures the influencing relations among agents, that is, when preferences (like utilities, beliefs, opinions) or choices of agents are affected by others [17]. Recently, in the fields of artificial intelligence, economics, and politics, the dynamics of influence have been extensively studied, especially in the framework of social networks [1,3–17,19–22], using the ties (links) between nodes (agents) to represent the influencing relationships between them [17].

2.1   Influence in Social Networks

Many of influence models assume a unidimensional value as utility or belief for each agent, which will be affected by his or her "neighbors" (namely, linked other agents) depending on the social structure [17]. Jackson [15] described a social network scenario where each agent's belief will be influenced (in the form of learning) by other agents' beliefs and also his or her own belief according to a weight allocation [17], where weights may be bigger or smaller to represent stronger or weaker influences among agents; and he [15] used a matrix composed of entries that each represent the weight of influence between two agents to express the influences in a social network [17]. Besides, [13] proposed a model combining opinion diffusion and social networks, assumed that each agent's binary opinion will be affected by his or her neighbors in the network according to the trust the agent has in them, namely, the more trust, the bigger weight of influence [17]. Some most recent works [1, 21] also discussed the influence and its impact on the evolution of experts' opinions in the framework of social networks.

2.2 Influence of Coalitions of Agents

Through comparing influence models and command games, Grabisch and Rusinowska [7–12] proposed a series of specific influence functions in social networks and systematically discussed the influence of a coalition upon an individual in depth, but in some sense they "skipped" the influence of individuals directly to the influence of a coalition, not distinguished the influence from a coalition as an extra effect in addition to that of individual influences. They mainly [7, 9–12] assumed that agents are to make a yes/no (acceptance/rejection) binary choice, where each has an inclination to say either. Due to the mutual-influence among agents, the choice of each agent may be different from his or her original inclination; they deem such a transformation from inclinations to choices as influence [17].

2.3 Social Choice Functions and Social Influence Functions

Most recently, Luo [17] discussed the simultaneous influence of more than one agent on another agent by both non-ordering and ordering approaches. He [17] extended the classic *social choice functions*, such as the Borda count and the Condorcet method, to signed and weighted *social influence functions* in the context of social networks, namely, the influence can be varied both in strength: stronger or weaker, and in polarity: positive or negative. Moreover, he [17] extended the KSB distance metric to a *matrix influence function*, defined the rule of how to transform each preference ordering into a matrix and then set a distance metric to compute the distance between any two *ordering matrices*; the preference that has the smallest weighted sum of the distances from all influencing agents' preferences will then be the resulting preference of the influenced agent [17].

In conclusion, previous work discussed the influence from multiple agents but mainly in an individual (independent) way, which assumed that all influencing agents exert their own influences independently from each other. Thus, the result for the influenced agent could be a simple linear weighted aggregation of all influencing agents' preferences or choices. However, these previous studies less discussed the coalitions of influencing agents (possessing the same beliefs, opinions or choices) and particularly ignored that the structures (namely, influencing relations) within influencing agents will also produce additional and extraordinary influencing effects.

3   Graphical and Mathematical Expressions of the Three Levels of Influence

If we look back at previous work on influence models, we could conclude that the discussions about the influence from coalitions and particularly the influence from structures actually advance and complete the system of influence studied. Now, we could propose a new analytical framework of influence as the *three levels of influence*, in which the influences discussed in previous work can be mainly classified as the *level I influence from independent agents (individual influence)*, while the *level II influence from coalitional agents (coalitional influence)* is less discussed and the *level III influence from structural agents (structural influence)* has been ignored. Only in the third level, the influence from the structures (namely, the influencing relations) themselves are considered in the process and computation of multiple influences as origins of influence, not just as paths or channels of influence anymore.

Definition 1. *(Group Decision-making Society with Mutual-Influence) Assume a society* $\mathbb{S} = \{\mathbb{N}, \mathbb{M}, \mathbb{P}, \mathbb{C}, \mathbb{W}\}$: $\mathbb{N} = \{1, 2, .., n\}$ *is the set of all agents (a general term for decision-makers, voters, game players, etc.);* $\mathbb{M} = \{o_1, o_2, ..., o_m\}$ *is the set of all alternatives (candidates);* $\mathbb{P} = \{P_{(1)}, P_{(2)}, ..., P_{(n)}\}$ *is the set of all agents' preferences (such as utilities, beliefs, decision-making probabilities, etc.);* $\mathbb{C} = \{C_{(1)}, C_{(2)}, ..., C_{(n)}\}$ *is the set of all agents' choices out of the set of alternatives;* $\mathbb{W}$ *is the matrix whose entries are the weights of influence between each of two agents,* $\mathbb{W} = [w_{(i,j)}]$ $(i, j \in \mathbb{N})$, *in which* $w_{(i,j)}$ *means the weight of influence from agent* i *to agent* j.

3.1   Level I Influence from Independent Agents

Definition 2. *(Individual Influence) is the influence just from agents as individuals. When an agent is simultaneously influenced by more than one agent but these influencing agents are independent from each other, separately exerting their own influences, the influences from different agents can be simply linearly summed just by their respective weights (of influence). This kind of influence from independent agents can be expressed as a directed line marked as* $\rightarrow_{x,y}$, *in which* x *indicates the influencing subject and* y *indicates the influenced object. The weight of this kind of influence can be defined as* $w_{x,y}$, *which indicates the weight of influence from* x *to* y.

Example 1. (A Graphical Expression of the Individual Influence) As in Fig. 1, we assume there are five agents 1–5, where agent 5 is simultaneously influenced by agents 1, 2, 3, 4 and 5 (himself or herself)[1], who all exert their own influences separately on agent 5. This is the basic hypothesis of most of previous work assumed so far. Thus, there are five *individual (independent) influences*, from 1 to 5 (marked as $\rightarrow_{1,5}$), from 2 to 5 ($\rightarrow_{2,5}$), from 3 to 5 ($\rightarrow_{3,5}$), from 4 to 5 ($\rightarrow_{4,5}$), and from 5 to 5 himself or herself ($\rightarrow_{5,5}$).

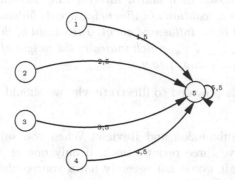

Fig. 1. Level I-influence from individual (Independent) agents

If the agents' preferences (such as utilities, beliefs, etc.) over an issue are expressed in a normalized value, then the resulting preference of the influenced agent will be a simple weighted sum of all influencing agents' preferences:

$$P'_5 = \frac{w_{1,5}P_1 + w_{2,5}P_2 + w_{3,5}P_3 + w_{4,5}P_4 + w_{5,5}P_5}{|w_{1,5}| + |w_{2,5}| + |w_{3,5}| + |w_{4,5}| + |w_{5,5}|}$$

Definition 3. *(Individual Influence Function) transforms multiple independent influencing agents' preferences to the influenced agent's preference. If this influence function considers the iteration of influence and multiperiod interactions, then:*

$$P_{(i)}(t+1) = \frac{\sum_{j\in\mathbb{N}} w_{(j,i)}P_{(j)}(t)}{\sum_{j\in\mathbb{N}} |w_{(j,i)}|} \qquad i \in \mathbb{N}$$

In which $P_{(i)}(t+1)$ represents the preference of agent i after t-th mutual influence, and $P_{(j)}(t)$ represents the preference of agent j at t-th mutual influence.

---

[1] The influence from oneself cannot be ignored. In reality, your current preference or choice over an issue is remarkably affected by your previous preferences or choices over the same or similar issue, which explains well why under identical influences from other people, some people can insist on their own preferences or choices while others change [17].

3.2   Level II Influence from Coalitional Agents

Definition 4. *(Coalitional Influence) considers (a portion of) influencing agents as a united coalition, especially those agents possessing the same or similar beliefs, opinions (like an opinion alliance), etc., which, under certain circumstances, will create an extra influencing effect (named the* coalitional influence*)[2]. Thus, the multiple influences felt by the influenced agent might not merely be a simple weighted sum of all influences from individuals. This* coalitional influence *can be expressed as a dotted directed line marked as* $--\rightarrow_{X,y}$*, where* $X = \{x_1, x_2, ..., x_c\}$ *is a coalition of influencing agents holding the same or similar preferences and y is the influenced agent. The weight of this kind of influence can be defined as* $w_{\{x_1,x_2,...,x_c\},y}$*, which indicates the weight of the influence from the coalition* $X = \{x_1, x_2, ..., x_c\}$ *to y.*

A daily example is provided to illustrate why we should consider the *coalitional influence.*

Example 2. (Paper Submission and Review) When you submit a paper to a conference and receive three reviews back, if only one or two reviewers give negative evaluations, it would not severely harm your confidence and feelings, and you might think they do not really understand your paper, but if three independent reviewers all judge your paper as "rubbish", then you would likely feel despair (the hard feeling would be more than three times that of one negative evaluation), which can be understood as a kind of coalitional effect produced by multiple influencing agents possessing the same or similar views.

Example 3. (A Graphical Expression of the Coalitional Influence) As in Fig. 2, again we assume that agent 5 is simultaneously influenced by agents 1, 2, 3, 4 and 5 (himself or herself), but agent 5 further observes or believes that agents 1, 2, and 3 have the same or similar preferences (like beliefs, opinions, etc.). Then, not only do the five agents 1, 2, 3, 4, and 5 all separately exert their own influences on agent 5, but also the coalition formed by agents 1, 2, and 3 (as an opinion alliance) would exert an extra *coalitional influence*, which is marked as $--\rightarrow_{\{1,2,3\},5}$.

If the agents' preferences are expressed in a normalized value, let $w_{\{1,2,3\},5}$ represents the weight of influence from agents $\{1,2,3\}$ as a uniform coalition to agent 5, and define χ as the function for the *coalitional influence*, then:

$$P'_5 = \frac{w_{1,5}P_1 + w_{2,5}P_2 + w_{3,5}P_3 + w_{4,5}P_4 + w_{5,5}P_5 + w_{\{1,2,3\},5}\chi\{P_1, P_2, P_3\}}{|w_{1,5}| + |w_{2,5}| + |w_{3,5}| + |w_{4,5}| + |w_{5,5}| + |w_{\{1,2,3\},5}|}$$

---

[2] A similar concept is the peer pressure.

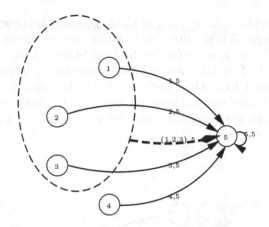

Fig. 2. Level II-influence from coalitional agents

Definition 5. *(Coalitional Influence Function) transforms multiple coalitional influencing agents' preferences to the influenced agent's preference. If it is expressed in a general multiperiod form, then:*[3]

$$P_{(i)}(t+1) = \frac{\sum_{j \in \mathbb{N}} w_{(j,i)} P_{(j)}(t) + \sum_{\mathfrak{c} \in \mathbb{C}_{[\mathbb{N}](i)}(t)} w_{(\mathfrak{c},i)} \chi[\mathfrak{c}]}{\sum_{j \in \mathbb{N}} |w_{(j,i)}| + \sum_{\mathfrak{c} \in \mathbb{C}_{[\mathbb{N}](i)}(t)} |w_{(\mathfrak{c},i)}|} \qquad i \in \mathbb{N}$$

In which $\mathbb{C}_{[\mathbb{N}](i)}(t)$ represents the set of all coalitions of agents with the same or similar preferences within the set of agents $\mathbb{N}$ from the perspective of agent i at t-th mutual influence, $\chi[\mathfrak{c}]$ ($\mathfrak{c} \in \mathbb{C}_{[\mathbb{N}](i)}(t)$) represents the coalitional influence produced by coalition $\mathfrak{c}$, and $w_{(\mathfrak{c},i)}$ represents the weight of influence from coalition $\mathfrak{c}$ to agent i.

3.3   Level III Influence from Structural Agents

Definition 6. *(Structural Influence) comes from the influencing relations among the influencing agents. The structures constituted by the influencing relations, which are represented as links in a social network, will also be perceived as origins of influence but not just channels of influence and then produce an extra influencing effect (named the structural influence). This structural influence can be expressed as: $\dashrightarrow_{x_1 \to x_2, y}$. The influencing subject discussed here is not an individual x_1 or x_2 or a coalition of x_1 and x_2 anymore but rather an influencing relation from x_1 to x_2. The weight of this kind of influence can be defined as $w_{x_1 x_2, y}$, which indicates the weight of the influence from the influencing relationship (x_1 influenced x_2) to y.*

---

[3] This is just a general expression of how to address the *coalitional influence*, one specific *coalitional influence function* can be found in the Sect. 4.

Example 4. (A Graphical Expression of the Structural Influence) As in Fig. 3, again we assume agent 5 is simultaneously influenced by agents 1, 2, 3, 4 and 5 (himself or herself), but agent 5 further finds or believes that among the three influencing agents 1, 2, and 3, there are two influencing relations from agent 1 to 2 and from agent 1 to 3. Thus, not only do the five agents 1, 2, 3, 4, and 5 all separately exert their own influence on agent 5, but also the two influencing relations among them would produce extra influencing effects, which are marked as $--\rightarrow_{1\rightarrow2,5}$ and $--\rightarrow_{1\rightarrow3,5}$.

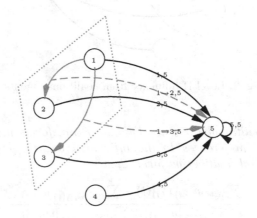

Fig. 3. Level III-influence from structural agents

If the agents' preferences are expressed in a normalized value, let $w_{12,5}$ and $w_{13,5}$ respectively represent the weights of influence from influencing relation $1 \rightarrow 2$ and influencing relation $1 \rightarrow 3$ to agent 5, and define φ as the function for the *structural influence*, then:

$$P_5' = \frac{w_{1,5}P_1 + w_{2,5}P_2 + w_{3,5}P_3 + w_{4,5}P_4 + w_{5,5}P_5 + w_{12,5}\varphi[P_1, P_2] + w_{13,5}\varphi[P_1, P_2]}{|w_{1,5}| + |w_{2,5}| + |w_{3,5}| + |w_{4,5}| + |w_{5,5}| + |w_{12,5}| + |w_{13,5}|}$$

Definition 7. *(Structural Influence Function) transforms multiple interacting influencing agents' preferences to the influenced agent's preferences. If it is expressed in a general multiperiod form, then:*[4]

$$P_{(i)}(t+1) = \frac{\sum_{j\in\mathbb{N}} w_{(j,i)} P_{(j)}(t) + \sum_{\mathfrak{s}\in\mathbb{S}_{[\mathbb{N}](i)}(t)} w_{(\mathfrak{s},i)}\varphi[\mathfrak{s}]}{\sum_{j\in\mathbb{N}} |w_{(j,i)}| + \sum_{\mathfrak{s}\in\mathbb{S}_{[\mathbb{N}](i)}(t)} |w_{(\mathfrak{s},i)}|} \quad i \in \mathbb{N}$$

In which $\mathbb{S}_{[\mathbb{N}](i)}(t)$ *represents the set of all structures (influencing relations) among the set of agents* $\mathbb{N}$ *from the perspective of agent* i *at* t*-th mutual influence,* $\varphi[\mathfrak{s}]$ *(*$\mathfrak{s} \in \mathbb{S}_{[\mathbb{N}](i)}(t)$*) represents the* structural influence *produced by influencing relation* $\mathfrak{s}$*, and* $w_{(\mathfrak{s},i)}$ *represents the weight of influence from influencing relation* $\mathfrak{s}$ *to agent* i.

---

[4] This is just a general expression of how to address the *structural influence*, one specific *structural influence function* will be discussed in the Section 4.

The Relations among the Three Levels of Influence

We have to admit that this work is just a mathematical modelling and graphical expression for the *individual, coalitional and structural influence*, and we have not evaluated and compared the relative merits of the above different influence models (functions). A critical reason is that the relations among the *three levels of influence* are not substitutional or competitive, but all of them together could fully describe the complicated features of influence in the real-world settings.

4 The Interplay Between the Coalitional Influence and the Structural Influence: A Probability-Based Approach

As the *three levels of influence* may work together in reality. To better understand the mechanisms of the *coalitional influence* and the *structural influence* and particularly their mixed effects, we provide a simple group decision-making example and use a probability-based choice approach to illustrate of how to address the *three levels of influence*:

Example 5. (The Mixed Effects of the Coalitional and Structural Influences) As in Fig. 4, we assume a group decision-making system with eight agents (1–8) making a choice with three alternatives: $\{a, b, c\}$. From the perspective of agent 8 at the bottom, he or she is simultaneously influenced by eight agents, with four agents (1, 2, 4, 6) choosing a, one agent (3) choosing b, and three agents (5, 7, and 8 himself or herself) choosing c. The weights of influence are marked on the links. What's more, the agent 8 observes or believes that there are two influencing relations among the three influencing agents saying a and one influencing relation between the two influencing agents saying b. Assume the influencing relations are specific as "following" (the influenced one follows what the influencing one says); and assume that all influence are positive.

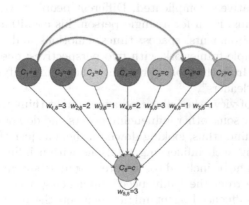

Fig. 4. An example of the mixed effects of the coalitional and structural influence

4.1   Individual Influences

If just considering the independent influences from all influencing agents but not the coalitions and structures among them, it is easy to form a linear weighted function to obtain the influence result.

Definition 8. *(Probability-based Individual Influence Function) Assume that $P_{o(i)}(t+1)$ is the preference of agent i for alternative o after t-th mutual influence, which can be expressed by a probability of agent i choosing alternative o. It will be influenced by other agents choosing alternative o at t-th mutual influence, and according to their weights of influence on agent i:*

$$P_{o(i)}(t+1) = \frac{\sum_{C_{(j)}(t)=o} w_{(j,i)}}{\sum_{j \in \mathbb{N}} w_{(j,i)}} \quad i \in \mathbb{N}, o \in \mathbb{M}$$

In which $C_{(j)}(t) = o$ means the choice of agent j at t-th mutual influence is alternative o.

In the Example 5, assume that $P_{a(8)}, P_{b(8)}, P_{c(8)}$ are respectively the probabilities of the influenced agent 8 choosing a, b, c after the mutual influence:

$$P_{a(8)} = \frac{w_{1,8} + w_{2,8} + w_{4,8} + w_{6,8}}{w_{1,8} + w_{2,8} + w_{3,8} + w_{4,8} + w_{5,8} + w_{6,8} + w_{7,8} + w_{8,8}} = \frac{8}{16} = 0.5000$$

$$P_{b(8)} = \frac{w_{3,8}}{w_{1,8} + w_{2,8} + w_{3,8} + w_{4,8} + w_{5,8} + w_{6,8} + w_{7,8} + w_{8,8}} = \frac{1}{16} = 0.0625$$

$$P_{c(8)} = \frac{w_{5,8} + w_{7,8} + w_{8,8}}{w_{1,8} + w_{2,8} + w_{3,8} + w_{4,8} + w_{5,8} + w_{6,8} + w_{7,8} + w_{8,8}} = \frac{7}{16} = 0.4375$$

4.2   Structural Influences

However, if the influencing effects from structures (influencing relations) are considered, there would be still different perspectives to understand and address, as human minds are natively complicated. Different people have varied personalities and value systems. Even for a single person, his cognition will be different under changing environments, spaces, times, emotions and other cases. Here, we just give one simple framework with two perspectives: one is weakening the weights of influence from the "followers", the other is intensifying the weights of influence from the "leaders".

From one angle of view, the influenced agent might think that some influencing agents just follow some other influencing agents and do not have independent ideas or their own mind (thus, looking down on them as just "followers"); therefore, after perceiving such influencing relations within influencing agents, the influenced agent might be inclined to "ignore", or more specifically, decrease the weights of influence from the "influenced" influencing agents. From the other angle of view, the influenced agent might focus on the influencers or uninfluenced ones but not the influenced ones within influencing agents, thinking that the reason for some influencing agents always insist on their ideas and even influenced more other agents is that they indeed have correct or better beliefs (like truth-holders) or that they are very influential, forceful or powerful (like opinion leaders or authorities), and deem that these "pure" influencing agents (like wise men or leaders)' preferences or choices might be safer or more beneficial; therefore, after perceiving such influencing relations, the influenced agent might be inclined to "empathize", or more specifically, increase the weights of influence from the "uninfluenced" influencing agents.

To realize this solution for the *structural influence*, first, we should distinguish the influencing agents with uninfluenced (original) preferences and influenced preferences (also influenced by other influencing ones). Thus, some more delicate variables should be defined to achieve such classification.

Definition 9. *(Probability-based Structural Influence Function) Assume that α, β is a pair of structural influence coefficients, which are timed respectively by the weights of "uninfluenced" influencing agents and the weights of "influenced" influencing agents. The structural influence coefficients should satisfy $\alpha + \beta = 1$ and $\alpha \geq \beta$ by common sense. Assume that $P^S_{o(i)}(t+1)$ is the probability of agent i choosing alternative o after t-th mutual influence considering the* Structural *influence, which will be influenced by other agents choosing o at t-th mutual influence according to their weights of influence, and also affected by these agents' "roles" in the influencing relationships within the influencing agents:*

$$P^S_{o(i)}(t+1) = \frac{\alpha \sum_{C_{(j)}(t)=o=C_{(j)}(t-1)} w_{(j,i)} + \beta \sum_{C_{(j)}(t)=o\neq C_{(j)}(t-1)} w_{(j,i)}}{\alpha \sum_{C_{(j)}(t)=C_{(j)}(t-1)} w_{(j,i)} + \beta \sum_{C_{(j)}(t)\neq C_{(j)}(t-1)} w_{(j,i)}} \quad i \in N, o \in M$$

In the Example 5, assume that $P^S_{a(8)}$ is the probability of the influenced agent 8 choosing a after the mutual influence combining the Structural influence:

$$P^S_{a(8)} = \frac{\alpha w_{1,8} + \alpha w_{2,8} + \beta w_{4,8} + \beta w_{6,8}}{\alpha w_{1,8} + \alpha w_{2,8} + \alpha w_{3,8} + \beta w_{4,8} + \alpha w_{5,8} + \beta w_{6,8} + \beta w_{7,8} + \alpha w_{8,8}}$$

$$= \begin{cases} \frac{0.5\times3+0.5\times2+0.5\times2+0.5\times1}{0.5\times3+0.5\times2+0.5\times1+0.5\times2+0.5\times3+0.5\times1+0.5\times1+0.5\times3} = 0.5000, & \alpha = 0.5, \beta = 0.5 \\ \frac{0.8\times3+0.8\times2+0.8\times1+0.2\times2+0.2\times1}{0.8\times3+0.8\times2+0.8\times1+0.2\times2+0.8\times3+0.2\times1+0.2\times1+0.8\times3} = 0.4423, & \alpha = 0.8, \beta = 0.2 \\ \frac{1\times3+1\times2+0\times2+0\times1}{1\times3+1\times2+1\times1+0\times2+1\times3+0\times1+0\times1+1\times3} = 0.4167, & \alpha = 1, \beta = 0 \end{cases}$$

When $\alpha = 0.5, \beta = 0.5$, the *structure influence* is actually not considered; when $\alpha = 1, \beta = 0$, the weights of influence from the "followers" are totally eliminated, namely, just ignoring the "followers"; and we also assume a relatively mild case $\alpha = 0.8, \beta = 0.2$. We can find that the probability of the influenced agent 8 choosing a is reduced after considering the *structural influence*, which makes sense as agent 8 finds or believes that half of influencing agents choosing a are just "followers", without independent ideas or their own mind.

4.3   Coalitional Influences

If the influencing effects from coalitions (coalitional agents) are also considered, the mixed effects with the influencing effects from structures (structural agents) should be addressed.

Definition 10. *(Probability-based Coalitional Influence Function)* *Assume that* $P_{o(i)}^{\text{SC}}(t)$ *is the probability of agent* i *choosing alternative* o *after both the* Structural *influence and the* Coalitional *influence. If the coalitional influence works in a way similar to the Majority rule, which means once the probability of agent* i *choosing* o *(namely, the weighted ratio of influencing agents choosing* o*) exceeds* $\frac{1}{2}$*, then the resulting choice will be* o *for sure; otherwise, unsure, but depending on the comparison of the probabilities of the influenced agent choosing different alternatives.*

$$P_{o(i)}^{\text{SC}}(t) = \begin{cases} 1, & P_{o(i)}^{\text{S}}(t) > \frac{1}{2} \\ P_{o(i)}^{\text{S}}(t), & \bigwedge_{o' \in \text{M}} P_{o'(i)}^{\text{S}}(t) \leq \frac{1}{2} \quad i \in \mathbb{N}, o \in \mathbb{M} \\ 0, & otherwise \end{cases}$$

As shown in Fig. 5, the two curves of different colors mean that when $P_{o(i)}^{\text{S}}(t) < \frac{1}{2}$*, there are two different possible outcomes for* $P_{o(i)}^{\text{SC}}(t)$*; while if* $P_{o(i)}^{\text{S}}(t)$ *reaches the majority, there will be 100% uniformly.*

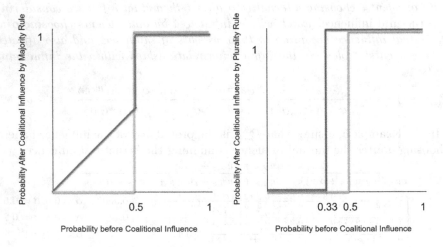

Fig. 5. Decision-making probability after the coalitional influence by the Majority rule and the Plurality rule

If the coalitional influence works in a way similar to the Plurality rule, which means that once the probability of agent i *choosing* o *exceeds* $\frac{1}{2}$*, then* o *will definitely be the resulting choice (as a majority is certainly a plurality); and once the probability of agent* i *choosing* o *falls short of* $\frac{1}{m}$*, then* o *will definitely not be the resulting choice (as less than* $\frac{1}{m}$ *is certainly not a plurality for* m *alternatives); otherwise, unsure, but depending on the comparison of the probabilities of agent* i *choosing different alternatives.*

$$P_{o(i)}^{SC}(t) = \begin{cases} 1, & P_{o(i)}^{S}(t) > \frac{1}{2} \\ 1, & \frac{1}{2} \geq P_{o(i)}^{S}(t) > \frac{1}{m} \bigwedge_{o' \in M \setminus \{o\}} P_{o(i)}^{S}(t) > P_{o'(i)}^{S}(t) \\ 0, & otherwise \end{cases} \quad i \in \mathbb{N}, o \in \mathbb{M}$$

As shown in Fig. 5, the two curves of different colors mean that when $P_{o(i)}^{S}(t) > \frac{1}{2}$ or $< \frac{1}{m}$, then there will be 100% or 0% uniformly; otherwise, there are two different possible outcomes for $P_{o(i)}^{SC}(t)$.

In the Example 5, when *structural influence coefficients* $\alpha = 0.8, \beta = 0.2$, we have the preference of the influenced agent 8 after the *structural influence* as: $P_{a(8)}^{S} = 0.4423, P_{b(8)}^{S} = 0.0769, P_{c(8)}^{S} = 0.4808$. If the *coalitional influence* works in a way similar to the Majority rule, then $P_{a(8)}^{SC} = P_{a(8)}^{S} = 0.4423, P_{b(8)}^{SC} = P_{b(8)}^{S} = 0.0769, P_{c(8)}^{SC} = P_{c(8)}^{S} = 0.4808$ as no one reaches the majority; while if the *coalitional influence* works in a way similar to the Plurality rule, then $P_{a(8)}^{SC} = 0, P_{b(8)}^{SC} = 0, P_{c(8)}^{SC} = 1$ as $P_{c(8)}^{S}$ is larger than both $P_{a(8)}^{S}$ and $P_{b(8)}^{S}$, thus the plurality.

5 Discussion, Conclusion and Future Work

We consider settings of group decision-making where agents' preferences or choices can be influenced by each other. To address the simultaneous influence of more than one agent on another agent, the previous work mainly discussed the multiple influences in an individual way, assuming that all influencing agents exert their own influences independently from each other and the resulting preference or choice of the influenced agent could be simply a linear weighted aggregation of all influencing agents' preferences or choices. Some previous work discussed the influence of coalitions of multiple agents. As for some influencing agents holding the same or similar beliefs, opinions, or choices, an extra influencing effect besides the separate individual influences should be considered. However, another important influencing effect, which could be named the *structural influence* has been ignored, in which the structure (namely, influencing relations) among influencing agents should be addressed to determine the result of the influenced agent. Actually, previous work just perceived the structure (represented as links or ties in social networks) among agents just as the path or channel of influence, while ignore the structure itself can also exert some extra influences.

However, the influencing effect from structures on an agent is not easy to be expressed and computed, as it considers two disparate categories of variables: the influencing subject is the influencing relationship among agents, and the influenced object is the preference or choice of one individual agent. In this paper, we provide an elementary model of how to address the influence from structures on an agent, accompanied by the influence from coalitions of agents. Based on previous work and new thoughts, we extend a new model of influence with three levels: the first level is the influence from independent agents (*individual influence*), the second level is the influence from coalitional agents (*coalitional influence*), and the third level is the influence from structural agents (*structural influence*).

The discussion of the *three levels of influence* has both theoretical values and practical meaning in a wide range of disciplines like (distributed) artificial intelligence, economics, and decision theory (particularly group decision, voting, and games), and even politics and international relations. However, our work is not yet sufficient, as other prospects remain:

- Though starting from common sense, we have not validated the models of the *three levels of influence* (particularly the *structural influence* that we proposed) to accurately characterize influences among agents in reality. In the future, a contribution uniting other disciplines like psychology, cognitive science and behavioural science will be much valuable.
- While addressing the *structural influence*, we introduce the *structural influence coefficients* α and β (the former for "uninfluenced" influencing agents and the latter for the "influenced" influencing agents) and provide different assignments of specific values to display the model and observe the effects. As the *structural influence coefficients* are quite subjective and may be diversified for different people, it is not easy to accurately estimate the coefficients. In the future, the *structural influence coefficients* could be systematically studied, such as what they actually are from the aspect of psychology, how they are measured and computed, and what their computational costs are.
- While addressing the interplay between the *structural influence* and the *coalitional influence*, we only provide one elementary model by a probability-based choice approach: decreasing the influenced one's weight of influence or increasing the uninfluenced one's weight of influence for the *structural influence*; amplifying or reducing the decision-making probability referring to some classical social choice functions for the *coalitional influence*. There might be many other meaningful analytical frameworks to be discussed in the future; indeed, as human minds are complicated, different people have varied value systems and their own judgements, and even for a single person, his cognition will change under different environments, spaces, time, emotions and so on. Besides, an ordering-based approach instead of the non-ordering approach for the *three levels of influence* could be tried in future work.

Acknowledgment. This study is supported by a National Natural Science Foundation of China Grant (71804006) and a National National Natural Science Foundation of China and European Research Council Cooperation and Exchange Grant (7161101045).

References

1. Capuano, N., Chiclana, F., Fujita, H., Herrera-Viedma, E., Loia, V.: Fuzzy group decision making with incomplete information guided by social influence. IEEE Trans. Fuzzy Syst. **26**(3), 1704–1718 (2018)
2. Degroot, M.H.: Reaching a consensus. J. Am. Stat. Assoc. **69**(345), 118–121 (1974)
3. Demarzo, P.M., Vayanos, D., Zwiebel, J.: Persuasion bias, social influence, and unidimensional opinions. Quart. J. Econ. **118**(3), 909–968 (2003)

4. Friedkin, N.E., Johnsen, E.C.: Social influence and opinions. J. Math. Soc. **15**(3–4), 193–206 (1990)
5. Friedkin, N.E., Johnsen, E.C.: Social positions in influence networks. Soc. Networks **19**(3), 209–222 (1997)
6. Golub, B., Jackson, M.O.: Naive learning in social networks and the wisdom of crowds. Am. Econ. J. Microecon. **2**(1), 112–149 (2010)
7. Grabisch, M., Rusinowska, A.: A model of influence in a social network. Theory Decis. **69**(1), 69–96 (2010)
8. Grabisch, M., Rusinowska, A.: A model of influence with an ordered set of possible actions. Theory Decis. **69**(4), 635–656 (2010)
9. Grabisch, M., Rusinowska, A.: Iterating influence between players in a social network. In: 16th Coalition Theory Network Workshop (2011)
10. Grabisch, M., Rusinowska, A.: Measuring influence in command games. Soc. Choice Welfare **33**(2), 177–209 (2009)
11. Grabisch, M., Rusinowska, A.: Influence functions, followers and command games. Games Econ. Behav. **72**(1), 123–138 (2011)
12. Grabisch, M., Rusinowska, A.: A model of influence based on aggregation function. Math. Soc. Sci. **66**(3), 316–330 (2013)
13. Grandi, U., Lorini, E., Perrussel, L.: Propositional opinion diffusion. In: International Conference on Autonomous Agents and Multiagent Systems, pp. 989–997 (2015)
14. Hoede, C., Bakker, R.: A theory of decisional power. J. Math. Soc. **8**, 309–322 (1982)
15. Jackson, M.O.: Social and Economic Networks. Princeton University Press, Princeton (2008)
16. Luo, H.: Agent-based modeling of the UN Security Council decision-making: with signed and weighted mutual-influence. In: 14th Social Simulation Conference (2018)
17. Luo, H.: How to address multiple sources of influences in group decision-making: from a non-ordering to an ordering approach. In: 19th International Conference on Group Decision and Negotiation (2019)
18. Luo, H., Meng, Q.: Multi-agent simulation of SC reform and national game. World Econ. Polit. **6**, 136–155 (2013). in Chinese
19. Maran, A., Maudet, N., Pini, M.S., Rossi, F., Venable, K.B.: A framework for aggregating influenced CP-nets and its resistance to bribery. In: Twenty-Seventh AAAI Conference on Artificial Intelligence (2013)
20. Maudet, N., Pini, M.S., Venable, K.B., Rossi, F.: Influence and aggregation of preferences over combinatorial domains. In: International Conference on Autonomous Agents and Multiagent Systems (2012)
21. Prez, L.G., Mata, F., Chiclana, F., Gang, K., Herrera-Viedma, E.: Modelling influence in group decision making. Soft. Comput. **20**(4), 1653–1665 (2016)
22. Salehi-Abari, A., Boutilier, C.: Empathetic social choice on social networks. In: International Conference on Autonomous Agents and Multi-Agent Systems, pp. 693–700 (2014)

Betweenness Spaces: Morphism and Aggregation Functions

Marta Cardin[(✉)] [ID]

Department of Economics, Ca' Foscari University of Venice,
Sestiere Cannaregio 873, Venice, Italy
mcardin@unive.it

Abstract. The notion of betweenness space or of a convex structure is an abstraction of the standard notion of convexity in a linear space. We first consider a ternary betweenness relation that gives rise to an interval space structure and then we propose a more general definition of betweenness. We study morphism between abstract convex spaces and we characterize aggregation function that are monotone with respect to a betweenness relation.

Keywords: Betweenness · Convexity · Aggregation function

1 Introduction

There are two ways of defining a convexity on a set: by intersection of a family of subset (for example half spaces in an Euclidean space) or by the property of being closed with a family of finitary functions (linear convex combinations in a vector space).

The present paper considers these two aspects of convexity, a very general definition of a betweenness relation and then of an abstract convex space. It is proved that the two definitions are equivalent. In this context it is important to note that in [1] and [2] many significant examples of betweenness spaces are detailed. The paper is organized as follows. In Sect. 2, we consider the case of ternary betweenness and of interval spaces. In Sect. 3 we propose a more general definition of betweenness while in Sect. 4 we study morphisms and aggregation functions in convex spaces.

2 Ternary Betweenness and Interval Spaces

The notion of a point lying between two given points on a geometric line or a totally ordered set has strong intuitive appeal, and has been generalized in a number of directions In all of these, betweenness is taken to be a ternary relation that satisfies certain conditions.

The ternary relation of betweenness comes up in different structures on a given set, reflecting intuitions that range from order-theorethic to the geometrical and topological settings with different meanings.

© Springer Nature Switzerland AG 2019
V. Torra et al. (Eds.): MDAI 2019, LNAI 11676, pp. 92–97, 2019.
https://doi.org/10.1007/978-3-030-26773-5_8

Modern axiomatic definition of betweenness is due to Hedlíková [3] who introduced the ternary representation of betweenness. These relations have been introduced in the context of abstract convexity in [7], in the context of property spaces (see for example [4] and [5]) as well as in graph theory (see [6]).

In a lattice L is defined a ternary betweenness relation

$$B = \{(x, z, y) \in L^3 : x \wedge y \leq z \leq x \vee y\}.$$

This ternary relation satisfies the following properties:

[B1] (Reflexivity) If $z \in \{x, y\}$ then $B(x, z, y)$

[B1](Symmetry) If $B(x, z, y)$ then $B(y, z, x)$

[B1](Transitivity) If $B(x, x', y)$, $B(x, y', y)$ and $B(x', z, y')$ then $B(x, z, y)$. Let us assume that these properties characterize a ternary betweenness relation. There is a close link between interval spaces defined below and the relation of betweenness. We introduce interval spaces $(X, I(x, y))$ (see [8] and the references therein) namely a set X and a function $I \colon X^2 \to \mathcal{P}(X)$ that satisfy the following properties:

[I1] (Extension) $\{x, y\} \subseteq I(x, y)$.

[I2] (Symmetry) $I(x, y) = I(y, x)$.

[I3] (Convexity) If $\{x', y'\} \subseteq I(x, y)$ then $I(x', y') \subseteq I(x, y)$.

We could consider interval spaces that also satisfies the following property:

[I4] (Idempotence) $I(x, x) = x$.

The following proposition can be easily proved.

Proposition 1. *If on the set X is defined a ternary betweenness the function*

$$I(x, y) = \{z \in X \mid B(x, z, y)\}$$

defines an interval space $(X, I(x, y))$.

If $(X, I(x, y))$ is an interval spaces the ternary relation

$$B(x, z, y) \quad \text{if and only if} \quad z \in I(x, y)$$

defines a ternary betweenness in X.

3   Betweenness Spaces as Abstract Convex Structures

Let us consider a more general definition of betweenness as a binary relation involving points and finite subsets of a non empty set X. We introduce this relation by considering for every $k \in \mathbb{N}$ a k-ary interval function $I_k \colon X^k \to \mathcal{P}(X)$ such that

F1 $x_i \in I_k(x_1, \ldots, x_k)$ for every $k, 1 \leq i \leq k$

F2 I_k is a symmetric function for every $k \in \mathbb{N}$

F3 if $y_1, \ldots, y_h \in I_k(x_1, \ldots, x_k)$ then $I_h(y_1, \ldots, y_h) \subseteq I_k(x_1, \ldots, x_k)$

The set $I_k(x_1, \ldots, x_k)$ is called a generalized k-interval and we say that an element x in X is between the elements $x_1, \ldots, x_k$ if $x \in I_k(x_1, \ldots, x_k)$. Then we assume that no point is between the empty set and we can describe also betweenness as binary relation involving points and finite subsets of a given set (see [7]). We say that an element x is between a finite set $\{x_1, \ldots, x_k\}$ of elements of X if $x \in I_k(x_1, \ldots, x_k)$

A non empty set X endowed with a family of k-ary functions $I_k \colon X^k \to \mathcal{P}(X)$ that satisfy properties F1–F3 is called a betweenness space.

Then we consider the general notion of abstract convexity structure studied in [7] and we refer to [7] for a general theory of convexity.

A *convex structure* consists of a set X and a collection $\mathcal{C}$ of subsets of X such that the following properties are satisfied: **C1** $\emptyset$ and X belong to $\mathcal{C}$

C2 if $C_i, \in \mathcal{C}$ then $\bigcap_i C_i \in \mathcal{C}$

C3 if $C_1 \subseteq C_2 \ldots \subseteq C_i \ldots$ are elements of $\mathcal{C}$ then $\bigcup_i C_i \in \mathcal{C}$

The elements of $\mathcal{C}$ are called convex sets of X and the pair $(X, \mathcal{C})$ is called a convex space. A convex set with a convex complement is called an half-space.

Moreover, the convexity notion allows us to define the notion of the convex hull operator, which is similar to that of the closure operator in topology. If X is a set with a convexity $\mathcal{C}$ and A is a subset of X, then the convex hull of $A \subseteq X$ is the set

$$\mathrm{co}(A) = \bigcap \{C \in \mathcal{C} : A \subseteq C\}. \tag{1}$$

This operator enjoys certain properties that are identical to those of usual convexity: for instance $\mathrm{co}(A)$ is the smallest convex set that contains set A. It is also clear that C is convex if and only if $\mathrm{con}(C) = C$.

The convex hull of a set $\{x_1, \ldots, x_k\}$ is called an k-interval and is denoted by $[x_1, \ldots, x_n]$. A 2-interval $[a, b]$ is called an interval or the segment joining a, b.

A convex structure is completely determined by its hull operator, or even by its effect on finite sets since if an element x belongs to $\mathrm{co}(A)$ then x belongs to $\mathrm{co}(F)$ where F is a finite subset of A (see Proposition 2.1 of [7]).

A convexity $\mathcal{C}$ is called n-ary ($n \in \mathbb{N}$) if $A \subseteq \mathcal{C}$ whenever $\mathrm{co}(F) \subseteq A$ for all $F \subseteq A$ where F has at most n elements.

Note that convex spaces are often derived from some mathematical structure (see for example [2]). The following result proves that a convex space is completely characterized by its betweenness relation.

Proposition 2. *Let X a non empty set and $\{I_k : k \in \mathbb{N}\}$ a family of functions $I_k \colon X^k \to \mathcal{P}(X)$ that satisfy properties F_1, F_2 and F_3. Then there exists a convexity $\mathcal{C}$ on X such that*

$$x \in I_k(x_1, \ldots, x_k) \iff [\text{ for all } C \in \mathcal{C} : x_1, \ldots, x_k \in C \implies x \in C]. \tag{2}$$

Conversely if $(X, \mathcal{C})$ is a convex space the functions $I_k \colon X^k \to \mathcal{P}(X)$ defined by

$$I_k(x_1, \ldots, x_k) = \mathrm{co}(\{x_1, \ldots, x_k\}) \tag{3}$$

satisfy properties F_1, F_2 and F_3.

Proof. Let $\{I_k : k \in \mathbb{N}\}$ a family of functions $I_k \colon X^k \to \mathcal{P}(X)$ that satisfy properties F_1, F_2, and F_3. Then we define a convexity $\mathcal{C}$ on X where $C \in \mathcal{C}$ if and only if for every k $x \in I_k(x_1, \ldots, x_k)$ and $x_1, \ldots, x_k \in C$ then $x \in C$.

We can easily prove that $\emptyset$ and X belong to $\mathcal{C}$ and that the intersection of elements of $\mathcal{C}$ is an element of $\mathcal{C}$.

Moreover if we consider a chain $C_1 \subseteq C_2 \ldots \subseteq C_i \ldots$ of elements of $\mathcal{C}$ when $\{x_1, \ldots, x_k\} \subseteq \bigcup_i C_i$ there exists j such that $\{x_1, \ldots, x_k\} \subseteq C_j$ then we can get that $I_k(x_1, \ldots, x_k) \subseteq C_j \subseteq \bigcup_i C_i$.

Now we have to prove that if for all $C \in \mathcal{C}$ if $x_1, \ldots, x_k \in C$ implies that $x \in C$ then $x \in I_k(x_1, \ldots, x_k)$. This can be proved by noting that by property F3 $I_k(x_1, \ldots, x_k)$ is an element of the convexity $\mathcal{C}$. The second part can be easily verified.

4   Morphisms and Aggregation Functions

The class of convexity spaces can be considered as a category and the morphisms are the convexity preserving functions.

If X and Y are betweenness space a map $f \colon X \to Y$ is a morphism if for every $x_1, \ldots, x_k \in X$

$$f(I_k(x_1, \ldots, x_k)) \subseteq I'_k(f(x_1), \ldots, f(x_k))$$

where I_k and I'_k are n-ary functions in X and Y respectively.

If X and Y are interval spaces with betweenness function B and B' respectively, a map $f \colon X \to Y$ is a morphism if for every $x, y, z \in X$

$$B(x, z, y) \quad \Longrightarrow \quad B'(f(x), f(z), f(y)).$$

The following proposition proves a very natural property of morphisms of abstract convex structures.

Proposition 3. *If X and Y are betweenness space, a map $f \colon X \to Y$ is a morphism if and only if for every convex set C in Y $f^{-1}(C)$ is convex in X.*

Proof. Let f be a morphism and C a convex set in Y. If $\{x_1, \ldots, x_k\} \subseteq f^{-1}(C)$ then $f(I_k(x_1, \ldots, x_k)) \subseteq I_k(f(x_1), \ldots, f(x_k)) \subseteq C$.

Then we get that $I_k(x_1, \ldots, x_k) \subseteq f^{-1}(C)$ and then we can prove that $f^{-1}(C)$ is convex.

Conversely note that $\{x_1, \ldots, x_k\} \subseteq f^{-1}I_k(f(x_1), \ldots, f(x_k))$ then $I_k(x_1, \ldots, x_k) \subseteq f^{-1}(I_k(f(x_1), \ldots, f(x_k))$ since $f^{-1}I_k(f(x_1), \ldots, f(x_k))$ is convex.

Therefore we can conclude that $f(I_k(x_1, \ldots, x_k)) \subseteq I_k(f(x_1), \ldots, f(x_k))$.

If $N = \{1, \ldots, n\}$ and X is a betweenness space, then an aggregation function is a map $F \colon X^n \to X$. Note that the space X^n is a convex space with the product convexity (see [7] for details).

We consider now some properties that an aggregation functional $F \colon X^n \to X$ may or may not satisfy.

Let F be an aggregation function, $F \colon X^n \to X$ acting on a betweenness space X F is said to be monotone if when $F(x_1, \ldots x_i \ldots, x_n) \in I_k(z_1, \ldots, z_k)$ for $z_1, \ldots, z_k \in X$ and $y_i \in I_k(z_1, \ldots, z_k)$, $F(x_1, \ldots y_i \ldots, x_n) \in I_k(z_1, \ldots, z_k)$.

F is idempotent if and only if for every $x \in X$, $F(x, x, \ldots, x) = x$.

Now we can easily prove the following statement.

Proposition 4. *If X is a betweenness space and $F \colon X^n \rightarrow \mathcal{X}$ an aggregation function then F is monotone if and only if for every $C \in \mathcal{C}$ if $F(x_1, \ldots x_i \ldots, x_n) \in C$ and $y_i \in C$ then $F(x_1, \ldots y_i \ldots, x_n) \in C$.*

We characterize the class of monotone aggregation functions.

A family of subset of N, $\mathcal{F} \subseteq \mathcal{P}(N)$ is called an upper set if when $A \subseteq B$ and $A \in \mathcal{F}$ then $B \in \mathcal{F}$.

Proposition 5. *Let X be a betweenness space and $F \colon X^n \rightarrow \mathcal{X}$ an idempotent aggregation function. Then F is monotone if and only if for every $C \in \mathcal{C}$ there exists an upper set $\mathcal{F}_C$ such that*

$$F(\mathbf{x}) \in \bigcap \{C : \{i \in N : x_i \in C\} \in \mathcal{F}_C\}. \tag{4}$$

Proof. Note that if $F \colon X^n \rightarrow \mathcal{X}$ is an idempotent aggregation function that satisfies (4) then if $F(x_1, \ldots, x_n) \in C$ and $(y_1, \ldots, y_n) \in X^n$ is such that $\{i \in N : x_i \in C\} \subseteq \{i \in N : y_i \in C\}$ then obviously $F(x_1, \ldots, x_n) \in C$.

If C is an element of $\mathcal{C}$ and $\mathbf{x}$ an element of X^n, let $A(\mathbf{x}, C) = \{i \in N : x_i \in C\}$. Being F a monotone and idempotent function we can prove that if we consider two elements $\mathbf{x}, \mathbf{y} \in X^n$ such that for every $i \in N$, $x_i \in C$ if and only if $y_i \in C$ then $F(\mathbf{x}) \in C$ if and only if $F(\mathbf{y}) \in C$.

We say that a set A is C-decisive if there exists $\mathbf{x} \in X^N$ such that $A(\mathbf{x}, C) = A$ and $F(\mathbf{x}) \in C$. Hence a set is C-decisive if and only if for every $\mathbf{x} \in X^n$ such that $A(\mathbf{x}, C) = A$, $F(\mathbf{x}) \in C$.

For every $C \in \mathcal{C}$ let $\mathcal{F}_C$ the family of C-decisive subsets of N. Then we get that for every $\mathbf{x} \in X^n$, $F(\mathbf{x}) \in C$ if and only if $A(\mathbf{x}, C) \in \mathcal{F}_C$. The family of subset $\mathcal{F}_C$ is non empty since F is idempotent. For every $C \in \mathcal{C}$ the family of subsets $\mathcal{F}_C$ is an upper set since F is a monotone function.

Now we consider the class of monotone aggregation functions that are also morphisms.

Proposition 6. *If X is a betweeness space and $F \colon X^n \rightarrow \mathcal{X}$ a monotone and idempotent aggregation function. Then F is a morphism if and only if for every $C \in \mathcal{C}$ there exists a totally ordered set $\mathcal{F}_C$ of elements of $\mathcal{P}(N)$ such that*

$$F(\mathbf{x}) \in \bigcap \{C : \{i \in N : x_i \in C\} \in \mathcal{F}_C\}. \tag{5}$$

Proof. If we consider a product of finitely many convex spaces all convex sets are product of convex sets (see Proposition 1.10.2 in [8]). Then F is a morphism if and only $F^{-1}(C)$ is a product of convex sets $C_i, 1 \leq i \leq n$ for every $C \in \mathcal{C}$.

Hence $F(x_1, \ldots, x_n) \in C$ if and only if $x_i \in C_i$ and then if and only $\mathcal{F}_C$ is a totally ordered subset of $\mathcal{P}(N)$.

5 Concluding Remarks

We have considered and characterized an abstract definition of betweenness. We have shown that this definition is equivalent to that of abstract convex space and we have considered also morphisms and aggregation functions in betweenness spaces. We plan to consider other classes of aggregation functions, and to find more applications of our results in future work.

References

1. Cardin, M.: Aggregation over property-based preference domains. In: Torra, V., Mesiar, R., De Baets, B. (eds.) AGOP 2017. AISC, vol. 581, pp. 130–137. Springer, Cham (2018). https://doi.org/10.1007/978-3-319-59306-7_13
2. Cardin, M.: Sugeno integral on property-based preference domains. In: Kacprzyk, J., Szmidt, E., Zadrożny, S., Atanassov, K.T., Krawczak, M. (eds.) IWIF-SGN/EUSFLAT -2017. AISC, vol. 641, pp. 400–407. Springer, Cham (2018). https://doi.org/10.1007/978-3-319-66830-7_36
3. Hedlíková, J.: Ternary spaces, media and Chebyshev sets. Czechoslovak Math. J. **33**, 373–389 (1983)
4. Nehring, K., Puppe, C.: The structure of strategy-proof social choice - part I: general characterization and possibility results on median spaces. J. Econ. Theory **135**(1), 269–305 (2007)
5. Nehring, K., Puppe, C.: Abstract Arrowian aggregation. J. Econ. Theory **145**, 467–494 (2010)
6. Rautenbach, D., Schäfer, P.M.: Strict betweennesses induced by posets as well as by graphs. Order **28**, 89–97 (2011)
7. van de Vel, M.L.J.: Theory of Convex Structures North-Holland Mathematical Library, vol. 50. Elsevier, Amsterdam (1993)
8. Vannucci, S.: Weakly unimodal domains, anti-exchange properties, and coalitional strategy-proofness of aggregation rules. Math. Soc. Sci **84**, 50–67 (2016)

On Idempotent n-ary Uninorms

Jimmy Devillet[1(✉)], Gergely Kiss[2], and Jean-Luc Marichal[1]

[1] University of Luxembourg, Mathematics Research Unit,
Esch-sur-Alzette, Luxembourg
{jimmy.devillet,jean-luc.marichal}@uni.lu
[2] Hungarian Academy of Sciences, Alfréd Rényi Institute of Mathematics,
Budapest, Hungary
kigergo57@gmail.com

Abstract. In this paper we describe the class of idempotent n-ary uninorms on a given chain. When the chain is finite, we axiomatize the latter class by means of the following conditions: associativity, quasitriviality, symmetry, and nondecreasing monotonicity. Also, we show that associativity can be replaced with bisymmetry in this new axiomatization.

1 Introduction

Let X be a nonempty set and let $n \geq 2$ be an integer. For a few decades, many classes of binary aggregation functions have been investigated due to their great importance in data fusion (see, e.g. [8] and the references therein). Among these classes, the class of binary uninorms plays an important role in fuzzy logic. Recently, the study of the class of n-ary uninorms gained an increasing interest (see, e.g. [9]).

This paper, which is a shorter version of [6][1], focuses on characterizations of the class of idempotent n-ary uninorms (Definition 3). In Sect. 2, we provide a characterization of these operations and show that they only depend on the extreme values of the variables (Proposition 1). We also provide a description of these operations as well as an alternative axiomatization when the underlying set is finite (Theorem 1). In particular, we extend characterizations of the class of idempotent binary uninorms obtained by Couceiro et al. [4, Theorems 12 and 17] to the class of idempotent n-ary uninorms. In Sect. 3, we investigate some subclasses of bisymmetric n-ary operations and derive several equivalences involving associativity and bisymmetry. More precisely, we show that if an n-ary operation has a neutral element, then it is associative and symmetric if and only if it is bisymmetric (Corollary 1). Also, we show that if an n-ary operation is quasitrivial and symmetric, then it is associative if and only if it is bisymmetric (Corollary 1). These observations enable us to replace associativity with bisymmetry in our axiomatization (Theorem 2).

We adopt the following notation throughout. We use the symbol X_k if X contains $k \geq 1$ elements, in which case we assume without loss of generality

---

[1] This paper is also an extended version of [7].

© Springer Nature Switzerland AG 2019
V. Torra et al. (Eds.): MDAI 2019, LNAI 11676, pp. 98–104, 2019.
https://doi.org/10.1007/978-3-030-26773-5_9

that $X_k = \{1, \ldots, k\}$. Finally, for any integer $k \geq 1$ and any $x \in X$, we set $k \cdot x = x, \ldots, x$ (k times). For instance, we have $F(3 \cdot x, 2 \cdot y) = F(x, x, x, y, y)$.

Recall that a binary relation R on X is said to be

- *total* if $\forall x, y$: xRy or yRx;
- *transitive* if $\forall x, y, z$: xRy and yRz implies xRz;
- *antisymmetric* if $\forall x, y$: xRy and yRx implies $x = y$.

Recall also that a *total ordering on X* is a binary relation $\leq$ on X that is total, transitive, and antisymmetric. The ordered pair $(X, \leq)$ is then called a *chain*.

Definition 1. *An operation $F: X^n \to X$ is said to be*

- idempotent *if $F(n \cdot x) = x$ for all $x \in X$;*
- quasitrivial *(or conservative) if $F(x_1, \ldots, x_n) \in \{x_1, \ldots, x_n\}$ for all $x_1, \ldots, x_n \in X$;*
- symmetric *if $F(x_1, \ldots, x_n)$ is invariant under any permutation of $x_1, \ldots, x_n$;*
- associative *if*

$$F(x_1, \ldots, x_{i-1}, F(x_i, \ldots, x_{i+n-1}), x_{i+n}, \ldots, x_{2n-1})$$
$$= F(x_1, \ldots, x_i, F(x_{i+1}, \ldots, x_{i+n}), x_{i+n+1}, \ldots, x_{2n-1})$$

for all $x_1, \ldots, x_{2n-1} \in X$ and all $i \in \{1, \ldots, n-1\}$;
- bisymmetric *if*

$$F(F(\mathbf{r}_1), \ldots, F(\mathbf{r}_n)) = F(F(\mathbf{c}_1), \ldots, F(\mathbf{c}_n))$$

for all $n \times n$ matrices $[\mathbf{c}_1 \cdots \mathbf{c}_n] = [\mathbf{r}_1 \cdots \mathbf{r}_n]^T \in X^{n \times n}$;
- nondecreasing *for some total ordering $\leq$ on X if $F(x_1, \ldots, x_n) \leq F(x'_1, \ldots, x'_n)$ whenever $x_i \leq x'_i$ for all $i \in \{1, \ldots, n\}$.*

Given a total ordering $\leq$ on X, the *maximum* (resp. *minimum*) *operation on X* for $\leq$ is the symmetric n-ary operation $\max_{\leq}$ (resp. $\min_{\leq}$) defined by $\max_{\leq}(x_1, \ldots, x_n) = x_i$ (resp. $\min_{\leq}(x_1, \ldots, x_n) = x_i$) where $i \in \{1, \ldots, n\}$ is such that $x_j \leq x_i$ (resp. $x_i \leq x_j$) for all $j \in \{1, \ldots, n\}$.

Definition 2. *Let $F: X^n \to X$ be an operation. An element $e \in X$ is said to be a* neutral element *of F if*

$$F((i-1) \cdot e, x, (n-i) \cdot e) = x$$

for all $x \in X$ and all $i \in \{1, \ldots, n\}$.

2   A First Characterization

In this section we provide a characterization of the n-ary operations on X that are associative, quasitrivial, symmetric, and nondecreasing for some total ordering $\leq$ on X. We will also show that in the case where X is finite these operations are exactly the idempotent n-ary uninorms.

Recall that a *uninorm* on a chain $(X, \leq)$ is a binary operation $U \colon X^2 \to X$ that is associative, symmetric, nondecreasing for $\leq$, and has a neutral element (see [5,11]). It is not difficult to see that any idempotent uninorm is quasitrivial.

The concept of uninorm can be easily extended to n-ary operations as follows.

Definition 3 (see [9]). *Let $\leq$ be a total ordering on X. An n-ary uninorm is an operation $F \colon X^n \to X$ that is associative, symmetric, nondecreasing for $\leq$, and has a neutral element.*

The next proposition provides a characterization of idempotent n-ary uninorms. In particular, since any idempotent uninorm is quasitrivial, it shows that an idempotent n-ary uninorm always outputs either the greatest or the smallest of its input values.

Proposition 1. *Let $\leq$ be a total ordering on X and let $F \colon X^n \to X$ be an operation. Then F is an idempotent n-ary uninorm if and only if there exists a unique idempotent uninorm $U \colon X^2 \to X$ such that*

$$F(x_1, \ldots, x_n) = U(\min_{\leq}(x_1, \ldots, x_n), \max_{\leq}(x_1, \ldots, x_n)), \qquad x_1, \ldots, x_n \in X.$$

In this case, the uninorm U is uniquely defined as $U(x, y) = F((n-1) \cdot x, y)$.

We now introduce the concept of single-peaked total ordering which first appeared for finite chains in social choice theory (see Black [2,3]).

Definition 4. *Let $\leq$ and $\preceq$ be total orderings on X. We say that $\preceq$ is single-peaked for $\leq$ if for any $a, b, c \in X$ such that $a < b < c$ we have $b \prec a$ or $b \prec c$.*

When X is finite, the single-peakedness property of a total ordering $\preceq$ on X for some total ordering $\leq$ on X can be easily checked by plotting a function, say $f_{\preceq}$, in a rectangular coordinate system in the following way. Represent the reference totally ordered set $(X, \leq)$ on the horizontal axis and the reversed version of the totally ordered set $(X, \preceq)$, that is $(X, \preceq^{-1})$, on the vertical axis. The function $f_{\preceq}$ is defined by its graph $\{(x, x) : x \in X\}$.[2] We then see that the total ordering $\preceq$ is single-peaked for $\leq$ if and only if $f_{\preceq}$ has only one local maximum.

Example 1. Consider $X = X_6$ endowed with the usual total ordering $\leq$ defined by $1 < 2 < 3 < 4 < 5 < 6$. Figure 1 gives the functions $f_{\preceq}$ and $f_{\preceq'}$ corresponding to the total orderings $3 \prec 4 \prec 2 \prec 5 \prec 1 \prec 6$ and $4 \prec' 2 \prec' 6 \prec' 1 \prec' 3 \prec' 5$,

---

[2] When $X = X_k$ for some integer $k \geq 1$, the graphical representation of $f_{\preceq}$ is then obtained by joining the points $(1, 1), \ldots, (k, k)$ by line segments.

respectively, on X_6. We see that $\preceq$ is single-peaked for $\leq$ since $f_\preceq$ has only one local maximum while $\preceq'$ is not single-peaked for $\leq$ since $f_{\preceq'}$ has three local maxima.

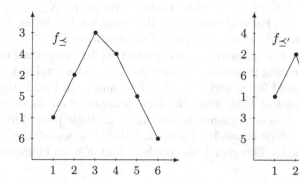

Fig. 1. $\preceq$ is single-peaked (left) while $\preceq'$ is not (right)

It is known (see, e.g., [1]) that there are exactly 2^{k-1} single-peaked total orderings on X_k for the usual total ordering $\leq$ defined by $1 < \ldots < k$.

The following theorem provides several characterizations of the class of associative, quasitrivial, symmetric, and nondecreasing operations $F\colon X^n \to X$. In particular, it provides a new axiomatization as well as a description of idempotent n-ary uninorms when the underlying set X is finite. In the latter case, it also extends characterizations of the class of idempotent uninorms obtained by Couceiro et al. [4, Theorems 12 and 17] to the class of idempotent n-ary uninorms.

Theorem 1. *Let $\leq$ be a total ordering on X and let $F\colon X^n \to X$ be an operation. The following assertions are equivalent.*

(i) F is associative, quasitrivial, symmetric, and nondecreasing for $\leq$.

(ii) There exists a quasitrivial, symmetric, and nondecreasing operation $G\colon X^2 \to X$ such that

$$F(x_1,\ldots,x_n) \;=\; G(\min{}_{\leq}(x_1,\ldots,x_n),\max{}_{\leq}(x_1,\ldots,x_n)), \qquad x_1,\ldots,x_n \in X.$$

(iii) There exists a total ordering $\preceq$ on X that is single-peaked for $\leq$ and such that $F = \max_\preceq$.

If $X = X_k$ for some integer $k \geq 1$, then any of the assertions $(i) - (iii)$ above is equivalent to the following one.

(iv) F is an idempotent n-ary uninorm.

Moreover, there are exactly 2^{k-1} operations $F\colon X_k^n \to X_k$ satisfying any of the assertions $(i) - (iv)$.

Now, let us illustrate Theorem 1 for binary operations. Recall that the contour plot of any operation $F \colon X_k^2 \to X_k$ is the undirected graph (X_k^2, E), where

$$E = \{\{(x,y),(u,v)\} \mid (x,y) \neq (u,v) \text{ and } F(x,y) = F(u,v)\}.$$

We can always represent the contour plot of any operation $F \colon X_k^2 \to X_k$ by fixing a total ordering on X_k. For instance, using the usual total ordering $\leq$ on X_6, in Fig. 2 (left) we represent the contour plot of an operation $F \colon X_6^2 \to X_6{}^3$. It is not difficult to see that F is quasitrivial and symmetric. To check whether F is associative and nondecreasing it suffices by Theorem 1 to find a total ordering $\preceq$ on X_6 that is single-peaked for $\leq$ and such that $F = \max_{\preceq}$. In Fig. 2 (right) we represent the contour plot of F by using the total ordering $\preceq$ on X_6 defined by $3 \prec 4 \prec 2 \prec 5 \prec 6 \prec 1$. It is not difficult to see that $\preceq$ is single-peaked for $\leq$. Also, we have $F = \max_{\preceq}$ which shows by Theorem 1 that F is associative and nondecreasing for $\leq$. Thus, by Theorem 1 we conclude that F is an idempotent uninorm.

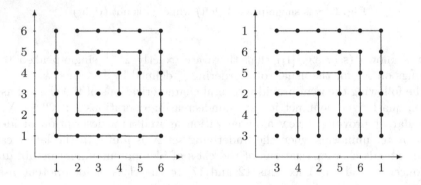

Fig. 2. An idempotent uninorm $F \colon X_6^2 \to X_6$

Remark 1. We observe that an alternative characterization of idempotent uninorms on chains was provided in [10]. Due to Proposition 1, we can extend this characterization to the class of idempotent n-ary uninorms.

3   An Alternative Characterization

In this section we investigate bisymmetric n-ary operations and derive several equivalences involving associativity and bisymmetry. More precisely, if an n-ary operation has a neutral element, then it is associative and symmetric if and only if it is bisymmetric. Also, if an n-ary operation is quasitrivial and symmetric, then it is associative if and only if it is bisymmetric. In particular, these observations enable us to replace associativity with bisymmetry in Theorem 1.

---

[3] To simplify the representation of the connected components, we omit edges that can be obtained by transitivity.

Definition 5. *We say that an operation* $F\colon X^n \to X$ *is* ultrabisymmetric *if*

$$F(F(\mathbf{r}_1), \ldots, F(\mathbf{r}_n)) = F(F(\mathbf{r}'_1), \ldots, F(\mathbf{r}'_n))$$

for all $n \times n$ *matrices* $[\mathbf{r}_1 \ \cdots \ \mathbf{r}_n]^T, [\mathbf{r}'_1 \ \cdots \ \mathbf{r}'_n]^T \in X^{n \times n}$, *where* $[\mathbf{r}'_1 \ \cdots \ \mathbf{r}'_n]^T$ *is obtained from* $[\mathbf{r}_1 \ \cdots \ \mathbf{r}_n]^T$ *by exchanging two entries.*

Ultrabisymmetry seems to be a rather strong property. However, as the next result shows, this property is satisfied by any operation that is bisymmetric and symmetric.

Proposition 2. *Let* $F\colon X^n \to X$ *be an operation. If* F *is ultrabisymmetric, then it is bisymmetric. The converse holds whenever* F *is symmetric.*

Proposition 3. *Let* $F\colon X^n \to X$ *be an operation. Then the following assertions hold.*

(a) If F *is quasitrivial and ultrabisymmetric, then it is associative and symmetric.*
(b) If F *is associative and symmetric, then it is ultrabisymmetric.*
(c) If F *is bisymmetric and has a neutral element, then it is associative and symmetric.*

Corollary 1. *Let* $F\colon X^n \to X$ *be an operation. Then the following assertions hold.*

(a) If F *is quasitrivial and symmetric, then it is associative if and only if it is bisymmetric.*
(b) If F *has a neutral element, then it is associative and symmetric if and only if it is bisymmetric.*

From Corollary 1 we immediately derive the following theorem, which is an important and surprising result.

Theorem 2. *In Theorem 1(i) we can replace associativity with bisymmetry. Also, in Theorem 1(iv) we can replace associativity and symmetry with bisymmetry.*

Acknowledgments. The first author is supported by the Luxembourg National Research Fund under the project PRIDE 15/10949314/GSM. The second author is also supported by the Hungarian National Foundation for Scientific Research, Grant No. K124749.

References

1. Berg, S., Perlinger, T.: Single-peaked compatible preference profiles: some combinatorial results. Soc. Choice Welf. **27**(1), 89–102 (2006)
2. Black, D.: On the rationale of group decision-making. J. Polit. Econ. **56**(1), 23–34 (1948)

3. Black, D.: The Theory of Committees and Elections. Kluwer Academic Publishers, Dordrecht (1987)
4. Couceiro, M., Devillet, J., Marichal, J.-L.: Characterizations of idempotent discrete uninorms. Fuzzy Sets Syst. **334**, 60–72 (2018)
5. De Baets, B.: Idempotent uninorms. Eur. J. Oper. Res. **118**, 631–642 (1999)
6. Devillet, J., Kiss, G., Marichal, J.-L.: Characterizations of quasitrivial symmetric nondecreasing associative operations. Semigroup Forum **98**(1), 154–171 (2019)
7. Devillet, J., Kiss, G., Marichal, J.-L.: Characterizations of idempotent n-ary uninorms. In: Proceedings of the 38th Linz Seminar on Fuzzy Set Theory (LINZ 2019), Linz, Austria, 4 p. 5–8 February 2019
8. Grabisch, M., Marichal, J.-L., Mesiar, J.-L., Pap, E.: Aggregation Functions: Encyclopedia of Mathematics and Its Applications, no. 127. Cambridge University Press, Cambridge (2009)
9. Kiss, G., Somlai, G.: A characterization of n-associative, monotone, idempotent functions on an interval that have neutral elements. Semigroup Forum **96**(3), 438–451 (2018)
10. Mesiarová-Zemánková, M.: A note on decomposition of idempotent uninorms into an ordinal sum of singleton semigroups. Fuzzy Sets Syst. **299**, 140–145 (2016)
11. Yager, R.R., Rybalov, A.: Uninorm aggregation operators. Fuzzy Sets Syst. **80**, 111–120 (1996)

On Aggregation of Risk Levels
Using T-Conorms

Māris Krastiņš[1,2(✉)]

[1] Department of Mathematics, University of Latvia, Jelgavas iela 3, Riga, Latvia
mk18032@lu.lv
[2] Institute of Mathematics and Computer Science, University of Latvia,
Raiņa bulvāris 29, Riga, Latvia
http://www.lumii.lv

Abstract. This paper deals with solutions for numeric evaluation of
risks containing several different risk factors assessed by experts. The proposed methods can be used to assess the risks and obtain the risk scores
in different industries, including financial industry, but they are also suitable for assessing risks in other areas, e.g. project management. While
risk is usually considered as a function of probability and impact with
strong quantitative background, there are many practical cases when
only qualitative risk assessment based on expert opinions can be used. At
the same time there are still requirements and needs for applying numerical values and mathematical models to such qualitative assessments. We
consider the options for aggregation of risk levels for corresponding risk
factors and obtaining consolidated risk level using transparent and self-
explanatory approach. The proposed models are constructed using maximum t-conorm and Łukasiewicz t-conorm. Practical example is provided
for calculation of consolidated risk score.

Keywords: Risk assessment · Aggregation operators ·
Maximum t-conorm · Łukasiewicz t-conorm

1 Introduction

The risk assessment methods and tools have been widely developed for different purposes over the last decades. Many industries consider risk assessment as
important part of their business, but for some of them the risk management
process is the milestone in their daily operations. The financial industry is one
of the most evident examples. Therefore we have based the main part of our
research on examples from this particular industry.

While such risks as the credit risk, the financial risk and the liquidity risk
have always had a solid statistic basis for numerical calculations and application

Partially supported by the project No LZP-2018/2-0338 'Development of fuzzy logic
based technologies for risk assessment by means of relation-grounded aggregation' of
the Latvian Council of Science.

© Springer Nature Switzerland AG 2019
V. Torra et al. (Eds.): MDAI 2019, LNAI 11676, pp. 105–112, 2019.
https://doi.org/10.1007/978-3-030-26773-5_10

of the probability theory, supervisory authorities have increased their expectations and demands for introducing mathematical models for assessment of the compliance risk, including the risk of money laundering, which lacks any solid mathematical background and can rely on expert opinions only. At the same time it should be noted that the number of subjects to Anti Money Laundering (AML) laws has significantly increased over the last decade. Therefore the models for scoring of money laundering risk often shall be implemented not only by financial institutions, but also by real estate brokers, gambling companies, notaries and other obliged entities pursuant to applicable laws.

The leading global AML software providers have developed different technological solutions to cope with increasing legal demands for scoring and monitoring of customers and their transactions. However, these solutions often have rather sophisticated underlying mathematical models, which are not openly disclosed even to the end users. It should be also admitted that so far there have been very few attempts in proposing the models for money laundering risk assessment in scientific publications. The review article [1], published in February 2018, summarizes all efforts used so far in finding the most appropriate approach for efficient handling of tasks related to prevention of money laundering and highlights importance of the suspicious transactions' detection. It is evident from the practical point of view that transactional patterns are just consequences from engaging into business with particular customers posing lower or higher risk of money laundering. A use case provided in [2] allows to cope with a very simple transactional pattern, but it is important to take into account the customer specifics as part of so called Know Your Customer (KYC) process as mentioned also in [1].

The main goal of the KYC process is to implement a robust solution allowing the obliged entities to assess the level of money laundering risk as part of the customer relationship establishment, often referred as an on-boarding process. The customers shall disclose different qualitative and quantitative data which can be afterwards evaluated by experts and transposed into risk factors with corresponding risk levels. Aggregation of these risk levels results in the risk scores. Different scales can be used for this purpose, but we will apply the fuzzy numbers and assign the values close to 0 for the lowest risks, and the values close to 1 for the highest risk. It shall be noted that some risk models are inverted by assigning the values close to 1 for the lowest risks, and the values close to 0 for the highest risk. Depending on the resulting money laundering risk scores obliged entities are required to apply risk mitigation actions, which include, but are not limited to regular monitoring of customer transactions, obtaining relevant documentation on customers' sources of wealth and funds etc.

The paper considers several options for risk level aggregation of such risk factors as customer residence, occupation or business for legal entities, customer reputation, estimated volumes and values of transactions and other similar factors as selected by the obliged entity or required by applicable laws and regulations. Section 2 provides an overview of aggregation principles which are further analysed by considering application of maximum t-conorm in Sect. 3 and

Łukasiewicz t-conorm in Sect. 4. Combination of these t-conorms is proposed in Sect. 5. A practical example of risk levels' aggregation based on expert evaluations is presented in the Sect. 6.

2 Conditions for Aggregation of Risk Factors in Qualitative Risk Assessment Model

The risk is usually considered as a function of probability and impact or likelihood and severity. Such definition is suitable for many industries as outlined, for example, in [3,4]. It also allows application of Mamdani-Type or Sugeno-Type fuzzy inference systems for obtaining the consolidated risk levels as described by [5,6]. This process is similar to different other practical applications provided in [7].

Contrary to the previous examples of risks, qualitative risk assessment often has rather limited quantitative basis for applying mathematical models. We will use the money laundering risk to explain our approach. This risk is rather specific and embraces comparably vague component of severity, which can be hardly characterised by any numeric value. It particularly applies to such indicators as reputational impact or business sustainability. At the same time AML laws and regulations require implementation of detailed KYC procedures and assessment of multiple risk factors for each customer. Therefore an expert opinion is among the most appropriate solutions for assigning the risk levels for corresponding notional risk factors. A simple example can be used to explain the need for a human decision in assessing the risk level. Let us consider two companies of different size and their estimated average transaction values. While EUR 100 000 payment would be treated as low risk indicator with value close to 0 for the large company, it would be definitely a high risk indicator with value close to 1 for the small company. Similar judgement is valid also for assessment of expected payment volumes which can be considered as another different risk factor.

Let us consider that all risk levels of corresponding risk factors $X_k, k \in \{1, ..., i\}$ and $k \in \mathbb{N}$ are expressed in the form of fuzzy set $\mu = (x_1, ..., x_i), i \in \{1, ..., n\}$ and $n \in \mathbb{N}$. In order to aggregate these risk levels we will use aggregation operator $\mathbf{A} : \bigcup_{n \in \mathbb{N}} [0,1]^n \to [0,1]$. An example of ten risk factors ($i = 10$) with corresponding risk levels for two sample customers is provided in Fig. 1. It is evident that customers can be different and with different risk levels for corresponding risk factors, which are notional and do not correspond to any particular values on x axis.

The use of risk level fuzzy sets, especially their graphical representation, provides a good preliminary overview of overall customer risk level. However, they do not encompass importance of each risk factor, and also do not provide a clear answer, if the obliged entity is facing high or low risk customer. Therefore we will explore the options for aggregation of risk levels using different aggregation operators in order to obtain the customer risk score. First of all we consider

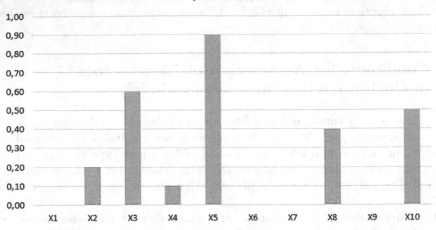

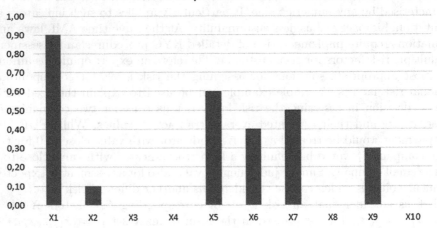

Fig. 1. Example of risk levels for two sample customers.

application of different average operators analysed by [8]. We define the fuzzy weighted average of risk levels $(x_1, ..., x_i)$ as follows:

$$W(x_1, ..., x_n) = \sum_{i=1}^{n} \omega_i x_i$$

where ω_i are weights for each corresponding x_i and $\sum_{i=1}^{n} \omega_i = 1$. At the first glimpse such approach could be considered as suitable since the risk score is the mean value of all weighted risk levels. However, due to specifics of money laundering risk there are many occasions when it is not reasonable to accept that the risk score is lower than value of the highest risk level in the fuzzy set $\mu = (x_1, ..., x_n)$. Therefore alternative aggregation operators should be considered.

3   Aggregation of Risk Levels with Maximum T-Conorm

As part of the risk level aggregation it can be assumed that for particular cases the total risk level of any sample customer cannot be lower than the highest (maximum) risk level of all risk factors. Therefore we can apply the maximum t-conorm $M(x_1, ..., x_n) = \max(x_1, ..., x_i)$, $i \in \{1, ..., n\}$ and $n \in \mathbb{N}$. It should be noted that all risk factors X_k may not be equally important for customer risk scoring. Therefore we apply fuzzy coefficients $a_i \in [0, 1], i \in \{1, ..., n\}$ and $n \in \mathbb{N}$ allowing to keep the values of certain initial risk levels or decrease them in similar way as proposed by [9]. In our model $\sum_{i=1}^{n} a_i \neq 1$, and fuzzy coefficients can be regarded as the indicators of risk appetite resulting in the fact that particular risk levels are decreased, if obliged entity considers them as less important. Consequently the maximum t-conorm can be expressed in the following format:

$$M(x_1, ..., x_n) = \max(a_1 x_1, ..., a_i x_i), \tag{1}$$

$i \in \{1, ..., n\}$ and $n \in \mathbb{N}$.

When applying such aggregation, the most important risk factors with corresponding risk levels are considered. However, other risk factors of lower importance with non-zero risk levels should not be disregarded as their aggregated impact could be more severe than the highest risk level of the most important risk factor. This implies that additional options for aggregation of the risk levels of non-critical risk factors are required.

4   Aggregation of Risk Levels with Łukasiewicz T-Conorm

Aggregation using arithmetic sum often results in values exceeding 1. If we consider the example provided in Fig. 1, it is evident that the sum of only two particular risk levels for both sample customers is greater than 1 while there are non-zero risk level values for four more risk factors. In order to overcome this problem, we apply Łukasiewicz t-conorm $L(x_1, ..., x_n) = \min(1, \sum_{i=1}^{n} x_i)$, $n \in \mathbb{N}$. As in the case of maximum t-conorm we note that the risk factors are not equally important. Therefore we apply the same fuzzy coefficients $a_i \in [0, 1], i \in \{1, ..., n\}$ and $n \in \mathbb{N}$ for calibration of the risk level values. The adjusted Łukasiewicz t-conorm is expressed as follows:

$$L(x_1, ..., x_n) = \min(1, \sum_{i=1}^{n} a_i x_i), \tag{2}$$

$n \in \mathbb{N}$. The formula (2) underlines the meaning of coefficients a_i as values expressing the risk appetite. It is evident that lower a_i means higher risk appetite, and more risk factors can be included in the total risk score unless it does not exceed the threshold of non-acceptable risk as internally set by the obliged entity.

While Łukasiewicz t-conorm and formula (2) provide a sound basis for suitable aggregation of money laundering risk levels into single value risk score, such aggregation may decrease the importance of some critical risk factors preserved in case of the maximum t-conorm. There is another scenario when the obliged entity may select some critical risk factors and benchmark their risk levels against aggregated risk level values of the remaining risk factors. Therefore construction of the combined aggregation operator is proposed in the Sect. 5.

5   Aggregation of Risk Levels with Combined T-Conorm

Let us assume that we have to deal with some critically important risk factors and less significant risk factors by splitting them in two groups and combining results using the following t-conorm: $C(x_1, ..., x_n) = \max(x_1, ..., x_k, \min(1, \sum_{j=k+1}^{n} x_j))$, $n \in \mathbb{N}$. Such t-conorm is commutative, monotone, and the number 1 acts as its identity element. However, this t-conorm is not associative as it basically consists of two independent parts. From the practical point of view, it is not critical that associativity does not hold. The t-conorm C is not unique in that sense, as there are similar examples of functions, like copulas and quasicopulas analysed in [10], where associativity does not hold.

When applying the combined t-conorm C we should consider the same principle of the importance of risk factors applied to maximum t-conorm and Łukasiewicz t-conorm. The resulting formula will be the following:

$$C(x_1, ..., x_n) = \max(a_1 x_1, ..., a_k x_k, \min(1, \sum_{j=k+1}^{n} a_j x_j)), \qquad (3)$$

$n \in \mathbb{N}$. In practice (3) will allow to select the critical risk factors out of the sum of remaining risk factors. Such calculation is two folded. First of all it secures that important high level risk factors are not missed, and secondly it allows to benchmark aggregated level of less important risk factors against critical risk factors.

6   Example of Aggregated Risk Score

The main purpose of the proposed aggregation model is to enable the end users, including obliged entities and supervisory authorities, to implement simple and clear solutions for obtaining the customer's money laundering risk score. In our example we will use four risk levels and apply them to each risk factor pursuant to the widely accepted international standards: low risk, medium risk, high risk, non-acceptable risk. Table 1 provides corresponding intervals for each risk level.

Table 1. Values of risk levels.

Risk level	Interval of risk score
Low	$[0, 0{,}30)$
Medium	$[0{,}30, 0{,}60)$
High	$[0{,}60, 0{,}95]$
Non-acceptable	$(0{,}95, 1]$

We will use a sample customer A from Fig. 1 with corresponding fuzzy set of 10 risk levels $\mu = (x_1, ..., x_{10}) = (0; 0, 2; 0, 6; 0, 1; 0, 9; 0; 0; 0, 4; 0; 0, 5)$. As the next step we will apply the following fuzzy coefficients characterising importance of corresponding risk factors: $a_1 = 1, a_2 = 1, a_3 = 1, a_4 = 0, 8, a_5 = 0, 7, a_6 = 0, 5, a_7 = 0, 5, a_8 = 0, 4, a_9 = 0, 3, a_{10} = 0, 2$. In such case (1) and (2) result in the following values: $M = 0, 63, L = \min(1, 1, 77) = 1$. It means that the maximum t-conorm returns the value slightly above the lowest value of the high risk, while application of Łukasiewicz t-conorm results in non-acceptable risk. Therefore intuitively these values could indicate that the most suitable level of aggregated risk score could be high, but certainly lower than non acceptable. In order to apply (3) we will select x_1, x_2, x_3, x_4 as the most critical risk factors. Such selection results in the following value of the combined t-conorm: $C = \max(0, 6; \min(1; 0, 89)) = 0, 89$. The result corresponding to high level risk (with risk score 0,89) is quite suitable for the obliged entity with rather low risk appetite. At the same time it should be admitted that the risk appetite can be increased by modifying fuzzy coefficients a_i, and a_5 in particular.

7   Conclusions

The proposed methods for aggregation of risk levels enable efficient calculation of risk scores. While simple t-conorms result in loss or overestimation of the importance of corresponding risk factors, combination of t-conorms provide a good basis for a simple risk scoring. Transparency of the combined t-conorm can allow supervisory authorities to identify the risk appetite of any obliged entity which has chosen to apply such aggregation model for obtaining money laundering risk scores. Further research will focus on the fine-tuning of the aggregation model and also introducing Mamdani-Type or Sugeno-Type fuzzy inference systems towards each risk factor of the money laundering risk.

References

1. Chen, Z., Khoa, L.D.V., Teoh, N.V., Nazir, A., Karuppiah, E.K., Lam, K.S.: Machine learning techniques for anti-money laundering (AML) solutions in suspicious transaction detection: a review. Knowl. Inf. Syst. **57**, 245–285 (2018)

2. Chen, Y.-T., Mathe, J.: Fuzzy computing applications for anti-money laundering and distributed storage system load monitoring. In: World Conference on Soft Computing (2011)
3. Choudhary, R., Raghuvanshi, A.: Risk assessment of a system security on fuzzy logic. Int. J. Sci. Eng. Res. 3(12), 1–4 (2012)
4. Hussin, H., Kaka, S., Majid, M.A.A.: A case study on fuzzy logic-based risk assessment in oil and gas industry. ARPN J. Eng. Appl. Sci. 11(5), 3049–3054 (2016)
5. Umoh, U.A., Udosen, A.A.: Sugeno-type fuzzy inference model for stock price prediction. Int. J. Comput. Appl. 103(3), 1–12 (2014)
6. Kaur, A.: Comparison of Mamdani-type and Sugeno-type fuzzy inference systems for air conditioning system. Int. J. Soft Comput. Eng. 2(01), 323–325 (2012)
7. Bojadziev, G., Bojadziev, M.: Fuzzy Sets, Fuzzy Logic, Applications. World Scientific Publishing Co., Pte. Ltd., Singapore (1998)
8. Calvo, T., Mayor, G., Mesiar, R.: Aggregation Operators. Physica-Verlag, Heidelberg (2002)
9. Peneva, V., Popchev, I.: Aggregation of fuzzy relations with fuzzy weighted coefficients. Adv. Stud. Contemp. Math. 15, 121–132 (2007)
10. Hájek, P., Mesiar, R.: On copulas, quasicopulas and fuzzy logic. Soft. Comput. 12, 1239–1243 (2008)

Towards an Adaptive Defuzzification:
Using Numerical Choquet Integral

Vicenç Torra[1][(✉)] and Joaquin Garcia-Alfaro[2]

[1] Hamilton Institute, Maynooth University, Maynooth, Ireland
vtorra@ieee.org
[2] Institut Polytechnique de Paris, CNRS SAMOVAR, Télécom SudParis, France
jgalfaro@ieee.org

Abstract. Fuzzy systems have been proven to be an effective tool for modeling and control in real applications. Fuzzy control is a well established area that is used in a large number of real systems. Fuzzy rule based systems are defined in terms of rules in which the concepts that define the rules (both in the antecedent and consequent) can be defined in terms of fuzzy sets. In applications, rules are fired and then a set of consequents need to be combined to make a final decision. This final decision is often computed by means of a defuzzification method. In this paper we discuss the defuzzification proces and propose the use of a Choquet integral for this process. In contrast with standard defuzzification methods which are based on mean operators (usually discrete), the Choquet integral permits us to have an output variable with values that have different importances and with interactions among the values themselves. To illustrate the approach, we use a numerical Choquet integral software for continuous functions that we have recently developed. We also position the application of the approach to handle the uncertainty associated to a mission-oriented Cyber-Physical System (CPS).

1 Introduction

Knowledge based systems are used in a large number of real-world applications. Among them, rule based systems stand out for their interpretability. As the name indicates, they are defined in terms of sets of rules, with each rule defined by an antecendent that establishes when the rule applies, and a consequent that establishes the conclusion when the rule applies. Different types of rules have been considered. We want to underline the case of rules that permit to represent some type of uncertainty.

Fuzzy rules [5,10] are the rules to be used when we need to consider vagueness and fuzziness in the concepts (either in the antecedent or the consequent). Recall that a concept on the reference set X is imprecise when different elements of X satisfy the concept (e.g., temperature below zero) and a concept is vague when there are values in X for which it is doubtful to affirm that they satisfy or not the concept (e.g., temperature is cold). Vague and imprecise concepts can be represented by means of fuzzy sets.

© Springer Nature Switzerland AG 2019
V. Torra et al. (Eds.): MDAI 2019, LNAI 11676, pp. 113–125, 2019.
https://doi.org/10.1007/978-3-030-26773-5_11

In contrast to standard (crisp) sets where characteristic functions are defined in terms of the Boolean sets $\{true, false\}$ or $\{0, 1\}$ (i.e., an element is either in a set A or not), fuzzy sets are defined in terms of membership functions that assign to each element a value in the [0,1] interval. So, for a set A on the reference set X, a characteristic function for A is typically of the form $\chi_A : X \to \{0, 1\}$ while a membership function is of the form $\mu_A : X \to [0, 1]$. Then, a value $\mu_A(x)$ of zero means no membership, a value of one means total membership, and values in $(0, 1)$ represent partial membership.

Fuzziness distinguishes from probability because while a fuzzy membership of e.g. 0.7 represents partial truth, probability of 0.7 is a measure of our certainty of being the fact completely true. To illustrate the difference [3] gives an example of two bottles A and B where A is marked with membership of 0.91 to be potable and B with probability 0.91 to be potable. The later means that B will be potable 91% of the trials, but 9% may be deadly. In contrast, the *fuzzy* bottle A will contain perfectly potable water, maybe not completely pure, but not deadly in any case.

Due to the fact that the rules are defined using fuzzy sets, and fuzzy sets can be partially satisfied, at a given time, several rules may apply. This is so because each of the rules may satisfy partially the conditions given in the antecedent. The degree of satisfaction of a rule (the truth value of the antecedent) is computed taking into account the fuzzy sets that define the rules, as well as appropriate operators (e.g., operators to model conjunction and disjunction of the concepts in the antecedent). Fuzzy systems usually fire all rules with positive degree of satisfaction. Then, this degree is propagated to the conclusion. This leads to a set of conclusions that need to be considered together, with each conclusion having the corresponding degree of satisfaction of the rule.

A variety of fuzzy rule based systems exist in the literature. A detailed discussion of their differences is beyond the interest of this paper. See, for instance, references [5,8–10,15,16,21] for a detailed description of some of them. We will focus on rule based systems in which the terms of both antecedents and consequents are described in terms of fuzzy sets. For simplicity, we will consider antecedents in which terms are only combined in a conjunctive way, and in which consequents have only one variable. Observe that this constraint is not relevant in our work because we focus on how to operate with the conclusions of the rules. Therefore, rules follow this pattern:

$$\textbf{If } V_1 \textbf{ is } T_1 \textbf{ and } V_2 \textbf{ is } T_2 \textbf{ and } \dots \textbf{and } V_2 \textbf{ is } T_2 \textbf{ then } V_O \textbf{ is } T_O$$

We will give a more accurate description of the rules in Sect. 2. Here, V_i represent variables, T_i terms that are represented by fuzzy sets, V_O is the output variable and T_O an output term.

In our study, the output of a rule will be a fuzzy set. This fuzzy set will be defined in terms of the fuzzy set T_O of a rule, and the degree of satisfaction of the antecedent. The collection of fuzzy sets obtained from all rules need to be combined (through a data fusion or aggregation process). We obtain in this way an aggregated fuzzy set. In order to obtain this set, we can proceed in different ways. In this work we will use the union of sets, as commonly used in the literature [5].

The combination of consequents of rules does not lead to an appropriate output. Most applications in control and modeling need as output a single value in the appropriate range of the output variable (e.g., the actual value to give to a controller). Therefore, a fuzzy set as system's output cannot be used without further processing. Defuzzification is the name of the process to select a single value from the aggregated fuzzy set. Different defuzzification strategies have been defined in the literature. The center of area is probably the most used one, but other solutions have been proposed.

Most defuzzification procedures can be seen as a two step process (see e.g. [6]). First, the aggregated fuzzy set is transformed into an appropriate distribution (e.g., a probability distribution via normalization or a distribution that eliminates inappropriate values). Then, an element is selected from the distribution (e.g., the expected value of the distribution).

In this paper, we propose the use of a Choquet integral [4] in the defuzzification process. This permits us to take into consideration some particularities of the domain of the output variable. For example, we may consider that different subdomains of the output variable have different relevance, and that different subdomains are incompatibles.

In a CPS, physical and software components are tightly connected. Real-time decisions are often needed, and systems need to adapt dynamically to their context. Rule-based approaches are appropriate for reactive systems. Changes in the environment cause changes in the input variables, that are quickly propagated into the output. Fuzzy rule based systems are an effective approach for implementing reactive systems.

Our model for defuzzification based on the Choquet integral permits us a dynamic modification of the focus of interest of the output variable. While keeping constant the fuzzy rule based system and the aggregation process, we will be able to adapt the center of interest of the output variable taking into account some additional variables of the environment. This will be done by means of a modification of the parameter of the Choquet integral (i.e., the fuzzy measure). As it is shown later, we consider an example with a simulated annealing-type approach where the defuzzification shifts from a less supported value to a more supported one (more conservative) when time progresses.

The structure of the paper is as follows. In Sect. 2 we give an overview of fuzzy measures and the Choquet integral. We will use these concepts later to formally define our approach. In Sect. 3 we describe the type of fuzzy rule based system we use in this work. In Sect. 4 we formalize our defuzzification method. In Sect. 5, we position the application of the approach to handle the uncertainty associated to a mission-oriented CPS. The paper finishes with some conclusions and lines for future work.

2   Fuzzy Measures and the Choquet Integral

Let us start reviewing the concept of fuzzy measure, also known as non-additive measure or monotonic game. Fuzzy measures generalize additive measures and probabilities replacing the additivity condition by a condition on monotonicity: monotonicity with respect to set inclusion.

Definition 1. *Let $(\Omega, \mathcal{F})$ be a measurable space. A set function μ defined on $\mathcal{F}$ is called a fuzzy measure if an only if*

- $0 \leq \mu(A) \leq \infty$ *for any $A \in \mathcal{F}$;*
- $\mu(\emptyset) = 0$;
- $\mu(\Omega) = 1$;
- *If $A_1 \subseteq A_2 \subseteq \mathcal{F}$ then*

$$\mu(A_1) \leq \mu(A_2)$$

The boundary condition $\mu(\Omega) = 1$ is not always required. In our context, it will be convenient. It corresponds to the condition that the probability of the whole reference set Ω is one.

When the set of reference is discrete, the Choquet integral of a function with respect to a fuzzy measure is defined as follows.

Definition 2. *Let X be a reference set, and let μ be a fuzzy measure on $X = \{x_1, \ldots, x_n\}$; then, the Choquet integral of a function $f : X \to \mathbb{R}^+$ with respect to μ is defined by*

$$(C) \int f d\mu = \sum_{i=1}^{n} [f(x_{s(i)}) - f(x_{s(i-1)})] \mu(A_{s(i)}), \tag{1}$$

where $f(x_{s(i)})$ indicates that the indices have been permuted so that $0 \leq f(x_{s(1)}) \leq \cdots \leq f(x_{s(n)}) \leq 1$, and where $f(x_{s(0)}) = 0$ and $A_{s(i)} = \{x_{s(i)}, \ldots, x_{s(n)}\}$.

When the reference set is not finite, the Choquet integral of f with respect to μ is defined with the following expression

$$(C) \int f d\mu := \int_0^\infty \mu_f(r) dr, \tag{2}$$

where $\mu_f(r) = \mu(\{x | f(x) \geq r\})$.

This definition generalizes the Lebesgue integral, and it reduces to the Lebesgue integral when μ is additive.

3   Fuzzy Rule Based Systems

We will consider fuzzy rules following the structure below. For the sake of simplicity, we consider that all rules have n input variables and a single output.

As we have indicated in the introduction, we only allow for conjunction to combine the variables and their terms in the antecedent.

R: If V_1 is t_{1a}^R and V_2 is t_{2b}^R and $\cdots$ V_n is t_{nz}^R then Y is t_{yo}^R

Here, V_i represents an input variable and X_i its domain, t_{ie}^R is a term for the ith input variable, Y represents an output variable and t_{yo}^R is a term for output variable. More specifically, for each variable X_i, we have a set of terms n_i terms denoted by $t_{i1}, \ldots, t_{in_i}$. Similarly, we have a set of n_o terms for the output variable Y. These terms are $t_{y1}, \ldots, t_{yn_o}$. Naturally, $t_{1a}^R \in \{t_{11}, \ldots, t_{1n_1}\}$, $t_{2b}^R \in \{t_{21}, \ldots, t_{2n_2}\}$, etc.

Each t_{ij} is described in terms of a fuzzy set defined on the domain of variable V_i. We denote this fuzzy set by μ_{ij}. Naturally, $\mu_{ij} : X_i \to [0,1]$.

Let us consider the rule below for controling the temperature of a device. We have two input variables ϵ and $\Delta\epsilon$ representing, respectively, the error (with the objective temperature) and error difference (change in the error in two consecutive time instants), and an output variable that controls the device. We have three terms, one for each variable. They are, positive, positive, and small-negative. Each term will be defined by a fuzzy set. Fuzzy sets will be defined in the range of the variable (this means that the term positive for variable ϵ may be different than the term positive for variable $\Delta\epsilon$).

Rule1: if ϵ is positive and $\Delta\epsilon$ is positive
           then control-variable is small-negative

Fuzzy inference for fuzzy rule based systems for this type of rules usually follows four steps. We give these fours steps formally below. We begin with an informal description based on the previous rule.

Step 1. Rules are fired when we have actual values for each of the input variables. E.g., we have that the error ϵ is 3 degrees ($\epsilon = 3$) and that the error has decreased 0.2 degrees ($\Delta\epsilon = -0.2$). Then, the degree of satisfaction for each variable is defined as the membership degree of the actual value of the variable using the appropriate membership function. In our case $\mu_{\epsilon,\text{positive}}(3)$.

In our rules, the only operator in the antecedent is **and**. Because of that, the membership degrees obtained for each variable in the antecedent are combined using an operator that models the conjunction. In fuzzy logic, t-norms play this role. The minimum $\min(a,b)$ and the product ab are examples of t-norms. This combination corresponds to the degree of satisfaction of the antecedent.

Step 2. The degree of satisfaction of the antecedent is propagated to the consequent. This is done clipping the membership function associated to the output variable using the degree of satisfaction of the antecedent. The process implies that when the antecendent is completely satisfied (i.e., degree equal to one), the output is just the fuzzy term. In contrast, when the antecedent is not at all satisfied (i.e., degree equal to zero), the output is just the empty set (i.e., all membership values are zero).

Step 3. All rules of our knowledge base (say KB) are fired applying the approach just described. We obtain in this way a collection of fuzzy sets (say μ_R for each rule R in KB). All these fuzzy sets are combined. It is usual to use the union of all fuzzy sets for this purpose. We obtain in this way an aggregated fuzzy set. From a mathematical point of view, this approach is to consider fuzzy rules in a disjunctive way. That is, either we apply R_1 or R_2 or ... or R_t. Under this interpretation, we take the union of the outputs (the output of R_1 or the output of R_2...). For crisp and disjoint rules, this would result into a single output. In the case of fuzzy rules, this step results into a fuzzy set that contains pieces of information of several rules. See e.g. Figure 1 that shows a typical output of Step 3. It corresponds to the union of two clipped fuzzy sets (one with maximum value at $y = 2$ and the other with maximum value at $y = 3$). Then, in Step 4 we obtain a kind of average of the outputs (average weighted by the degree of satisfaction of the rule).

In a previous work [18] we showed that the process of combination can be expressed in terms of a Sugeno integral [14]. Understanding the fusion of consequents in this way, we generalize the usual approach introducing a model in which rules do not need to be independent. A similar idea is present in [2,11] where the Choquet integral is used to combine outcomes from rules, also to permit non independent rules.

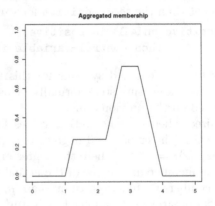

Fig. 1. Example of aggregated fuzzy set obtained in Step 3.

Step 4. A value is obtained from the aggregated set. This step is called defuzzification. The center of area is one of this defuzzification processes. This method is the one given in Step 4 below.

We give now a more formal definition of this process.

Step 1. Degree of satisfaction of the antecedent of each rule R

$$\alpha^R = T(\mu_{1a}^R(x_1), \mu_{2b}^R(x_2), \dots, \mu_n^R(x_{nz}))$$

for a t-norm T.

Step 2. Application of all rules R and computation of their consequents

$$\mu^R(x) = \min(\alpha^R, \mu_{yo}^R).$$

Step 3. Application of a set of rules and computation of the corresponding consequent. In terms of the membership function, it is the following:

$$\tilde{\mu} = \cup_{R \in KB} \mu^R.$$

If we consider memberships at given y in the space of the output Y, this is

$$\tilde{\mu}(y) = \cup_{R \in KB} \mu^R(y)$$

for all $y \in Y$. We often use a t-conorm (e.g., the maximum) to compute the union.

Step 4. Defuzzification

$$y^* = \frac{\sum_y \tilde{\mu}(y) \cdot y}{\sum_z \tilde{\mu}(z)} \text{ or } y^* = \frac{\int y\tilde{\mu}(y)dy}{\int \tilde{\mu}(z)dz}. \tag{3}$$

4 Defuzzification Based on the Choquet Integral

Equation 3 is one of the existing approaches to select a value from the aggregated fuzzy set. This approach is known as center of area or gravity defuzzification. Driankov et al. [5] describe five other methods: center of sums, center of largest area, first of maxima, middle of maxima, and height defuzzification.

As briefly explained in the introduction, following [6], most defuzzification procedures can be seen from a two step process perspective. The first step transforms the fuzzy set $\tilde{\mu}$ into an appropriate distribution, and the second step is about selecting an element from the distribution. In the center of area, the transformation is about building the following probability distribution from the fuzzy set $\tilde{\mu}$: $p(y) = \tilde{\mu}(y)/\sum_z \tilde{\mu}(z)$, or $p(y) = \tilde{\mu}(y)/\int \tilde{\mu}dz$ in the continuous case. Selection is defined as the expected value of the distribution p. First of maxima sets to zero all memberships that are not maximal and then selects the first y with a maximal membership (i.e., let $\alpha = \sup_z \tilde{\mu}(z)$ then $\tilde{\mu}'(y) = \alpha$ if and only if $\tilde{\mu}(y) = \alpha$ and select $\min_y\{y|\tilde{\mu}'(y) = \alpha\}$). Middle of maxima follows the same approach but selects $0.5(\min_y\{y|\tilde{\mu}'(y) = \alpha\} + \max_y\{y|\tilde{\mu}'(y) = \alpha\})$.

Mathematically, the two step process for the center of area corresponds to (i) building the probability distribution and then (ii) computing the expectation of this distribution. We use the definition of the expectation as the Lebesgue integral of the function $f(y) = y$ with respect to the probability distribution. That is, considering a continuous membership function $\tilde{\mu}$:

$$y^* = \int y dp$$

where $p(y) = \tilde{\mu}(y)/\int \tilde{\mu}(z)dz$.

Our contribution is to use in this process the Choquet integral of the function $f(y) = y$ with respect to a measure ν built from the fuzzy set $\tilde{\mu}$. That is, our proposal is to defuzzify using

$$y^* = (C) \int d\nu \tag{4}$$

where ν is built from $\tilde{\mu}$. Therefore, we need to consider ways to construct fuzzy measures ν defined for all the sets $[x, \infty]$ (note that the values of $\nu([x, \infty])$ are the only ones actually considered in the integration process). We give some examples below. In the examples we use $p(y) = \tilde{\mu}(y)/\int \tilde{\mu}(y)dz$ as above.

Example 1. If we define $\nu([x, \infty]) = \int_x^\infty p(x)dx$ then the Choquet integral corresponds to the Lebesgue integral, and the defuzzification is the center of area.

Example 2. If we define $\nu([x, \infty]) = Q(\int_x^\infty p(x)dx)$ for a distortion function Q (i.e., a function $Q(x)$ such that $Q(0) = 0$, $Q(1) = 1$ and monotonic with respect to x), we result into the continuous WOWA operator for defuzzification (see [17] for details). Note that measures of this form correspond to distorted probabilities. I.e., $\nu = QoP$.

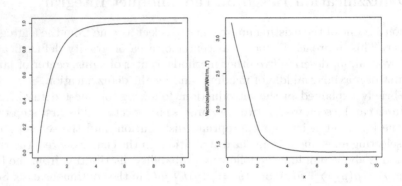

Fig. 2. Function $1 - e^{-k}$ used to define the distortion function Q_k, and the resulting defuzzified values.

Let us consider the continuous piece-wise linear functions $Q_{(a,b)}(x)$ defined as 0 for $x < a$, 1 for $x > b$, and linear between a and b. This is to model that the values that accumulate probabilities in (a, b) (as in quantile functions) are the most relevant in the defuzzification process. Let us consider a system where we can consider risky alternatives (with less suport) at the initial times, but more conservative alternatives (with more support) at later times. To model this situation, we consider $t_k = e^k$ (a popular cooling schedule – the exponential one – in simulated annealing). We then consider the intervals (a_k, b_k) defined as $(max(0, 1 - e^k - 0.1), min(1, 1 - e^k + 0.1))$ to build the corresponding distortions Q_k. The function $1 - e^k$ as well as the outcome of the defuzzification with respect to time k is given in Fig. 2 (left and right, respectively).

We can see in Fig. 2 (right) how the defuzzified value tend to the minimum when time increases.

Example 3. For a given fuzzy measure ν', we define the measure ν by set $\tilde{\mu}$:

$$\nu([x,\infty]) = \frac{(C)\int_{[x,\infty]} pd\nu'}{K}, \tag{5}$$

where $p(y)$ is defined as before $p(y) = \tilde{\mu}(y)/\int \tilde{\mu}(z)dz$, and with normalizing factor $K = (C)\int_{[-\infty,+\infty]} pd\nu'$. Note that in this equation, we can use equivalently either $\tilde{\mu}$ or p because we have the normalizing factor K. For $\nu' = \lambda$ (i.e., ν' is the Lebesgue measure), this definition results into the measure above $\nu([x,\infty]) = \int_x^\infty p(x)dx$. For fuzzy measures ν', this results into a fuzzy measure ν.

To illustrate this example, we have considered two cases: measures ν_1' and ν_2' which result using Expression 5 into two measures ν_1 and ν_2. They are the following ones:

- $\nu_1'([a,b]) = (l_1/5) + (l_2/5)^2 + 0.2(l_1/5)(l_2/5)^4$
- $\nu_2'([a,b]) = (l_1/5) + \sqrt{l_2/5} + 0.2(l_1/5)\sqrt{\sqrt{l_2/5}}$

where $l_1 = \lambda([a,b] \cap [0,3])$ and $l_2 = \lambda([a,b] \cap [3,6])$ (here λ is the Lebesgue measure).

The first measure gives more relevance for elements $x < 3$ than the second measure that have more relevance for elements $x > 3$. Compare the measure of $\nu_1([2,3]) = 0.6373673$ and $\nu_1([3,4]) = 0.08365445$ vs. $\nu_2([2,3]) = 0.2400694$ and $\nu_2([3,4]) = 0.6262801$. This implies that the defuzzification using ν_1 is smaller than the defuzzification using ν_2. Experiments show that this is true, as in the first case we obtain a defuzzified value of 2.306174 while in the second case it is 2.576696.

Example 3 shows that defuzzification using a Choquet integral provides additional flexibility, and that we can model situations in which different regions of the domain have different relevance. We can exploit non-additivity of the measure to reduce or increase the relevance of conservative or risky values. In addition, we can also consider measures in line with Example 2 where the measure is time-dependent, and, thus permits us to shift the focus of the system over time even when its inputs do not change.

The outcomes of these examples have been computed considering the membership functions as continuous functions and computing the Choquet integrals numerically. The numerical Choquet integral has been computed using the software in R we provided in [20]. In Example 3, we need to compute the Choquet integral of $f(x) = x$ with respect to ν in the interval [0,6]. ν is the Choquet integral of $\tilde{\mu}$ with respect to ν. The integration of f requires the computation of $\nu([a,5])$ for several values of a (i.e., several numerical Choquet integrals). This makes the process costly. E.g., the defuzzification of the fuzzy set using ν_2 takes 43 s in a Laptop (Intel(R) Core(TM) i7-8550U CPU @ 1.80 GHz).

5   Applying the Approach to Cyber-Physical Systems

We present in this section the application of the defuzzification approach presented in this paper, to handle the uncertainty associated to a mission-oriented CPS[1]. We assume a CPS in which entities integrate computation, communication and physical processes [7]. Examples include robotics and autonomous vehicles. Our CPS scenario assumes the existence of a decentralized process that computes corrective control actions based on our defuzzification approach. Mathematically, the CPS is modeled as a spatially distributed system whose control loops are closed by a wireless communication network. The communication network connects the different components of the CPS, assumed to be a series of mobile agents that exchange messages to complete a mission-oriented problem. The scenario follows previous work presented in [1,12] (cf. http://j. mp/scavesim for further details and some related media), in order to address classical theoretical problems studied in the CPS literature, such as stabilization of mobile systems and control-theoretic techniques addressing uncertainty. A quick summary of the scenario is presented next.

5.1   Trajectory Search Scenario

A series of mobile agents (e.g., unmanned aerial vehicles) must accomplish a mission. The mission relies on a trajectory search scenario. The agents must cross different segments of, e.g., city blueprints, by physically identifying an unknown number of intermediary trajectory points in each segment. Identifying and visiting all the trajectory points, as well as reaching the final destination, it is crucial to label the mission as accomplished. The agents collect and deliver information at each trajectory point, such as taking pictures and exchanging messages between them, in order to discover the way to reaching the following trajectory point. The mission is considered as accomplished when all the intermediate points specified in the trajectory are successfully visited by all the agents. The mission fails when at least one of the agents fails at identifying or visiting one of the points in the trajectory.

A trajectory is represented as an unknown number of intermediate points. At every trajectory point, the agents must collectively determine the next step by solving a search problem [1]. We define the *trajectory search problem* as a collective solution in which a bounded number of agents follow a trajectory from s (starting point) to t (terminal point), as depicted in Fig. 3.

The agents must discover the trajectory points instructed by a decentralized CPS process. The process is to collect the information from all the agents, and apply the defuzzification approach presented in this paper in the reasoning process. Together, the agents travel from a trajectory point to any other. With the trajectory points identified to vertices, they form a complete graph. We assume

---

[1] The term CPS, coined in 2006 by H. Gill at the National Science Foundation [13], refers to next generation embedded ICT systems, which include monitoring and control technologies in charge of physical components for pervasive applications.

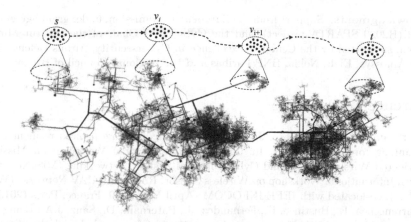

Fig. 3. Group of mobile agents traveling from s (initial point of a trajectory) to t (terminal point of the trajectory). Segments (v_i, v_{i+1}) represent the connection of two trajectory points (v_i and v_{i+1}). They cooperate to navigate from s to t.

that the agents are unaware of the terrain and locations of the trajectory points. However, their CPS process has the capability to correlate data collectively collected by the overall system and take decisions. The agents communicate and exchange information between them. The mission of the agents is assumed to be accomplished if all the agents successfully find all the trajectory points within the segments that start in s and end in t. See [22] for further details.

6   Concluding Remarks

In this paper we have proposed the use of Choquet integrals in the defuzzification of fuzzy rule based systems. We have shown with some examples that our definition permits to take into account the relevance of different regions. We have also considered using the proposed approach for the accomplishment of critical missions under uncertain conditions of a mission-oriented CPS. We have presented a CPS scenario in which a series of mobile agents are instructed to solve a trajectory search problem. The agents must discover and physically visit a series of unknown trajectory points. To successfully accomplish the mission, all the agents must collectively share information at each trajectory point, such as taking pictures or exchanging messages between them, in order to discover the following steps towards the following step of the trajectory. The mission is said to have been accomplished when all the intermediate segments specified in the trajectory are successfully visited by all the agents. The mission fails when one of the agents fails at identifying or visiting one of the trajectory points. Future work includes embedding the approach in a fuzzy system, and develop more efficient solutions to compute the defuzzified values using the Choquet integral.

Acknowledgments. Support from the European Commission, under grant agreement 830892 (H2020 SPARTA project), and the Cyber CNI Chair of the Institut Mines-Télécom, supported by the Center of excellence in Cybersecurity, Airbus Defence and Space, Amossys, EDF, Nokia, BNP Paribas and the Regional Council of Brittany.

References

1. Barbeau, M., Garcia-Alfaro, J., Kranakis, E.: Geocaching-inspired resilient path planning for drone swarms. In: Joint 7th International Workshop on Mission-Oriented Wireless Sensor and Cyber-Physical System Networking (MiSeNet) and 12th International Workshop on Wireless Sensor, Robot and UAV Networks (WiSARN), co-located with IEEE INFOCOM: April-May 2019, France, Paris (2019)
2. Barrenechea, E., Bustince, H., Fernandez, J., Paternain, D., Sanz, J.A.: Using the choquet integral in the fuzzy reasoning method of fuzzy rule-based classification systems. Axioms **2**, 208–223 (2013)
3. Bezdek, J.C.: Fuzzy models - what are they, and why? IEEE Trans. Fuzzy Syst. **1**, 1–1993 (1993)
4. Choquet, G.: Theory of capacities. Ann. Inst. Fourier **5**, 131–295 (1953/54)
5. Driankov, D., Hellendoorn, H., Reinfrank, M.: An Introduction to Fuzzy Control. Springer, Heidelberg (1993). https://doi.org/10.1007/978-3-662-11131-4
6. Filev, D., Yager, R.R.: A generalized defuzzification method under BADD distribution. Int. J. Intel. Syst. **6**(7), 687–697 (1991)
7. Kim, K., Kumar, P.: Cyber-physical systems: a perspective at the centennial. Proc. IEEE **100**, 1287–1308 (2012)
8. Magdalena, L.: Hierarchical fuzzy control of a complex system using meta-knowledge. In: Proceedings of the IPMU 2000 Conference, pp. 630–637 (2000)
9. Magdalena, L.: On the role of context in hierarchical fuzzy controllers. Int. J. Intell. Syst. **17**(5), 471–493 (2002)
10. Ross, T.J.: Fuzzy Logic with Engineering Applications, 4th edn. Wiley, Hoboken (2016)
11. Sanz, J., Lopez-Molina, C., Cerrón, J., Mesiar, R., Bustince, H.: A new fuzzy reasoning method based on the use of the Choquet integral. In: Proceedings of the EUSFLAT (2013)
12. Shi, W., Garcia-Alfaro, J., Corriveau, J.-P.: Searching for a black hole in interconnected networks using mobile agents and tokens. J. Parall. Distrib. Comput. **74**(1), 1945–1958 (2014)
13. Song, H., Rawat, D., Jeschke, S., Brecher, C.: Cyber-Physical Systems: Foundations, Principles and Applications. Elsevier, Amsterdam (2016)
14. Sugeno, M.: Theory of Fuzzy Integrals and its Applications, Ph.D. Dissertation, Tokyo Institute of Technology, Tokyo, Japan (1974)
15. Takahagi, E.: On fuzzy integral representation in fuzzy switching functions, fuzzy rules and fuzzy control rules. In: Proceedings of the 8th IFSA Conference, pp. 289–293 (1999)
16. Torra, V.: A review on the construction of hierarchical fuzzy systems. Int. J. Intell. Syst. **17**(5), 531–543 (2002)
17. Torra, V., Godo, L.: Continuous WOWA operators with application to defuzzification. In: Calvo, T., Mayor, G., Mesiar, R. (eds.) Aggregation Operators, pp. 159–176. Physica-Verlag, Heidelberg (2002). https://doi.org/10.1007/978-3-7908-1787-4_4

18. Torra, V., Narukawa, Y.: The interpretation of fuzzy integrals and their application to fuzzy systems. Int. J. Approx. Reasoning **41**(1), 43–58 (2006)
19. Torra, V., Narukawa, Y., Miyamoto, S.: Modeling decisions for artificial intelligence: theory, tools and applications. In: Torra, V., Narukawa, Y., Miyamoto, S. (eds.) MDAI 2005. LNCS (LNAI), vol. 3558, pp. 1–8. Springer, Heidelberg (2005). https://doi.org/10.1007/11526018_1
20. Torra, V., Narukawa, Y.: Numerical integration for the Choquet integral. Inf. Fusion **31**, 137–145 (2016)
21. Tunstel, E., Oliveira, M.A.A., Berman, S.: Fuzzy behavior hierarchies for multi-robot control. Int. J. Intell. Syst. **17**, 449–470 (2002)
22. http://www-public.imtbs-tsp.eu/~garcia_a/web/prototypes/scave/

Uninorms and Nullnorms and their Idempotent Versions on Bounded Posets

Martin Kalina[✉][iD]

Faculty of Civil Engineering, Slovak University of Technology,
Radlinského 11, Sk-810 05, Bratislava, Slovakia
kalina@math.sk

Abstract. The paper deals with uninorms and nullnorms as basic operations which are associative, commutative and monotone (increasing). These operations were first introduced on the unit interval and later generalized to bounded lattices. In this contribution we show how it is possible to generalize them to bounded posets. We will study their existence, property and conditions under which it is possible to construct their idempotent versions, since these are the most important operations from the point of view of possible applications.

Keywords: Bounded Poset · Associative operation ·
Idempotent uninorm · Idempotent nullnorm · Uninorm · Nullnorm

1 Introduction

Uninorms and nullnorms on the unit interval are special types of aggregation functions since, due to their associativity they can be straightforwardly extended to n-ary operations for arbitrary $n \in \mathbb{N}$. They are important in various fields of applications, e.g., neuron nets, fuzzy decision making and fuzzy modelling. They are interesting also from a theoretical point of view. Important is, among others, the class of idempotent operations, studied, e.g., in [1,2,9,15]. Recently they have been studied on bounded lattices (see, e.g., [4–6,9–11,22]).

Uninorms were introduced by Yager and Rybalov [27]. Special types of associative, commutative and monotone operations with neutral elements had been already studied in [12,16,17]. Deschrijver [13,14] has shown that on the lattice $L^{[0,1]}$ of closed subintervals of the unit interval there exist uninorms which are neither conjunctive nor disjunctive (i.e., whose annihilator is different from both, **0** and **1**). Particularly, he constructed uninorms having the neutral element $\mathbf{e} = [e, e]$, where $e \in]0, 1[$. In [22] the authors have shown that on arbitrary bounded lattice L it is possible to construct a uninorm regardless which element of L is chosen to be the neutral one. A different type of construction of uninorms on bounded lattices was presented in [4].

Supported by the VEGA grant agency, grant No. 2/0069/16 and 1/0006/19, and by the Science and Technology Assistance Agency under contract No. APVV-18-0052.

© Springer Nature Switzerland AG 2019
V. Torra et al. (Eds.): MDAI 2019, LNAI 11676, pp. 126–137, 2019.
https://doi.org/10.1007/978-3-030-26773-5_12

In [20] construction of a uninorm for arbitrary pair (e, a) of incomparable elements such that e is the neutral element and a the absorbing one, was presented. In [19] the author showed that on some special bounded lattices, one can construct operations which are both, proper uninorms and nullnorms, meaning that their neutral, as well as annihilators are different from both, 0 and 1.

T-operators were defined on $[0, 1]$ by Mas et al. in 1999. In 2001, Calvo et al. introduced the notion of nullnorms, also on $[0, 1]$. Both of these operations were defined as generalizations of t-norms and t-conorms. As Mas et al. in 2002 pointed out, t-operators and nullnorms coincide on $[0, 1]$. Particularly, under constrained that Op : $[0, 1]^2 \to [0, 1]$ is a commutative, associative and monotone operation then properties

(a) 0 and 1 are idempotent elements of Op and functions $\mathrm{Op}(0, \cdot)$ and $\mathrm{Op}(1, \cdot)$ are continuous,
(b) there exists $a \in [0, 1]$ such that 0 is a partial neutral element of Op on $[0, a]$, and 1 is a partial neutral element of Op on $[a, 1]$,

are equivalent to each other. Afterwards, only nullnorms were studied and later generalized as operations on bounded lattices [21]. In [6], the authors have pointed out some differences between nullnorms and t-operators on bounded lattices. In this paper we will deal only with nullnorms.

2   Basic Notations and Some Known Facts and Notions

In the whole paper, $(P, 0, 1, \leq)$ will denote a bounded poset, where $P \neq \emptyset$ is a given set and $0, 1$ its two distinguished elements such that $0 \leq x \leq 1$ for all $x \in P$. If it will cause no confusion, by P we will denote also the poset itself. First, we recall some basic properties of binary operations.

Definition 1. *Let P be a bounded poset and $*$ be a binary commutative operation on P. Then*

(i) *element $c \in P$ is said to be* idempotent *if $c * c = c$,*
(ii) *element $e \in P$ is said to be* neutral *if $e * x = x$ for all $x \in P$,*
(iii) *element $a \in P$ is said to be* annihilator *if $a * x = a$ for all $x \in P$.*

Lemma 1. *Let $*$ be a commutative and associative operation on L. Further, let c be an idempotent element. Assume that there exist elements $x, y \in L$ such that $x * c = y$. Then also $y * c = y$.*

Schweizer and Sklar [26] introduced the notion of a triangular norm (t-norm for brevity).

Definition 2 ([26]). *An operation $T \colon [0, 1]^2 \to [0, 1]$ is a t-norm if it is associative, commutative, monotone, and 1 is its neutral element.*

T-norms and t-conorms are dual to each other. If $T\colon [0,1]^2 \to [0,1]$ is a t-norm, then

$$S(x,y) = 1 - T(1-x, 1-y)$$

is the dual t-conorm to T. For details on t-norms and t-conorms see, e.g., [23].

As a generalization of both t-norms and t-conorms Yager and Rybalov [27] proposed the notion of uninorm.

Definition 3 ([27]). *An operation $U\colon [0,1]^2 \to [0,1]$ is a uninorm if it is associative, commutative, monotone, and if it possesses a neutral element $e \in [0,1]$.*

A uninorm U is *proper* if its neutral element $e \in {]}0,1{[}$.

Every uninorm has an annihilator. A uninorm with the annihilator 0 is conjunctive, and a uninorm with annihilator 1 is disjunctive.

Lemma 2 ([27]). *Let $U\colon [0,1]^2 \to [0,1]$ be a uninorm whose neutral element is e. Then its dual operation*

$$U^d(x,y) = 1 - U(1-x, 1-y)$$

is a uninorm whose neutral element is $1-e$. Moreover, U is conjunctive if and only if U^d is disjunctive.

Results in paper [27] imply the following assertion.

Lemma 3. *Let $U\colon [0,1]^2 \to [0,1]$ be a uninorm whose neutral element is e. Then there exists a t-norm $T_U\colon [0,1]^2 \to [0,1]$ and a t-conorm $S_U\colon [0,1]^2 \to [0,1]$ such that*

$$(\forall x, y \in [0,e]^2)(U(x,y) = e \cdot T_U(\tfrac{x}{e}, \tfrac{y}{e})),$$
$$(\forall x, y \in [e,1]^2)(U(x,y) = e + (1-e) \cdot S_U(\tfrac{x-e}{1-e}, \tfrac{y-e}{1-e})).$$

Lemma 4 ([27]). *Assume U is a uninorm with neutral element e. Then:*

1. *for any x and all $y > e$ we get $U(x,y) \geq x$,*
2. *for any x and all $y < e$ we get $U(x,y) \leq x$.*

Definition 4 ([7]). *An operation $V : [0,1]^2 \to [0,1]$ is said to be a nullnorm if V is associative, commutative, monotone, and moreover if there exists an element $a \in [0,1]$ such that*

(1b) $V(0,x) = x$ *for all $x \in [0,a]$,*
(2b) $V(1,x) = x$ *for all $x \in [a,1]$.*

A nullnorm is *proper* if $a \notin \{0,1\}$.

Remark 1.

(a) It is well-known that the element a in Definition 4 is the absorbing element of nullnorm V. Further, 0 is a partial neutral element of V on the interval $[0,a]$ and 1 is a partial neutral element of V on the interval $[a,1]$. Particularly, we have

$$V(0,0) = 0, \quad V(1,1) = 1. \tag{1}$$

(b) Setting $a = 0$ ($a = 1$) the nullnorm V becomes a t-norm (t-conorm). For properties of t-norms and t-conorms see, e.g., the monograph [23].

Definition 5. *Let $a \in P$ be a fixed element. Then $\|_a$ denotes the set of all elements of P that are incomparable with a.*

For more information on associative (and monotone) operations on $[0, 1]$ refer to the monographs [8, 18, 23].

3 Existence of Uninorms and Nullnorms on Bounded Posets

In this part we will show how it is possible to construct a proper uninorm and a proper nullnorm on every bounded poset P with at least 3 elements. Important is that on every bounded poset P there exists at least one t-norm T_D and one t-conorm S_D, namely

$$T_D(x, y) = \begin{cases} x & \text{if } y = 1, \\ y & \text{if } x = 1, \\ 0 & \text{otherwise,} \end{cases} \qquad S_D(x, y) = \begin{cases} x & \text{if } y = 0, \\ y & \text{if } x = 0, \\ 1 & \text{otherwise.} \end{cases} \qquad (2)$$

Theorem 1. *For arbitrary bounded poset P with at least 3 elements, and arbitrarily chosen element $e \notin \{0, 1\}$ there exist at least two uninorms U_c and U_d whose neutral element is e. The annihilator of U_c is 0 and of U_d is 1.*

Proof. Based on the t-norm T_D and t-conorm S_D we construct a conjunctive and a disjunctive uninorm:

$$U_c(x, y) = \begin{cases} x & \text{if } y = e, \text{ or if } x < e \text{ and } y > e, \\ & \quad \text{or if } x > e \text{ and } y \parallel e, \\ y & \text{if } x = e, \text{ or if } y < e \text{ and } x > e, \\ & \quad \text{or if } y > e \text{ and } x \parallel e, \\ 1 & \text{if } x > e \text{ and } y > e, \\ 0 & \text{otherwise,} \end{cases}$$

$$U_d(x, y) = \begin{cases} x & \text{if } y = e, \text{ or if } x > e \text{ and } y < e, \\ & \quad \text{or if } x < e \text{ and } y \parallel e, \\ y & \text{if } x = e, \text{ or if } y > e \text{ and } x < e, \\ & \quad \text{or if } y < e \text{ and } x \parallel e, \\ 0 & \text{if } x < e \text{ and } y < e, \\ 1 & \text{otherwise.} \end{cases}$$

Monotonicity and commutativity (symmetry) of both, U_c as well as U_d are obvious. We skip a detailed proof of the associativity of these operations. It is enough to realize that $U_c(x, y) = 0$ and $U_d(x, y) = 1$ if $(x, y) \in (\|_e)^2$. $\qquad \square$

Theorem 2. *For arbitrary bounded poset P with at least 3 elements, and arbitrarily chosen element $a \notin \{0,1\}$ there exists a nullnorm V whose annihilator is a.*

Proof. Similarly to the proof of Theorem 1, we will construct a nullnorm V based on the t-norm T_D and t-conorm S_D:

$$V(x,y) = \begin{cases} x & \text{if } x \leq a \text{ and } y = 0, \\ & \text{or if } x \geq a \text{ and } y = 1, \\ y & \text{if } y \leq a \text{ and } x = 0, \\ & \text{or if } y \geq a \text{ and } x = 1, \\ a & \text{otherwise.} \end{cases}$$

$\square$

Example 1. Consider poset P whose Hasse diagram is given on Fig. 1.

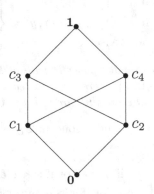

Fig. 1. Hasse diagram of the poset P

The poset P is not a lattice, but for any c_i, the intervals $[0, c_i]$ and $[c_i, 1]$ are lattices. If we choose c_3 to be the neutral element of a uninorm, there exist 4 t-norms on $[0, c_3]$ and the only t-conorm on $[c_3, 1]$. For every of the four t-norms we can choose from one of 3 possibilities for the annihilator, namely any element of $\{0, c_4, 1\}$. Using the constructions in proof of Theorem 1, we have the following uninorms U_c and U_d, respectively:

$$U_c(x,y) = \begin{cases} x & \text{for } (x,y) \in P \times \{c_3\}, \text{ for } (x,y) \in \{0, c_1, c_2\} \times \{c_4, 1\}, \\ & \text{and for } (x,y) \in \{1\} \times \{c_4\}, \\ y & \text{for } (x,y) \in \{c_3\} \times P, \text{ for } (x,y) \in \{c_4, 1\} \times \{0, c_1, c_2\}, \\ & \text{and for } (x,y) \in \{c_4\} \times \{1\}, \\ 0 & \text{otherwise,} \end{cases}$$

$$U_d(x,y) = \begin{cases} x & \text{for } (x,y) \in P \times \{c_3\} \text{ and } (x,y) \in \{0, c_1, c_2\} \times \{c_4\}, \\ y & \text{for } (x,y) \in \{c_3\} \times P \text{ and } (x,y) \in \{c_4\} \times \{0, c_1, c_2\}, \\ 0 & \text{for } (x,y) \in \{0, c_1, c_2\}^2, \\ 1 & \text{otherwise.} \end{cases}$$

Except of uninorms U_c and U_d, we can construct another uninorm U_3 with the neutral element c_3, whose annihilator is equal to c_4. If we choose meet and join for the respective underlying t-norm and t-conorm on the intervals $[0, c_3]$ and $[c_3, 1]$, we get the following

$$U_3(x,y) = \begin{cases} x \wedge y & \text{for } (x,y) \in [0, c_3]^2, \\ x \vee y & \text{for } (x,y) \in \{c_3, 1\}^2, \\ c_4 & \text{otherwise.} \end{cases}$$

When constructing a proper nullnorm on P, we can choose any of the elements c_i to serve as the annihilator. For the annihilator equal to c_3 we have 4 t-conorms on $[0, c_i]$ and 1 t-norm on $[c_i, 1]$. When we want to use the construction from the proof of Theorem 2, we get the following:

$$V(x,y) = \begin{cases} x & \text{for } (x,y) \in [0, c_3] \times \{0\} \text{ and } (x,y) \in \{c_3, 1\} \times \{1\}, \\ y & \text{for } (x,y) \in \{0\} \times [0, c_3] \text{ and } (x,y) \in \{1\} \times \{c_3, 1\}, \qquad (3) \\ c_3 & \text{otherwise.} \end{cases}$$

4   Idempotent Uninorms and Nullnorms

Consider again Example 1 and the operations constructed there. The uninorms U_1, U_2, U_3 are idempotent, while the nullnorm V is not idempotent.

Lemma 5. *Let P be the poset whose Hasse diagram is sketched on Fig. 1. There exists no idempotent nullnorm on P.*

Proof. We will not consider all cases. Similarly to the nullnorm V given by formula (3), assume that c_3 is the annihilator of nullnorm V. Then the following holds

$$V(0, 1) = c_3 \geq V(c_4, 0) \geq \begin{cases} c_1 = V(0, c_1), \\ c_2 = V(0, c_2) \end{cases}.$$

This gives $V(c_4, 0) = c_3 \leq V(c_4, c_4)$. I.e., there are only two possibilities for the value $V(c_4, c_4)$, namely c_3 and 1. This gives that V is not idempotent. All the other cases could be proven in a similar way.   □

To give conditions under which there exist idempotent uninorms and/or nullnorms on a bounded poset P, first we provide the following Theorems 3 and 4 which are necessary conditions for the existence of an idempotent uninorm and nullnorm, respectively.

Theorem 3. *Let P be a bounded poset with at least three elements and $e \notin \{0, 1\}$. There exists an idempotent t-norm on $[0, e]$ if and only if $([0, e], \leq)$ is a meet semi-lattice.*

There exists an idempotent t-conorm on $[e, 1]$ if and only if $([e, 1], \leq)$ is a join semi-lattice.

Proof. We show that a t-norm $T : [0, e]^2 \to [0, e]$ is idempotent if and only if $T(x, y) = x \wedge y$. Assume for some $(x, y) \in [0, e]^2$ $T(x, y) = z \neq x \wedge y$. Then there exists s that is bounded from above by both, x and y and either $s > z$ or $s \parallel z$. Monotonicity of T implies $T(s, s) \leq T(x, y) \neq s$ and this implies that T is not idempotent.

Similarly we could show that a t-conorm S on $[e, 1]$ is idempotent if and only if $S(x, y) = x \vee x$.

In the same way as we proven Theorem 3 we could prove the following assertion.

Theorem 4. *Let P be a bounded poset with at least three elements and $a \notin \{0, 1\}$. There exists the idempotent t-conorm on $[0, a]$ if and only if $([0, a], \leq)$ is a join semi-lattice.*

There exists the idempotent t-norm on $[a, 1]$ if and only if $([a, 1], \leq)$ is a meet semi-lattice.

The following theorems state some sufficient conditions under which it is possible to construct an idempotent uninorm on P.

Theorem 5. *Let $(P, 0, 1, \leq)$ be a bounded poset with at least three elements and $e \notin \{0, 1\}$. Assume $[0, e]$ is a meet semi-lattice, $[e, 1]$ is a join semi-lattice. Further assume $(\|_e, \leq)$ is either a meet semi-lattice such that $x < e$ and $y \in \|_e$ imply $x < y$, or $(\|_e, \leq)$ is a join semi-lattice such that $x > e$ and $y \in \|_e$ imply $y < x$. Then there exists an idempotent uninorm on P whose neutral element is e.*

Proof. Assume $\|_e$ is a meet semi-lattice fulfilling that $x < e$ and $y \in \|_e$ imply $x < y$. Then the following operation is a conjunctive uninorm:

$$
U_c(x, y) = \begin{cases}
x \wedge y & \text{for } (x, y) \in [0, e]^2 \text{ and for } (x, y) \in \|_e^2, \\
x \vee y & \text{for } (x, y) \in [e, 1]^2, \\
x & \text{if } y \in [e, 1] \text{ and } x < e \text{ or } x \in \|_e, \\
& \quad \text{or if } x < e \text{ and } y \in \|_e, \\
y & \text{if } x \in [e, 1] \text{ and } y < e \text{ or } y \in \|_e, \\
& \quad \text{or if } y < e \text{ and } x \in \|_e .
\end{cases}
$$

In case $\|_e$ is a join semi-lattice fulfilling $x > e$ and $y \in\|_e$ imply $y < x$, the following operation is a disjunctive uninorm:

$$U_d(x,y) = \begin{cases} x \wedge y & \text{for } (x,y) \in [\mathbf{0}, e]^2, \\ x \vee y & \text{for } (x,y) \in [e, \mathbf{1}]^2 \text{ and for } (x,y) \in\|_e^2, \\ x & \text{if } y \in [\mathbf{0}, e] \text{ and } x > e \text{ or } x \in\|_e, \\ & \text{or if } x > e \text{ and } y \in\|_e, \\ y & \text{if } x \in [\mathbf{0}, e] \text{ and } y > e \text{ or } y \in\|_e, \\ & \text{or if } y > e \text{ and } x \in\|_e. \end{cases}$$

Concerning associativity of U_c, this is implied by the fact that both, meet and join, are associative operations and we assume their existence on the respective parts of P, and further, elements of $[e, \mathbf{1}]$ are partial neutral element for the remaining part of P and elements of $\|_e$ are partial neutral elements for values $x < e$.

Similarly we could show the associativity of U_d.   □

Theorem 6. *Let $(P, \mathbf{0}, \mathbf{1}, \leq)$ be a bounded poset with at least three elements and $e \notin \{\mathbf{0}, \mathbf{1}\}$. Assume $[\mathbf{0}, e[\cup \|_e$ is a meet semi-lattice, $[e, \mathbf{1}]$ is a join semi-lattice ($[\mathbf{0}, e]$ is a meet semi-lattice and $]e, \mathbf{1}] \cup \|_e$ is a join sem-lattice). Then there exists an idempotent uninorm on P whose neutral element is e.*

Proof. We construct an idempotent uninorm in the case $[\mathbf{0}, e[\cup \|_e$ is a meet semi-lattice.

$$U_{c_1} = \begin{cases} x \vee y & \text{for } (x,y) \in [e, \mathbf{1}]^2, \\ x & \text{for } (x,y) \in ([\mathbf{0}, e[\cup \|_e) \times [e, \mathbf{1}], \\ y & \text{for } (x,y) \in [e, \mathbf{1}] \times ([\mathbf{0}, e[\cup \|_e), \\ x \wedge y & \text{otherwise.} \end{cases}$$

□

The following example illustrates that the constraints in Theorems 5 and 6 are really just sufficient.

Example 2. Assume a poset P_1 given by the following Hasse diagram (Fig. 2) Set $c_3 = e$. As can be easily checked, $\|_e = \|_{c_3} = \{c_4, c_5, c_6\}$ and the set $\{c_4, c_5, c_6\}$ is neither a meet nor a join semi-lattice and moreover, there exists neither join nor meet for any pair of different elements from $\|_{c_3}$, i.e., P_1 fulfils neither the constraints of Theorem 5 nor of Theorem 6.

Define a preference among elements of $\|_{c_3}$ by $c_4 \prec c_5 \prec c_6$, and

$$c_i \curlywedge c_j \quad \text{if } c_i \preceq c_j \text{ for } (c_i, c_j) \in (\|_{c_3})^2.$$

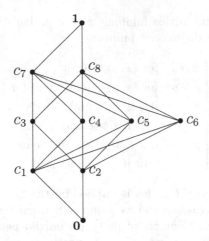

Fig. 2. Hasse diagram of the poset P_1

We construct now a conjunctive uninorm on P_1 whose neutral element is c_3.

$$
U_{c_2} = \begin{cases}
x \vee y & \text{for } (x,y) \in [c_3, \mathbf{1}]^2, \\
x & \text{for } (x,y) \in (\{\mathbf{0}, c_1, c_2\} \cup \{c_7, c_8, \mathbf{1}\}) \times (\{c_3\} \cup \ \|_{c_3}), \\
& \text{and for } (x,y) \in \{\mathbf{0}, c_1, c_2\} \times \{c_7, c_8, \mathbf{1}\} \cup \ \|_{c_3} \times \{c_3\}, \\
y & \text{for } (x,y) \in (\{c_3\} \cup \ \|_{c_3}) \times (\{\mathbf{0}, c_1, c_2\} \cup \{c_7, c_8, \mathbf{1}\}) \\
& \text{and for } (x,y) \in \{c_7, c_8, \mathbf{1}\} \times \{\mathbf{0}, c_1, c_2\} \cup \{c_3\} \times \ \|_{c_3}, \\
x \curlywedge y & \text{for } (x,y) \in (\|_{c_3})^2, \\
x \wedge y & \text{otherwise}
\end{cases}
$$

The uninorm U_{c_2} is idempotent.

A necessary condition the set $\|_e$ has to fulfil to yield the existence of an idempotent uninorm, is formulated in the next theorem.

Theorem 7. *Let* $(P, \mathbf{0}, \mathbf{1}, \leq)$ *be a bounded poset with at least three elements and* $e \notin \{\mathbf{0}, \mathbf{1}\}$. *Assume there exists an eight-truple* $\{c_1, c_2, c_3, c_4, c_5, c_6, c_7, c_8\} \in \|_e$ *having Hasse diagram depicted in Fig. 3, whose join of* (c_1, c_2), (c_3, c_4), (c_5, c_6), *and meet of* (c_3, c_4), (c_5, c_6), (c_7, c_8), *are not defined. Then there exists no idempotent uninorm* U *on* P *whose neutral element is* e.

Proof. Denote $C = \{c\|_1, c_2, c_3, c_4, c_5, c_6, c_7, c_8\}$. Since U is an idempotent uninorm, for arbitrary $(c_i, c_j) \in C^2$ such that $c_i \leq c_j$, we have

$$
c_i = U(c_i, c_i) \leq U(c_i, c_j) \leq U(c_j, c_j) = c_j.
$$

Let us define the value of $U(c_3, c_4)$. By monotonocity of U we get $U(c_3, c_4) \in C$. Because of the symmetry of Hasse diagram in Fig. 3 it is enough to check three possibilities, $U(c_3, c_4) \in \{c_1, c_3, c_5\}$. Set $U(c_3, c_4) = c_5$. Then

$$
c_5 = U(c_3, c_4) \leq U(c_3, c_6) \in [c_3, c_6],
$$

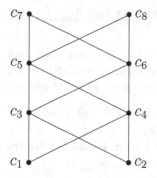

Fig. 3. Hasse diagram of the eight-tuple $\{c_1, c_2, \ldots, c_8\}$

which is a contradiction. Similarly, for $U(c_3, c_4) = c_1$ we get the follwing contradiction

$$c_1 = U(c_3, c_4) \geq U(c_2, c_4) \in [c_2, c_4].$$

The last possibility is $U(c_3, c_4) = c_3$. Then

$$c_3 = U(c_3, c_4) \leq U(c_5, c_4) \in [c_4, c_5], \quad c_3 = U(c_3, c_4) \leq U(c_6, c_4) \in [c_4, c_6].$$

The above formulas imply $U(c_5, c_6) \geq c_5$ and $U(c_5, c_6) \geq c_6$. Set $U(c_5, c_6) = c_7$. In this case we get the following contradiction

$$c_7 = U(c_5, c_6) \leq U(c_6, c_8) \in [c_6, c_8]. \qquad \square$$

The following theorem states conditions under which it is possible to construct an idempotent nullnorm on P.

Theorem 8. *Let $(P, 0, 1, \leq)$ be a bounded poset with at least three elements and $a \notin \{0, 1\}$. Assume $[0, a]$ is a join semi-lattice, $[a, 1]$ is a meet semi-lattice. Further assume that there exist $c_1 \in [0, a]$ and $c_2 \in [a, 1]$ such that $\|_a \subset [c_1, c_2]$ and $x \parallel y$ for all $x \in \|_a$ and $y \in \,]c_1, a] \cup [a, c_2[\cup \|_{c_1} \cup \|_{c_2}$, and moreover, $(\|_a \cup \{c_1, c_2\}, \leq)$ is either a meet or a join semi-lattice. Then there exists an idempotent nullnorm on P whose annihilator is a.*

Proof. We provide a construction of an idempotent nullnorm for the case $\|_a$ is a meet semi-lattice.

$$V(x, y) = \begin{cases} x \vee y & \text{for } (x, y) \in [0, a]^2, \\ x \wedge y & \text{for } (x, y) \in [a, 1]^2 \text{ and for } (x, y) \in \|_a^2, \\ x & \text{if } y \in \|_a \text{ and } x \in [c_1, a] \cup [a, c_2] \cup \|_{c_1} \cup \|_{c_2}, \\ y & \text{if } x \in \|_a \text{ and } y \in [c_1, c_2], \\ c_1 & \text{if } (x, y) \in [0, c_1] \times \|_a \cup \|_a \times [0, c_1], \\ c_2 & \text{if } (x, y) \in [c_2, 1] \times \|_a \cup \|_a \times [c_2, 1], \\ a & \text{otherwise.} \end{cases}$$

$\square$

Finally, we provide a necessary condition for the existence of an idempotent nullnorm on P.

Theorem 9. *Let $(P, 0, 1, \leq)$ be a bounded poset with at least three elements and $a \notin \{0, 1\}$. Assume $[0, a]$ is a join semi-lattice, $[a, 1]$ is a meet semi-lattice. If there exists an idempotent nullnorm V on P whose annihilator is a, then:*

(i) *for arbitrary pair $(c_1, c_2) \in [0, a]^2$ and arbitrary $c_3 \in \|_a$ such that $c_3 \geq c_1$ and $c_3 \geq c_2$, we have that $V(c_1, c_2) \leq c_3$,*

(ii) *for arbitrary pair $(c_4, c_5) \in [a, 1]^2$ and arbitrary $c_6 \in \|_a$ such that $c_6 \leq c_4$ and $c_6 \leq c_5$, we have that $V(c_4, c_5) \geq c_6$.*

Proof. We show the case (i). Assume $(c_1, c_2) \in [0, a]^2$ and arbitrary $c_3 \in \|_a$ such that $c_3 \geq c_1$ and $c_3 \geq c_2$, and $V(c_1, c_2) \not\leq c_3$. Since V is idempotent, $V(c_1, c_3) \in [c_1, c_3]$ and because of monotonicity of V we get $V(c_1, c_3) \geq V(c_1, c_2) \notin [c_1, c_3]$, which is a cintradiction. □

5   Conclusions

In this contribution we have introduced uninorms and nullnorms on bounded posets. We have shown that on every bounded poset with at least three elements it is possible to construct a conjunctive and a disjunctive proper uninorm regardless which element we choose to serve as the neutral one. Further we have shown that on every bounded poset with at least three elements it is possible to construct a proper nullnorm. Further, we have provided some sufficient and some necessary conditions under which it is possible to construct idempotent uninorms and/or nullnorms.

References

1. De Baets, B.: Idempotent uninorms. Eur. J. Oper. Res. Int. **118**, 631–642 (1999)
2. De Baets, B., Fodor, J., Ruiz-Aguilera, D., Torrens, J.: Idempotent uninorms on finite ordinal scales. Int. J. Uncertainty Fuzziness Knowl. Based Syst. **17**(1), 1–14 (2009)
3. Birkhoff, G.: Lattice Theory. American Mathematical Society Colloquium Publishers, Providence (1967)
4. Bodjanova, S., Kalina, M.: Construction of uninorms on bounded lattices. In: SISY 2014, IEEE 12th International Symposium on Intelligent Systems and Informatics, Subotica, Serbia, pp. 61–66 (2014)
5. Bodjanova, S., Kalina, M.: Uninorms on bounded lattices – recent development. In: Kacprzyk, J., Szmidt, E., Zadrożny, S., Atanassov, K.T., Krawczak, M. (eds.) IWIFSGN/EUSFLAT -2017. AISC, vol. 641, pp. 224–234. Springer, Cham (2018). https://doi.org/10.1007/978-3-319-66830-7_21
6. Bodjanova, S., Kalina, M.: Nullnorms and T-operators on bounded lattices: coincidence and differences. In: Medina, J., et al. (eds.) IPMU 2018. CCIS, vol. 853, pp. 160–170. Springer, Cham (2018). https://doi.org/10.1007/978-3-319-91473-2_14

7. Calvo, T., De Baets, B., Fodor, J.: The functional equations of Frank and Alsina for uninorms and nullnorms. Fuzzy Sets Syst. **120**, 385–394 (2001)
8. Calvo, T., Kolesárová, A., Komorníková, M., Mesiar, R.: Aggregation operators: properties, classes and construction methods. In: Calvo, T., Mayor, G., Mesiar, R. (eds.) Aggregation Operators, pp. 3–104. Physica-Verlag GMBH, Heidelberg (2002)
9. Çayli, G.D., Drygaś, P.: Some properties of idempotent uninorms on a special class of bounded lattices. Inf. Sci. **422**, 352–363 (2018)
10. Çayli, G.D., Karaçal, F.: Construction of uninorms on bounded lattices. Kybernetika **53**, 394–417 (2017)
11. Çayli, G.D., Karaçal, F., Mesiar, R.: On a new class of uninorms on bounded lattices. Inf. Sci. **367–368**, 221–231 (2016)
12. Czogała, E., Drewniak, J.: Associative monotonic operations in fuzzy set theory. Fuzzy Sets Syst. **12**, 249–269 (1984)
13. Deschrijver, G.: A representation of t-norms in interval valued L-fuzzy set theory. Fuzzy Sets Syst. **159**, 1597–1618 (2008)
14. Deschrijver, G.: Uninorms which are neither conjunctive nor disjunctive in interval-valued fuzzy set theory. Inf. Sci. **244**, 48–59 (2013)
15. Devillet, J., Kiss, G., Marichal, J.-L.: Characterizations of quasitrivial symmetric nondecreasing associative operations. Semigroup Forum **98**(1), 154–171 (2019)
16. Dombi, J.: Basic concepts for a theory of evaluation: the aggregative operator. Eur. J. Oper. Res. **10**, 282–293 (1982)
17. Dombi, J.: A general class of fuzzy operators, the DeMorgan class of fuzzy operators and fuzziness measures induced by fuzzy operators. Fuzzy Sets Syst. **8**, 149–163 (1982)
18. Grabisch, M., Marichal, J.-L., Mesiar, R., Pap, E.: Aggregation functions. In: Encyclopedia of Mathematics and its Applications, vol. 127. Cambridge University Press, Cambridge (2009)
19. Kalina, M.: On uninorms and nullnorms on direct product of bounded lattices. Open Phys. **14**(1), 321–327 (2016)
20. Kalina, M., Král, P.: Uninorms on interval-valued fuzzy sets. In: Carvalho, J.P., Lesot, M.-J., Kaymak, U., Vieira, S., Bouchon-Meunier, B., Yager, R.R. (eds.) IPMU 2016. CCIS, vol. 611, pp. 522–531. Springer, Cham (2016). https://doi.org/10.1007/978-3-319-40581-0_42
21. Karaçal, F., Ince, M.A., Mesiar, R.: Nullnorms on bounded lattices. Inf. Sci. **325**, 227–236 (2015)
22. Karaçal, F., Mesiar, R.: Uninorms on bounded lattices. Fuzzy Sets Syst. **261**, 33–43 (2015)
23. Klement, E.P., Mesiar, R., Pap, E.: Triangular Norms. Springer-Verlag, Heidelberg (2000)
24. Mas, M., Mayor, G., Torrens, J.: t-operators. J. Uncertainty Fuzziness Knowl. Based Syst. **7**, 31–50 (1999)
25. Mas, M., Mayor, G., Torrens, J.: t-operators and uninorms in a finite totally ordered set. Int. J. Intell. Syst. **14**(9), 909–922 (1999)
26. Schweizer, B., Sklar, A.: Probabilistic Metric Spaces. North Holland, New York (1983)
27. Yager, R.R., Rybalov, A.: Uninorm aggregation operators. Fuzzy Sets Syst. **80**, 111–120 (1996)

Derivative for Discrete Choquet Integrals

Yasuo Narukawa[1] and Vicenç Torra[2]($\boxtimes$)

[1] Tamagawa University, 6-1-1 Tamagawagakuen, Machida, Tokyo 194-8610, Japan
nrkwy@eng.tamagawa.ac.jp
[2] Hamilton Institute, Maynooth University, Eolas building, North Campus,
Maynooth, Ireland
vtorra@ieee.org

Abstract. In this paper we study necessary and sufficient conditions for the existence of the derivative for fuzzy measures when we are considering the Choquet integral. Results apply to discrete domains. The main result is based on the definition we introduce of compatible permutation for two pairs of measures (μ, ν).

As an application of the main result, we present the conditions for possibility measures.

1 Introduction

Choquet integral [2] permits to integrate a function with respect to a fuzzy measure. Fuzzy measures [3,12], also known as capacities and non-additive measures, generalize standard measures replacing the additivity condition by a monotonicity one. Then, when a fuzzy measure is additive, Choquet integral reduces to the Lebesgue integral.

The Radon-Nikodym derivative is a very important concept related to the Lebesgue integral. The Radon-Nikodym theorem establishes that we can express one additive measure with respect to another one under some conditions. In particular, the condition of absolute continuity between two measures plays a pivotal role.

In addition to its intrinsic mathematical interest, the Radon-Nikodym derivative is useful in practical applications. More particularly, it has been used to define distances and divergences between pairs of measures. In particular, f-divergences, which are defined in terms of Radon-Nikodym derivatives, are extensively used in statistics and information theory. Recall that the Hellinger distance, the Kullback-Leibler divergence, the Rényi distance and the variation distance are all examples of f-divergences. They are used to compare probability distributions, and the Kullback-Leibler divergence can also be used to define the entropy.

Because of its theoretical and applied interest, the problem of defining Radon-Nikodym-like derivatives for fuzzy measures is a relevant question. In the continuous case, Graf, Mesiar and Sipos, Nguyen, Rébillé, and Sugeno [5–9] have studied the existence and computation of a Radon-Nikodym like derivative for

© Springer Nature Switzerland AG 2019
V. Torra et al. (Eds.): MDAI 2019, LNAI 11676, pp. 138–147, 2019.
https://doi.org/10.1007/978-3-030-26773-5_13

non-additive measures in the context of Choquet integrals. We have considered the problem ourselves in the context of defining f-divergence for fuzzy measures. We have considered both discrete and continuous case. See e.g. [11, 13] (for f-divergence and Hellinger distances) and [10] (for the definition of the entropy). This derivative has also been used in [1] to define an alternative expression for f-divergence.

In this paper we consider the problem of existence of the derivative when the reference set is finite. More particularly, we consider the problem of finding necessary and sufficient conditions on the existence of the derivative.

The structure of the paper is as follows. In Sect. 2 we review the concepts that are needed in the paper. In Sect. 3 we present the main results. In Sect. 4, as the application of the main result, we present the conditions for possibility measures.

2   Preliminaries

Let us consider the universal set $X := \{x_1, x_2. \cdots, x_n\}$. Let us review the definitions of fuzzy measure and Choquet integral.

Definition 1. *A set function μ such that $\mu(\emptyset) = 0$ and that is monotonic with respect to the set inclusion (i.e., $\mu(A) \le \mu(B)$ when $A \subset B$) is called a fuzzy measure, non-additive measure, capacity or monotonic game.*

It is often also required that μ satisfies $\mu(X) = 1$. We do not require this condition in this paper.

Definition 2. *A fuzzy measure μ on $(X, 2^X)$ is called a possibility measure, if $\mu(A \cup B) = \mu(A) \vee \mu(B)$ for $A, B \in 2^X$. Here $\vee$ is understood as the maximum.*

Definition 3. *Let μ be a fuzzy measure and f be a function $f : X \to [0, \infty)$. The Choquet integral of the function f with respect to the fuzzy measure μ is defined by*

$$(C) \int f d\mu = \int_0^\infty \mu(\{x | f(x) \ge \alpha\}) d\alpha \tag{1}$$

Let $A \subset X$. The Choquet integral of the function f over A with respect to the fuzzy measure μ is defined by

$$(C) \int_A f d\mu = \int_0^\infty \mu(\{x | f(x) \ge \alpha\} \cap A) d\alpha \tag{2}$$

When μ is additive, this expression corresponds to the classical Lebesgue integral. Using Eq. 2 we can consider defining measures in terms of other measures. That is, we can define a measure μ from another measure ν as follows

$$\mu(A) = (C) \int_A f d\nu = \int_0^\infty \nu(\{x | f(x) \ge \alpha\} \cap A) d\alpha \tag{3}$$

Given μ and ν in Eq. 3, we can consider the problem of finding the function f. When the measures are additive, this corresponds to the Radon-Nikodym derivative, as the Choquet integral reduces to the Lebesgue integral.

3   Condition for the Existence of a Derivative

While for additive fuzzy measures the Radon-Nikodym derivative exist when the measures are absolutely continuous, this is not the case for fuzzy measures. Because of that, it is rellevant to study when the derivative exist. We give some conditions for its existence in this section. From now on, we will consider measures ν such that $\nu(\{x_i\}) \neq 0$ for all x_i.

Definition 4. *Let μ, ν be fuzzy measures on $(X, 2^X)$. We say that μ and ν are compatible if there exists a permutation σ on $\{1, 2, \ldots, n\}$ such that*

$$\frac{\mu(\{x_{\sigma(1)}\})}{\nu(\{x_{\sigma(1)}\})} \geq \frac{\mu(\{x_{\sigma(2)}\})}{\nu(\{x_{\sigma(2)}\})} \geq \cdots \geq \frac{\mu(\{x_{\sigma(n)}\})}{\nu(\{x_{\sigma(n)}\})}.$$

A permutation σ satisfying this condition is said to be a compatible permutation for (μ, ν).

From the definition of compatible permutation for a pair of measures (μ, ν), it is easy to prove the following proposition.

Proposition 1. *Let σ be a compatible permutation for (μ, ν). Then, we have*

$$\left| \begin{matrix} \mu(\{x_{\sigma(k)}\}) & \nu(\{x_{\sigma(k)}\}) \\ \mu(\{x_{\sigma(k+1)}\}) & \nu(\{x_{\sigma(k+1)}\}) \end{matrix} \right| \geq 0$$

for $k = 1, 2, \ldots, n-1$.

We will now give the main theorem of this paper. Let us now consider the following. Let $x_k \in X$ and $f(x_k) := \frac{\mu(\{x_k\})}{\nu(\{x_k\})}$ for $k = 1, 2, \cdots, n$.
Since

$$(C) \int_{\{x_k\}} f d\nu = f(x_k)\nu(\{x_k\}),$$

we have

$$\mu(\{x_k\}) = (C) \int_{\{x_k\}} f d\nu.$$

Let $A_2 := \{x_{i_1}, x_{i_2}\}$. Suppose that

$$\mu(A_2) = (C) \int_{A_2} f d\nu.$$

with $f(x_{\sigma(i_1)}) \geq f(x_{\sigma(i_2)})$
Since

$$(C) \int_{A_2} f d\nu = f(x_{\sigma(i_2)})[\nu(\{x_{\sigma(i_1)}, x_{\sigma(i_2)}\}) - \nu(\{x_{\sigma(i_1)}\})] + f(x_{\sigma(i_1)})\nu(\{x_{\sigma(i_1)}\})$$

$$= \frac{\mu(\{x_{\sigma(i_2)}\})}{\nu(\{x_{\sigma(i_2)}\})}[\nu(\{x_{\sigma(i_1)}, x_{\sigma(i_2)}\}) - \nu(\{x_{\sigma(i_1)}\})] + \frac{\mu(\{x_{\sigma(i_1)}\})}{\nu(\{x_{\sigma(i_1)}\})}\nu(\{x_{\sigma(i_1)}\})$$

$$= \frac{\mu(\{x_{\sigma(i_2)}\})}{\nu(\{x_{\sigma(i_2)}\})}[\nu(\{x_{\sigma(i_1)}, x_{\sigma(i_2)}\}) - \nu(\{x_{\sigma(i_1)}\})] + \mu(\{x_{\sigma(i_1)}\}),$$

we have

$$\mu(\{x_{\sigma(i_1)}, x_{\sigma(i_2)}\}) - \mu(\{x_{\sigma(i_1)}\}) = \frac{\mu(\{x_{\sigma(i_2)}\})}{\nu(\{x_{\sigma(i_2)}\})}[\nu(\{x_{\sigma(i_1)}, x_{\sigma(i_2)}\}) - \nu(\{x_{\sigma(i_1)}\})],$$

that is,

$$\nu(\{x_{\sigma(i_2)}\})[\mu(\{x_{\sigma(i_1)}, x_{\sigma(i_2)}\}) - \mu(\{x_{\sigma(i_1)}\})] = \mu(\{x_{\sigma(i_2)}\})[\nu(\{x_{\sigma(i_1)}, x_{\sigma(i_2)}\}) - \nu(\{x_{\sigma(i_1)}\})] \quad (4)$$

Let $A_3 := \{x_{i_1}, x_{i_2}, x_{i_3}\}$. Suppose that Eq. 4 and

$$\mu(A_3) = (C)\int_{A_3} f d\nu.$$

with $f(x_{\sigma(i_1)}) \geq f(x_{\sigma(i_2)}) \geq f(x_{\sigma(i_3)})$.

Then, we have

$$(C)\int_{A_3} f d\nu = f(x_{\sigma(i_3)})[\nu(\{x_{\sigma(i_1)}, x_{\sigma(i_2)}, x_{\sigma(i_3)}\}) - \nu(\{x_{\sigma(i_1)}, x_{\sigma(i_2)}, \})]$$

$$+ f(x_{\sigma(i_2)})[\nu(\{x_{\sigma(i_1)}, x_{\sigma(i_2)}\}) - \nu(\{x_{\sigma(i_1)}\})] + f(x_{\sigma(i_1)})\nu(\{x_{\sigma(i_1)}\})$$

$$= \frac{\mu(\{x_{\sigma(i_3)}\})}{\nu(\{x_{\sigma(i_3)}\})}[\nu(\{x_{\sigma(i_1)}, x_{\sigma(i_2)}, x_{\sigma(i_3)}\}) - \nu(\{x_{\sigma(i_1)}, x_{\sigma(i_2)}, \})]$$

$$+ \frac{\mu(\{x_{\sigma(i_2)}\})}{\nu(\{x_{\sigma(i_2)}\})}[\nu(\{x_{\sigma(i_1)}, x_{\sigma(i_2)}\}) - \nu(\{x_{\sigma(i_1)}\})] + \frac{\mu(\{x_{\sigma(i_1)}\})}{\nu(\{x_{\sigma(i_1)}\})}\nu(\{x_{\sigma(i_1)}\})$$

$$= \frac{\mu(\{x_{\sigma(i_3)}\})}{\nu(\{x_{\sigma(i_3)}\})}[\nu(\{x_{\sigma(i_1)}, x_{\sigma(i_2)}, x_{\sigma(i_3)}\}) - \nu(\{x_{\sigma(i_1)}, x_{\sigma(i_2)}, \})]$$

$$+ \mu(\{x_{\sigma(i_1)}, x_{\sigma(i_2)}\}) - \mu(\{x_{\sigma(i_1)}\}) + \mu(\{x_{\sigma(i_1)}\})$$

$$= \frac{\mu(\{x_{\sigma(i_3)}\})}{\nu(\{x_{\sigma(i_3)}\})}[\nu(\{x_{\sigma(i_1)}, x_{\sigma(i_2)}, x_{\sigma(i_3)}\}) - \nu(\{x_{\sigma(i_1)}, x_{\sigma(i_2)}, \})] + \mu(\{x_{\sigma(i_1)}, x_{\sigma(i_2)}\})$$

Then we have

$$\nu(\{x_{\sigma(i_3)}\})[\mu(\{x_{\sigma(i_1)}, x_{\sigma(i_2)}, x_{\sigma(i_3)}\}) - \mu(\{x_{\sigma(i_1)}, x_{\sigma(i_2)}\})]$$

$$= \mu(\{x_{\sigma(i_3)}\})[\nu(\{x_{\sigma(i_1)}, x_{\sigma(i_2)}, x_{\sigma(i_3)}\}) - \nu(\{x_{\sigma(i_1)}, x_{\sigma(i_2)}\})]$$

Let $A_k := \{x_{i_1}, x_{i_2}, \ldots, x_{i_k}\}$. Then, by induction, if

$$\mu(A_k) = (C)\int_{A_k} f d\nu,$$

then we have

$$\nu(\{x_{\sigma(i_k)}\})[\mu(\{x_{\sigma(i_1)}, x_{\sigma(i_2)}, \ldots, x_{\sigma(i_k)}\}) - \mu(\{x_{\sigma(i_1)}, x_{\sigma(i_2)} \ldots, x_{\sigma(i_{k-1})}\})]$$

$$= \mu(\{x_{\sigma(i_3)}\})[\nu(\{x_{\sigma(i_1)}, x_{\sigma(i_2)}, \ldots, x_{\sigma(i_k)}\}) - \nu(\{x_{\sigma(i_1)}, x_{\sigma(i_2)} \ldots, x_{\sigma(i_{k-1})}\})].$$

Therefore we have the next theorem

Theorem 1. *Let μ, ν be fuzzy measures on $(X, 2^X)$ and σ be a compatible permutation for (μ, ν). Then, there exists a function f on X such that*

$$\mu(A_k) = (C) \int_{A_k} f \, d\nu$$

for all $A_k = \{x_{i_1}, x_{i_2}, \cdots, x_{i_k}\} \subset X$ if and only if

$$\begin{vmatrix} 1 & \mu(\{x_{\sigma(i_1)}, x_{\sigma(i_2)} \ldots, x_{\sigma(i_{k-1})}\}) & \nu(\{x_{\sigma(i_1)}, x_{\sigma(i_2)} \ldots, x_{\sigma(i_{k-1})}\}) \\ 1 & \mu(\{x_{\sigma(i_1)}, x_{\sigma(i_2)}), \ldots, x_{\sigma(i_k)}\}) & \nu(\{x_{\sigma(i_1)}, x_{\sigma(i_2)}), \ldots, x_{\sigma(i_k)}\}) \\ 0 & \mu(\{x_{\sigma(i_k)}\}) & \nu(\{x_{\sigma(i_k)}\}) \end{vmatrix} = 0 \quad (5)$$

for all $k = 2, \ldots, n$.

Here, $x_{\sigma(i_k)}$ will be the element with smallest $\mu(\{x_{\sigma(i_k)}\})/\nu(\{x_{\sigma(i_k)}\})$ in the set $x_{\sigma(i_1)}, x_{\sigma(i_2)}, \ldots, x_{\sigma(i_k)}$.

We illustrate this theorem with an example. We give two measures on a reference set of three elements that are compatible. We show that the determinants of Theorem 1 are zero and thus, there exists a derivative of one measure with respect to the other one.

Example 1. Let $X := \{x_1, x_2, x_3\}$ and let μ and ν two non-additive measures defined as in Table 1.

Table 1. Two measures μ and ν that are compatible.

A	$\{x_1\}$	$\{x_2\}$	$\{x_3\}$	$\{x_1, x_2\}$	$\{x_2, x_3\}$	$\{x_1, x_3\}$	$\{x_1, x_2, x_3\}$
$\mu(A)$	0.2	0.3	0.4	0.5	0.6	0.55	0.8
$\nu(A)$	0.1	0.3	0.8	0.4	0.9	0.8	1

We can observe that

$$\frac{\mu(\{x_1\})}{\nu(\{x_1\})} > \frac{\mu(\{x_2\})}{\nu(\{x_2\})} > \frac{\mu(\{x_3\})}{\nu(\{x_3\})}.$$

Let us now check that the determinants are zero for all $A_k \subseteq X$. We need to consider only $k = 2$ and $k = 3$ as there are only 3 elements in X.

Let us begin with $k = 2$, and we need to consider the sets $\{x_1, x_2\}$, $\{x_1, x_3\}$, $\{x_2, x_3\}$. Then, for the first set we obtain the following determinant that is equal to zero:

$$\begin{vmatrix} 1 & \mu(\{x_1\}) & \nu(\{x_1\}) \\ 1 & \mu(\{x_1, x_2\}) & \nu(\{x_1, x_2\}) \\ 0 & \mu(\{x_2\}) & \nu(\{x_2\}) \end{vmatrix} = \begin{vmatrix} 1 & 0.2 & 0.1 \\ 1 & 0.5 & 0.4 \\ 0 & 0.3 & 0.3 \end{vmatrix} = 0.$$

For the second set we obtain the following determinant that is also equal to zero:

$$\begin{vmatrix} 1 & \mu(\{x_1\}) & \nu(\{x_1\}) \\ 1 & \mu(\{x_1,x_3\}) & \nu(\{x_1,x_3\}) \\ 0 & \mu(\{x_3\}) & \nu(\{x_3\}) \end{vmatrix} = \begin{vmatrix} 1 & 0.2 & 0.1 \\ 1 & 0.55 & 0.8 \\ 0 & 0.4 & 0.8 \end{vmatrix} = 0.$$

Similarly, for the third set we obtain the following determinant that is also equal to zero:

$$\begin{vmatrix} 1 & \mu(\{x_2\}) & \nu(\{x_2\}) \\ 1 & \mu(\{x_2,x_3\}) & \nu(\{x_2,x_3\}) \\ 0 & \mu(\{x_3\}) & \nu(\{x_3\}) \end{vmatrix} = \begin{vmatrix} 1 & 0.3 & 0.3 \\ 1 & 0.6 & 0.9 \\ 0 & 0.4 & 0.8 \end{vmatrix} = 0.$$

Then, for $k = 3$ we need to consider the only set with 3 elements. That is, $\{x_1, x_2, x_3\}$. In this case we have the following determinant that is also equal to zero.

$$\begin{vmatrix} 1 & \mu(\{x_1,x_2\}) & \nu(\{x_1,x_2\}) \\ 1 & \mu(\{x_1,x_2,x_3\}) & \nu(\{x_1,x_2,x_3\}) \\ 0 & \mu(\{x_3\}) & \nu(\{x_3\}) \end{vmatrix} = \begin{vmatrix} 1 & 0.5 & 0.4 \\ 1 & 0.8 & 1 \\ 0 & 0.4 & 0.8 \end{vmatrix} = 0.$$

Therefore, Theorem 1 implies that defining

$$f(x_1) = \frac{\mu(\{x_1\})}{\nu(\{x_1\})}, f(x_2) = \frac{\mu(\{x_2\})}{\nu(\{x_2\})}, f(x_3) = \frac{\mu(\{x_3\})}{\nu(\{x_3\})},$$

or, more specifically, with

$$f(x_1) = 2.0, f(x_2) = 1.0 \text{ and } f(x_3) = 0.5$$

we have

$$\mu(A) = (C)\int_A f d\nu,$$

for all A. This last equation can be checked with straightforward computation.

4   Possibility Measures

Let us consider two possibility measures μ and ν. We will reconsider for this type of measures Theorem 1 and make the condition for the existence of the derivative simpler.

Definition 5. *Let μ and ν be compatible fuzzy measures on $(X, 2^x)$ and σ be a compatible permutation on (μ, ν). Then, μ (resp. ν) is said to be weakly monotone decreasing for σ if*

$$\mu(\{x_{\sigma(1)}\}) \geq \mu(\{x_{\sigma(2)}\}) \geq \cdots \geq \mu(\{x_{\sigma(n)}\})$$

(resp. $\nu(\{x_{\sigma(1)}\}) \geq \nu(\{x_{\sigma(2)}\}) \geq \cdots \geq \nu(\{x_{\sigma(n)}\})$).

Suppose that μ and ν are weakly monotone increasing for σ
Since

$$\mu(\{x_{\sigma(i_1)}, x_{\sigma(i_2)} \ldots, x_{\sigma(i_{k-1})}\}) = \mu(\{x_{\sigma(i_1)}, x_{\sigma(i_2)}, \ldots, x_{\sigma(i_k)}\}) = \mu(\{x_{\sigma(i_1)}\})$$

and

$$\nu(\{x_{\sigma(i_1)}, x_{\sigma(i_2)} \ldots, x_{\sigma(i_{k-1})}\}) = \nu(\{x_{\sigma(i_1)}, x_{\sigma(i_2)}, \ldots, x_{\sigma(i_k)}\}) = \nu(\{x_{\sigma(i_1)}\}),$$

for any $k = 2, \ldots, n$ we have that we can prove that the following equality holds

$$\begin{vmatrix} 1 & \mu(\{x_{\sigma(i_1)}, \ldots, x_{\sigma(i_{k-1})}\}) & \nu(\{x_{\sigma(i_1)}, \ldots, x_{\sigma(i_{k-1})}\}) \\ 1 & \mu(\{x_{\sigma(i_1)}, \ldots, x_{\sigma(i_k)}\}) & \nu(\{x_{\sigma(i_1)}, \ldots, x_{\sigma(i_k)}\}) \\ 0 & \mu(\{x_{\sigma(i_k)}\}) & \nu(\{x_{\sigma(i_k)}\}) \end{vmatrix}$$

and

$$= \begin{vmatrix} 1 & \mu(\{x_{\sigma(i_1)}\}) & \nu(\{x_{\sigma(i_1)}\}) \\ 1 & \mu(\{x_{\sigma(i_1)}\}) & \nu(\{x_{\sigma(i_1)}\}) \\ 0 & \mu(\{x_{\sigma(i_k)}\}) & \nu(\{x_{\sigma(i_k)}\}) \end{vmatrix} = 0$$

for all $k = 2, \ldots, n$.

Therefore, as this implies that Eq. 5 holds for all $k = 2, \ldots, n$, applying
Theorem 1, we have the next theorem.

Theorem 2. Let μ and ν be compatible possibility measures on $(X, 2^x)$ and σ
be a compatible permutation on (μ, ν).
If μ and ν are weakly monotone increasing for σ, there exists a function f
on X such that

$$\mu(A_k) = (C) \int_{A_k} f d\nu$$

for all $A_k = \{x_{i_1}, x_{i_2}, \cdots, x_{i_k}\} \subset X$.

Let us consider a special case, let ν be a 0-1 possibility measure such that
$\nu(A) = 1$, if $A \neq \emptyset$, $\nu(A) = 0$, if $A \neq \emptyset$.
Then for every possibility measure μ, μ and ν are compatible.
Let σ be a compatible permutation for (μ, ν). Then we have

$$\frac{\mu(\{x_{\sigma(1)}\})}{1} \geq \frac{\mu(\{x_{\sigma(2)}\})}{1} \geq \cdots \geq \frac{\mu(\{x_{\sigma(n)}\})}{1}.$$

μ and ν are both weakly monotone decreasing.
Therefore we have the next corollary.

Corollary 1. Let ν be a 0-1 possibility measure such that $\nu(A) = 1$, if $A \neq \emptyset$,
$\nu(A) = 0$, if $A \neq \emptyset$. For every possibility measure μ, there exists a function f on
X such that

$$\mu(A_k) = (C) \int_{A_k} f d\nu$$

for all $A_k = \{x_{i_1}, x_{i_2}, \cdots, x_{i_k}\} \subset X$.

If μ and ν have some strict condition, we have the converse of Theorem 2.

Definition 6. *Let μ, ν be fuzzy measures on $(X, 2^X)$. We say that μ and ν are strict compatible if there exists a permutation σ on $\{1, 2, \ldots, n\}$ such that*

$$\frac{\mu(\{x_{\sigma(1)}\})}{\nu(\{x_{\sigma(1)}\})} > \frac{\mu(\{x_{\sigma(2)}\})}{\nu(\{x_{\sigma(2)}\})} > \cdots > \frac{\mu(\{x_{\sigma(n)}\})}{\nu(\{x_{\sigma(n)}\})}.$$

Suppose that possibility measures μ and ν are strict compatible and μ is not weakly monotone increasing. Then, there exist $l, m (1 \le l < m \le n)$ such that $\mu(\{x_{\sigma(i_l)}\}) < \mu(\{x_{\sigma(i_m)}\})$.

Let $A_m = \{x_{\sigma(i_l)}, x_{\sigma(i_m)}\}$, and let us define $\alpha_1, \alpha_2, \beta_1, \beta_2$ and D as follows:

$$\mu(\{x_{\sigma(i_l)}\}) = \alpha_1, \mu(\{x_{\sigma(i_m)}\}) = \alpha_2, \nu(\{x_{\sigma(i_l)}\}) = \beta_1, \nu(\{x_{\sigma(i_m)}\}) = \beta_2$$

and

$$D = \begin{vmatrix} 1 & \mu(\{x_{\sigma(i_l)}\}) & \nu(\{x_{\sigma(i_l)}\}) \\ 1 & \mu(\{x_{\sigma(i_l)}, x_{\sigma(i_m)}\}) & \nu(\{x_{\sigma(i_l)}, x_{\sigma(i_m)}\}) \\ 0 & \mu(\{x_{\sigma(i_m)}\}) & \nu(\{x_{\sigma(i_m)}\}) \end{vmatrix}.$$

Observe that from these definitions it follows $\alpha_1 < \alpha_2$. Then, we have for D the following:

$$D = \begin{vmatrix} 1 & \alpha_1 & \beta_1 \\ 1 & \alpha_2 & \beta_1 \vee \beta_2 \\ 0 & \alpha_2 & \beta_2 \end{vmatrix} = \beta_2(\alpha_2 - \alpha_1) - \alpha_2(\beta_1 \vee \beta_2 - \beta_1)$$

Then, if $\beta_1 \ge \beta_2$, we have

$$D = \beta_2(\alpha_2 - \alpha_1) - \alpha_2(\beta_1 - \beta_1) = \beta_2(\alpha_2 - \alpha_1) > 0,$$

and if $\beta_1 < \beta_2$ we have

$$D = \beta_2(\alpha_2 - \alpha_1) - \alpha_2(\beta_2 - \beta_1) = \alpha_2\beta_1 - \alpha_1\beta_2.$$

Since (μ, ν) is strict and as σ is a compatible permutation for (μ, ν) (i.e., $\alpha_1/\alpha_2 \le \beta_1/\beta_2$), we have that

$$D = \beta_1\beta_2\left(\frac{\alpha_2}{\beta_2} - \frac{\alpha_1}{\beta_1}\right) < 0.$$

In any case, we have $D \ne 0$.

Therefore we have the next proposition.

Proposition 2. *Let μ, ν be fuzzy measures on $(X, 2^X)$ which are strict compatible, and let σ be a compatible permutation for (μ, ν).*

Suppose that there exists a function f on X such that

$$\mu(A_k) = (C)\int_{A_k} f d\nu$$

for all $A_k = \{x_{i_1}, x_{i_2}, \cdots, x_{i_k}\} \subset X$, $k = 2, \ldots, n$. Then μ and ν are both weakly monotone decreasing.

Table 2. Possibility measures μ and ν defined by the measures on the singletons.

A	$\{x_1\}$	$\{x_2\}$	$\{x_3\}$	$\{x_4\}$
$\mu(A)$	0.9	0.8	0.6	0.4
$\nu(A)$	0.8	0.8	0.7	0.5

Example 2. Let $X := \{x_1, x_2, x_3, x_4\}$ and possibility measures defined as in Table 2.

Then (μ, ν) are strictly compatible, and μ and ν are weakly monotone.

From Proposition 2, it follows that there exists a function f on X such that

$$\mu(A_k) = (C) \int_{A_k} f d\nu$$

for all $A_k = \{x_{i_1}, x_{i_2}, \cdots, x_{i_k}\} \subset X$, $k = 2, \ldots, n$.

5   Conclusion

In this paper we have studied the problem of existence of Radon-Nikodym-like derivatives for fuzzy measures. We have proven a theorem on the necessary and sufficient conditions based on the definition of compatible permutation for pairs of measures. We have introduced this definition. We have also shown how these results apply to possibility measures.

References

1. Agahi, H.: A modified Kullback-Leibler divergence for non-additive measures based on Choquet integral. Fuzzy Sets Syst. **367**, 107–117 (2019)
2. Choquet, G.: Theory of capacities. Ann. Inst. Fourier 5, 131–295 (1953/1954)
3. Denneberg, D.: Non Additive Measure and Integral. Kluwer Academic Publishers, Dordrecht (1994)
4. Denneberg, D.: Conditional expectation for monotone measures, the discrete case. J. Math. Econ. **37**, 105–121 (2002)
5. Graf, S.: A Radon-Nikodym theorem for capacities, Journal für die reine und angewandte Mathematik, 192-214 (1980)
6. Mesiar, R., Šipoš, J.: Radon-Nikodym-like theorem for fuzzy measures. Journal of Fuzzy Mathematics, 1, 873–878
7. Nguyen, H.T.: An Introduction to Random Sets. Chapman and Hall, CRC Press, London (2006)
8. Rébillé, R.: A Super Radon-Nikodym derivative for almost subadditive set functions. Int. J. Unc. Fuzziness and Knowl.-Based Syst. **21**(03), 347–365 (2013)
9. Sugeno, M.: A note on derivatives of functions with respect to fuzzy measures. Fuzzy Sets Syst. **222**, 1–17 (2013)
10. Torra, V.: Entropy for non-additive measures in continuous domains. Fuzzy Sets Syst. **324**, 49–59 (2017)

11. Torra, V., Narukawa, Y., Sugeno, M.: On the f-divergence for non-additive measures. Fuzzy Sets Syst. **292**, 364–379 (2016)
12. Torra, V., Narukawa, Y., Sugeno, M. (eds.) Non-Additive Measures: Theory and Applications, Springer (2013)
13. Torra, V., Narukawa, Y., Sugeno, M., Carlson, M.: Hellinger distance for fuzzy measures. In: Proceedings EUSFLAT conference (2013)

Data Science and Data Mining

A Non-Negative Matrix Factorization for Recommender Systems Based on Dynamic Bias

Wei Song[(✉)] and Xuesong Li

School of Information Science and Technology,
North China University of Technology, Beijing 100144, China
songwei@ncut.edu.cn

Abstract. Recommender systems help individuals in a community to find information or items that are most likely to meet their needs. In this paper, we propose a new recommendation model called non-negative matrix factorization for recommender systems based on dynamic bias (NMFRS-DB). As well as the two factor matrices, the proposed method incorporates two bias matrices, which improve the interpretability of the recommendations by expressing differences between observed and estimated ratings. First, the relevant probabilistic distributions are modeled, and then the factor matrices and bias matrices are calculated. Finally, the algorithm is described and explained. To evaluate the proposed method, we conduct experiments on three real-world datasets. The experimental results demonstrate the effectiveness of the NMFRS-DB model.

Keywords: Recommender system · Matrix factorization · Probability distribution · Rating matrix · Dynamic bias

1 Introduction

Recommender systems (RS) are an effective tool for filtering information [1]. Various types of recommendation algorithms have been proposed, including collaborative filtering [7], content-based recommendation [9], and hybrid RS [5].

As the Netflix Prize competition has demonstrated, matrix factorization (MF) models are very suitable for generating product recommendations [4]. In MF, both the users and items are modeled by vectors of factors inferred from item ratings. This produces two low-rank matrices, representing the relation between users and factors and the relation between items and factors, respectively. The product of these two matrices is used to predict users' future preferences. As a variant of MF, non-negative matrix factorization (NMF) is one of the most effective decomposition tools for extracting factors from user–item rating matrix. NMF factorizes and reconstructs matrices whose elements are all greater than or equal to zero. Thus, a dense matrix, similar to the original one, can be constructed iteratively.

Hernando et al. [2] proposed an NMF recommender system based on a Bayesian probabilistic model. In their approach, each element in two non-negative factor matrices

© Springer Nature Switzerland AG 2019
V. Torra et al. (Eds.): MDAI 2019, LNAI 11676, pp. 151–163, 2019.
https://doi.org/10.1007/978-3-030-26773-5_14

is within the range [0, 1], which provides an understandable probabilistic meaning. However, this approach has a high training cost, and is prone to overfitting.

In this paper, we propose an improved version of the model described in [2]. The contributions of our non-negative matrix factorization for recommender systems based on dynamic bias (NMFRS-DB) are as follows. First, two matrices composed of dynamic biases are incorporated into NMF, allowing the differences between observed ratings and estimated ratings to be calculated explicitly. This strategy improves the interpretability and quality of recommendations. Second, for elements modeling the relationship between items and factors, evidence of both liking and disliking are considered using two parameters, rather than the single parameter used in [2]. This strategy improves the accuracy of the predictions. Third, the calculation of the probability that the rating given by user u_j for item i_k belongs to a certain group is simplified. Because this parameter is critical for computing the estimated ratings, this simplification reduces the burden of model learning.

2   Problem Description

RS is typically described by a *rating matrix* (RM). Let $\mathcal{U} = \{u_1, u_2, \ldots, u_M\}$ be a set of users and $\mathcal{I} = \{i_1, i_2, \ldots, i_N\}$ be a set of items. The associated RM R is an $M \times N$ matrix. Each entry $r_{j,k}$ of R corresponds to user u_j's $(1 \leq j \leq M)$ preference for item i_k $(1 \leq k \leq N)$. If $r_{j,k} \neq 0$, then u_j has rated i_k; otherwise, the item has not been rated by this user.

MF factorizes the original RM into two rank-L matrices A and B, where A is an $M \times L$ matrix, B is an $N \times L$ matrix, and $L < <\min\{M, N\}$. A is interpreted as the *user–factor matrix*, and the row vector $\vec{a}_j$ is a *user-specific latent factor vector*. B is interpreted as the *item–factor matrix*, and the row vector $\vec{b}_k$ is an *item-specific latent factor vector*. This factorization process is implemented by minimizing the cost function measuring the difference between R and the estimate AB^T. Specifically, each entry $r_{j,k}$ of R is estimated as $r_{j,k}^* = \vec{a}_j \cdot \vec{b}_k$. The goal of MF is:

$$ min \sum_{j=1}^{M} \sum_{k=1}^{N} (r_{j,k} - \vec{a}_j \cdot \vec{b}_k) \tag{1} $$

where $r_{j,k}$ is the observed rating given by user u_j to item i_k.

The main difference between NMF and MF is that NMF carries out the factorization process subject to a non-negative constraint, i.e., $A \geq 0$ and $B \geq 0$.

3   Related Work

The basic idea of MF-based RS is to identify the unknown L latent factors that allow us to estimate the ratings of users by solving an optimization problem. This approach is successful because it does not consider missing ratings to be equivalent to zero.

Once the factor matrices have been learned, MF-based RS can estimate the users' ratings efficiently. Thus, MF has become a popular methodology for RS [4], and several powerful algorithms have been proposed.

Recently, NMF has been shown to be useful in RS. NMF is applied when certain non-negativity constraints exist; these constraints make the results easier to explain, as it is natural to consider users producing non-negative ratings. Several NMF-based recommendation methods have been proposed, including regularized single-element-based NMF [6] and social similarity-based NMF [10].

In [3, 8], the problem of RS is studied from the perspective of probabilistic MF. In these studies, it was assumed that the row vectors of the two factor matrices follow a Gaussian distribution. The problem is that the ratings in RS are discrete, and the Gaussian distribution is not a discrete distribution. Thus, probabilistic interpretations of these methods are unintuitive. Using a Bayesian probabilistic model, Hernando et al. [2] proposed an NMF-based RS. In this method, the different roles of the users, items, and ratings are assumed to follow different probability distributions, and all the elements of the two factor matrices are probabilities within the range [0, 1]. The meaning of these matrices can be understood and explained, and the reported experimental results are good. However, the computational cost of their model is high, and the interpretability of the model training process is not sufficiently clear.

4 Proposed Method

4.1 Basic Model

Let R be a rating matrix and each rating $r_{j,k}$ in R be a real number in the range $[1, L]$, where L is the number of groups of users that the algorithm is going to identify. Here, we assume that the number of groups of users is the same as the maximal rating. Each user u_j could belong to one or more groups, and each item i_k could be liked by users in one or more groups. Besides the two matrices A and B described in Sect. 2, two bias matrices are also considered. The *estimated rating matrix* is calculated as:

$$R^* = AB^T + \frac{1}{2}C + \frac{1}{2}D \tag{2}$$

where C is the *bias matrix* of A, recording the mean values of the difference between the observed ratings and estimated ratings of users; the row vector of C is represented by $\vec{c}_j = (c_{j1}, c_{j2}, \ldots, c_{jN})$, where $c_{j1} = c_{j2} = \ldots = c_{jN}$, and each element can also be denoted by c_j. Similarly, D is the *bias matrix* of B, recording the mean values of the difference between the observed ratings and estimated ratings of items; the row vector of D is represented by $\vec{d}_k = (d_{k1}, d_{k2}, \ldots, d_{kN})$, where $d_{k1} = d_{k2} = \ldots = d_{kN}$, and each element can also be denoted by d_k.

Note that we use probabilistic MF in our model, so all elements in A, B, C, and D are in the range [0, 1]. As in [2], $a_{j,l}$ represents the probability that user u_j belongs to group l of users, such that $\sum_{l=1}^{L} a_{j,l} = 1$, and $b_{l,k}$ represents the probability that users in group l like item i_k. To fit this scenario, the elements of the rating matrix $\boldsymbol{R}$ are regularized in the range [0, 1] during the pre-processing phase.

4.2 Probabilistic Distribution

In this sub-section, we formally describe the ratings that users make over the items by means of the following probabilistic distributions. Here, we consider the problem in terms of the estimated rating r^* and the observed rating r. We assume that r^* is related to users, and r is related to items.

According to [2], $r^*_{j,k}$ follows a categorical distribution. That is:

$$r^*_{j,k} \sim Cat(\vec{a}_j + \vec{c}_j) \tag{3}$$

where $\vec{a}_j$ is the jth row vector of matrix $\boldsymbol{A}$ and $\vec{c}_j$ is the jth row vector of matrix $\boldsymbol{C}$. From [2], $\vec{a}_j$ is known to follow a Dirichlet distribution:

$$\vec{a}_j \sim Dir(\vec{\mu}_j) \tag{4}$$

where $\vec{\mu}_j$ is an L-dimensional vector and each of its elements $\mu_{j,s}$ is a parameter to be learned. According to [2], $r_{j,k}$ follows a binomial distribution:

$$r_{j,k} \sim Bin(L, b_{k,s}) \tag{5}$$

and each entry of matrix $\boldsymbol{B}$ follows a beta distribution:

$$b_{k,s} \sim Beta(\alpha_{k,s}, \beta_{k,s}) \tag{6}$$

where $\alpha_{k,s}$ and $\beta_{k,s}$ are parameters to be learned. $\alpha_{k,s}$ is related to the amount of evidence that the algorithm requires to deduce that user u_k likes item i_s, and $\beta_{k,s}$ is related to the amount of evidence that the algorithm requires to deduce that user u_k dislikes item i_s.

4.3 Factor Matrices

In this sub-section, we describe how to calculate entries in the factor matrices. As in [2], we adopt the variational inference technique (or variational Bayes) to calculate a distribution approaching the real posterior distribution.

Let us suppose that:

$$p(r^*_{j,k} = s) = \lambda_{j,k,s} \tag{7}$$

where $r^*_{j,k}$ is the estimated rating of user u_j for item i_k in $\boldsymbol{R^*}$, and $p(r^*_{j,k} = s)$ is the probability that $r^*_{j,k}$ is equal to s.

According to Eq. 2:

$$\lambda_{j,k,s} \propto \exp(\mathbb{E}_{q(\vec{a}_j),q(\vec{b}_k),q(r_{j,k}),q(c_j),q(d_k)} \ln p(\vec{a}_j, \vec{b}_k, r_{j,k}, r_{j,k}^*, c_j, d_k)) \tag{8}$$

where $p(\cdot)$ and $q(\cdot)$ denote the actual probability and posterior probability, respectively. From Eqs. 3 and 4, the estimated rating is related to μ, and:

$$a_{j,s} \propto \frac{\mu_{j,s}}{\sum_{f=1}^{L} \mu_{jf}} \tag{9}$$

where the sth element of vector $\vec{\mu}_j$ can be approximated as:

$$\mu_{j,s} = \gamma + c_j + \sum_{f=1}^{|\mathcal{I}|} \lambda_{jf,s} \tag{10}$$

in which $\gamma \in (0, +\infty)$ is the bias parameter of the Dirichlet distribution—the higher the value of γ, the more diverse the user's taste—and $|\mathcal{I}|$ is the number of elements in the set $\mathcal{I}$, i.e., the number of items. We can see from Eq. 10 that each entry $\mu_{j,s}$ ($1 \le s \le L$) of $\vec{\mu}_j$ can be approximated by the bias parameter, elements from the same row of matrix C, and items for which the estimated ratings of user u_j are s.

According to Eqs. 9 and 10:

$$a_{j,s} \propto c_j \tag{11}$$

Thus, we assume that the estimated ratings are related to the elements of matrix A. According to Eq. 6:

$$b_{k,s} \propto \frac{\alpha_{k,s}}{\alpha_{k,s} + \beta_{k,s}} \tag{12}$$

We approximate these two parameters by:

$$\alpha_{k,s} = \delta + \sum_{f=1}^{|\mathcal{U}|} r_{f,k} \lambda_{f,k,s} + d_k \tag{13}$$

$$\beta_{k,s} = \varepsilon + \sum_{f=1}^{|\mathcal{U}|} (1 - r_{f,k}) \lambda_{f,k,s} - d_k \tag{14}$$

where $\delta \in (0, +\infty)$ is a bias parameter indicating the degree to which user u_k likes item i_s and $\varepsilon \in (0, +\infty)$ is a bias parameter indicating the degree to which user u_k dislikes item i_s. That is, the higher the value of δ, the greater the degree to which u_k likes i_s; the higher the value of ε, the greater the degree to which u_k dislikes i_s. We can see from Eqs. 13 and 14 that each entry $b_{k,s}$ of matrix B can be approximated by the bias parameters, users' observed ratings on i_k, users whose estimated ratings for item i_k are s, and elements from the same row of matrix D.

According to Eqs. 13 and 14:

$$b_{k,s} \propto d_k \tag{15}$$

Thus, we assume that the observed ratings are related to the elements of matrix $\boldsymbol{B}$. Based on the above analysis, Eq. 8 can be approximated as:

$$\lambda_{j,k,s} \propto \exp(\mathrm{E}_{q(\vec{a}_j),q(\vec{b}_k)} (\ln p(r^*_{j,k} = s \,|\, \vec{a}_j) + \ln p(r_{j,k} = s \,|\, \vec{b}_k))) \tag{16}$$

Equation 16 can be further transformed as:

$$\lambda_{j,k,s} \propto \exp(\mathrm{E}_{q(\vec{a}_j)} (\ln p(r^*_{j,k} = s \,|\, \vec{a}_j)) + \mathrm{E}_{q(\vec{b}_k)} (\ln p(r_{j,k} = s \,|\, \vec{b}_k))) \tag{17}$$

According to Eq. 3, $r^*_{j,k}$ follows a categorical distribution, and so:

$$p(r^*_{j,k} = s \,|\, \vec{a}_j) = a_{j,s} \tag{18}$$

From Eq. 5, $r_{j,k}$ follows a binomial distribution. Thus, $r_{j,k} = s$ is only related to element $b_{k,s}$, and we have:

$$\begin{aligned} p(r_{j,k} = s \,|\, \vec{b}_k) &= p(r_{j,k} = s \,|\, b_{k,s}) \\ &= \binom{L}{r_{j,k}} b_{k,s}^{r_{j,k}} (1 - b_{k,s})^{(L - r_{j,k})} \propto b_{k,s}^{r_{j,k}} (1 - b_{k,s})^{(L - r_{j,k})} \end{aligned} \tag{19}$$

Substituting Eqs. 18 and 19 into Eq. 17, we have:

$$\lambda_{j,k,s} \propto \exp(\mathrm{E}_{q(\vec{a}_j)} (\ln a_{j,s}) + r_{j,k} \mathrm{E}_{q(\vec{b}_k)} (\ln b_{k,s}) + (L - r_{j,k}) \mathrm{E}_{q(\vec{b}_k)} (\ln(1 - b_{k,s}))) \tag{20}$$

According to the probability distribution of $a_{j,s}$, we have:

$$\mathrm{E}_{q(\vec{a}_j)} (\ln a_{j,s}) \propto \psi(\mu_{j,s}) - \psi(\textstyle\sum_{f=1}^{L} \mu_{j,f}) \tag{21}$$

and from the probability distribution of $b_{k,s}$, we can deduce that:

$$\mathrm{E}_{q(\vec{b}_k)} (\ln b_{k,s}) \propto \psi(\alpha_{k,s}) - \psi(\alpha_{k,s} + \beta_{k,s}) \tag{22}$$

$$\mathrm{E}_{q(\vec{b}_k)} (\ln(1 - b_{k,s})) \propto \psi(\beta_{k,s}) - \psi(\alpha_{k,s} + \beta_{k,s}) \tag{23}$$

Substituting Eqs. 21–23 into Eq. 20, we have:

$$\lambda_{j,k,s} \propto \exp(\psi(\mu_{j,s}) - \psi(\sum_{f=1}^{L} \mu_{j,f}) + r_{j,k}\psi(\alpha_{k,s}) + (L - r_{j,k})\psi(\beta_{k,s}) - L\psi(\alpha_{k,s} + \beta_{k,s}))$$

$$(24)$$

where $\psi(x)$ is the digamma function defined as:

$$\psi(x) = \frac{\Gamma'(x)}{\Gamma(x)} = \ln x - \frac{1}{2x} - \sum_{n=1}^{\infty} \frac{B_{2n}}{2nx^{2n}} \tag{25}$$

in which $\Gamma(x)$ is the gamma function and B_{2n} is the $2n$-th Bernoulli number. Thus, we have:

$$\exp(\psi(x)) \propto x \tag{26}$$

Using Eq. 26, we can simplify Eq. 24 as:

$$\lambda_{j,k,s} \propto \frac{\mu_{j,s} \cdot (\alpha_{k,s})^{r_{j,k}} \cdot (\beta_{k,s})^{(L-r_{j,k})}}{\sum_{f=1}^{L} \mu_{j,f} \cdot (\alpha_{k,s} + \beta_{k,s})^{L}} \tag{27}$$

4.4 Bias Matrices

In our method, the bias matrix C is assumed to be related to users, so we define c_j, the value of the jth row of C, as:

$$c_j = \frac{\sum_{f=1}^{|\mathcal{I}|} \left(r_{jf} - r_{jf}^*\right)}{|\mathcal{I}|} \tag{28}$$

Similarly, d_k, the value of the kth row of D, is defined as:

$$d_k = \frac{\sum_{f=1}^{|\mathcal{U}|} \left(r_{f,k} - r_{f,k}^*\right)}{|\mathcal{U}|} \tag{29}$$

4.5 Algorithm Description

Combining the above discussion, the proposed NMFRS-DB algorithm is described in Algorithm 1, which trains two factor matrices and two bias matrices and uses them to provide recommendations to the target users.

Algorithm 1	NMFRS-DB
Input	Rating matrix R, maximal number of iterations *max_iter*, number of groups of users L, bias parameters γ, δ, ε
Output	Matrices A, B, C, and D

```
1   Initialization( );
2   times=1;
3   while times<=max_iter do
4      Calculate C using Eq.28, and calculate D using Eq.29;
5      for j=1 to M do
6         for s=1 to L do
7            Calculate μ_{j,s} using Eq.10;
8         end for
9         for s=1 to L do
10            Approximate a_{j,s} using Eq.9;
11         end for
12      end for
13      for k=1 to N do
14         for s=1 to L do
15            Calculate α_{k,s} using Eq.13, and calculate β_{k,s} using Eq.14;
16         end for
17         for s=1 to L do
18            Approximate b_{k,s} using Eq.12;
19         end for
20      end for
21      for j=1 to M do
22         for k=1 to N do
23            for s=1 to L do
24               Approximate λ_{j,k,s} using Eq.27;
25            end for
26            Normalize λ such that ∑_{s=1}^{L} λ_{j,k,s} = 1;
27         end for
28      end for
29      Calculate R* using Eq.2;
30      times++;
31   end while
```

Step 1 of Algorithm 1 calls the Initialization procedure (see Algorithm 2). Step 2 then initializes the iteration number to 1. The main loop (Steps 3–31) constructs the recommender model. The bias matrices C and D are calculated in Step 4. The loop in Steps 5–12 updates matrix A, whereas the loop in Steps 13–20 updates matrix B. Note that Eqs. 9 and 12 are used to "approximate" the elements in A and B. That is, "=" is used to replace "$\propto$". Equation 27 in Step 24 is used in the same way. The values of λ are then calculated and normalized in the loop from Step 21–28. R^* is calculated in

Step 29. The number of iterations is then incremented by one in Step 30. Note that, after we have calculated the estimated ratings, we need to transform them to obtain non-normalized values. For example, when the typical rating scale $\{1, 2, 3, 4, 5\}$ is used, the calculated values should be multiplied by 5 before they are used to estimate user ratings.

Algorithm 2   Procedure Initialization( )
1 **for** j=1 to M **do**
2   Initialize $\overline{\mu_j}$ at random;
3   **for** s=1 to L **do**
4   Approximate $a_{j,s}$ using Eq.9;
5   **end for**
6 **end for**
7 **for** k=1 to N **do**
8   **for** s=1 to L **do**
9   Initialize $\alpha_{k,s}$ and $\beta_{k,s}$ at random;
10   Approximate $b_{k,s}$ using Eq.12;
11   **end for**
12 **end for**
13 **for** j=1 to M **do**
14   **for** k=1 to N **do**
15   **for** s=1 to L **do**
16   Approximate $\lambda_{j,k,s}$ using Eq.27;
17   **end for**
18   Normalize λ such that $\sum_{s=1}^{L} \lambda_{j,k,s} = 1$;
19   **end for**
20 **end for**
21 Initialize matrices C and D at random;
22 Calculate R^* using Eq.2;

In Algorithm 2, the factor matrix A is initialized randomly in the loop from Step 1–6. Similarly, B is initialized randomly in the loop from Step 7–12. Using these initial values, the values of λ are initialized and normalized in the loop from Step 13–20. Step 21 initializes the two bias matrices C and D at random. R^* is then calculated in Step 22. In Algorithm 2, Eqs. 9, 12 and 27 are used to compute approximations.

5 Performance Evaluation

We compare our NMFRS-DB algorithm with that of a latent factor model (LFM)-based recommender algorithm, denoted as LFM-RS [4], and a Bayesian probabilistic model-based recommender algorithm, denoted as BPM-RS [2].

5.1 Datasets

Three datasets were used for the evaluation. MovieLens 100K and MovieLens 1M can be downloaded from https://grouplens.org/datasets/movielens/; the ratings in these two datasets are real values ranging from 1.00–5.00. Jester Dataset 2 can be downloaded from http://eigentaste.berkeley.edu/dataset/; this dataset consists of real values ranging from −10.00–10.00, and the values are transformed to 1.00–5.00 in our experiments. Thus, L, the number of user groups, is set to 5. The datasets were divided into two parts, with 80% used as the training set T and 20% used as the testing set S.

5.2 Evaluation Metrics

We study the accuracy of the estimations given by our model through the Mean Absolute Error (MAE). This metric measures the average absolute error between the observed values and the estimated values, and is calculated as:

$$MAE = \frac{\sum_{(j,k) \in S} |r_{j,k} - r_{j,k}^*|}{|S|} \tag{30}$$

We can see that smaller values of MAE indicate more precise estimations.

We also evaluate the quality of the recommendations through the F1-measure:

$$F1 - measure = \frac{2 \times Precision \times Recall}{Precision + Recall} \tag{31}$$

where

$$Prcesion = \frac{|ES \cap AS|}{|ES|} \tag{32}$$

$$Recall = \frac{|ES \cap AS|}{|AS|} \tag{33}$$

in which ES is the set of items not rated by each user that have been estimated with the highest values (greater than 4), and AS is the set of items rated by each user with high values. Overall, the higher the value of the F1-measure, the greater the likelihood that the recommended items will be rated highly by users.

5.3 Parameter Settings

In our model, the bias parameters γ, δ, ε must be set to appropriate values. We first outline the approximate range of these parameters, and then determine their optimal values by progressive refinement. Tables 1, 2 and 3 illustrate the change in MAE with respect to γ, δ, and ε, respectively, on MovieLens 100K.

According to the results in Tables 1, 2 and 3, we set $\gamma = 0.01$, $\delta = 2.01$, and $\varepsilon = 1.61$ when using MovieLens 100K. Similarly, we determined values of $\gamma = 0.01$, $\delta = 2.20$, $\varepsilon = 1.70$ for MovieLens 1M and $\gamma = 0.10$, $\delta = 1.20$, $\varepsilon = 0.60$ for Jester Dataset 2.

Table 1. Change of MAE with respect to γ on MovieLens 100K

	0.01	0.21	0.41	0.61	0.81	1.01
MAE	0.663	0.668	0.671	0.674	0.672	0.677

Table 2. Change of MAE with respect to δ on MovieLens 100K

	0.01	1.01	2.01	3.01	4.01	5.01
MAE	0.678	0.672	0.665	0.687	0.684	0.701

Table 3. Change of MAE with respect to ε on MovieLens 100K

	0.01	0.81	1.61	2.41	3.21	4.01
MAE	0.700	0.691	0.663	0.684	0.699	0.721

5.4   Experimental Results

We first compare the difference between the observed values and the estimated values using MAE. The results are presented in Table 4.

Table 4. Comparison of MAE results

Dataset	LFM-RS	BPM-RS	NMFRS-DB
MovieLens 100 K	0.691	0.764	0.662
MovieLens 1 M	0.683	0.723	0.665
Jester Dataset 2	0.801	0.842	0.778

From Table 4, it is clear that the proposed NMFRS-DB model can estimate the users' ratings more accurately than the comparative methods on all three datasets. This is because evidence of both liking and disliking items is considered in NMFRS-DB. Thus, elements in the item–factor matrix can be approximated more accurately. As a result, the estimated ratings also have a higher accuracy.

A comparison of the F1-measure under different numbers of recommendations on the three datasets is presented in Table 5.

We can see from Table 5 that the F1-measure values from NMFRS-DB are superior to those of the other two methods. The reason is that the interpretability of NMFRS-DB is improved by modeling the observed and estimated ratings in bias matrices. Thus, the quality of recommendations can also be improved.

Table 5. Comparison results of F1-measure

Dataset	Algorithm	Number of recommendations			
		10	30	50	70
MovieLens 100 K	LFM-RS	0.018	0.044	0.063	0.074
	BPM-RS	0.021	0.062	0.068	0.070
	NMFRS-DB	0.073	0.088	0.091	0.094
MovieLens 1 M	LFM-RS	0.011	0.034	0.043	0.047
	BPM-RS	0.068	0.079	0.086	0.091
	NMFRS-DB	0.082	0.089	0.096	0.100
Jester Dataset 2	LFM-RS	0.083	0.103	0.102	0.095
	BPM-RS	0.088	0.104	0.100	0.097
	NMFRS-DB	0.095	0.110	0.110	0.110

6   Conclusions

Using dynamic bias, we have presented a recommendation model based on NMF called NMFRS-DB. In NMFRS-DB, two bias matrices are used alongside the user–factor matrix and item–factor matrix. From a probabilistic perspective, we analyzed how to approximate elements in the four matrices, and described the algorithms in detail. Experimental results show that the NMFRS-DB model is more precise than two other recommendation methods.

Acknowledgments. This work was partially supported by the Great Wall Scholar Program (CIT&TCD20190305), High Innovation Program of Beijing (2015000026833ZK04), and Beijing Urban Governance Research Center.

References

1. Alharthi, H., Inkpen, D., Szpakowicz, S.: A survey of book recommender systems. J. Intell. Inf. Syst. **51**(1), 139–160 (2018)
2. Hernando, A., Bobadilla, J., Ortega, F.: A non negative matrix factorization for collaborative filtering recommender systems based on a Bayesian probabilistic model. Knowl.-Based Syst. **97**, 188–202 (2016)
3. Hofmann, T.: Latent semantic models for collaborative filtering. ACM Trans. Inf. Syst. **22** (1), 89–115 (2004)
4. Koren, Y., Bell, R.M., Volinsky, C.: Matrix factorization techniques for recommender systems. IEEE Comput. **42**(8), 30–37 (2009)
5. Liu, J., Wang, D., Ding, Y.: PHD: a probabilistic model of hybrid deep collaborative filtering for recommender systems. In: ACML, pp. 224–239 (2017)
6. Luo, X., Zhou, M., Xia, Y., Zhu, Q.: An efficient non-negative matrix-factorization-based approach to collaborative filtering for recommender systems. IEEE Trans. Ind. Inform. **10** (2), 1273–1284 (2014)

7. Shi, Y., Larson, M., Hanjalic, A.: Collaborative filtering beyond the user-item matrix: a survey of the state of the art and future challenges. ACM Comput. Surv., **47**(1), (2014). https://dl.acm.org/citation.cfm?id=2556270
8. Salakhutdinov, R., Mnih, A.: Probabilistic matrix factorization. In: NIPS, pp. 1257–1264 (2007)
9. Wang, D., Liang, Y., Xu, D., Feng, X., Guan, R.: A content-based recommender system for computer science publications. Knowl.-Based Syst. **157**, 1–9 (2018)
10. Zhang, G., He, M., Wu, H., Cai, G.: Non-negative multiple matrix factorization with social similarity for recommender systems. In: BDCAT, pp. 280–286 (2016)

Forecasting Water Levels
of Catalan Reservoirs

Raúl Parada[✉], Jordi Font, and Jordi Casas-Roma

Faculty of Computer Science, Multimedia and Telecommunications,
Universitat Oberta de Catalunya Barcelona, Barcelona, Spain
{rparada,jordifm,jcasasr}@uoc.edu

Abstract. Reservoirs are largely natural or artificial lakes used as a source of water supply for society daily applications. However, reservoirs are limited natural resources which water levels vary according to annual rainfalls and other natural events. Therefore, prediction techniques are helpful to manage the water used more efficiently. This paper compares state-of-the-art methods to predict the water level in Catalan reservoirs comparing two approaches: using the water level uniquely, uni-variant, and adding meteorological data, multi-variant. With respect to relate works, our contribution includes a longer times series prediction keeping a high precision. The results return that combining Support Vector Machine and the multi-variant approach provides the highest precision with an R^2 value of 0.99.

Keywords: Forecasting · Reservoir · Time series analysis

1 Introduction

Since the beginning of time, water is a fundamental resource for existence. Thereby, its management is key to assure efficient water use worldwide. One way to manage the flow of water for society use, dams and reservoirs were built. Reservoirs are largely natural or artificial lakes used as a source of water supply to the population. The *"Agència Catalana de l'Aigua"* (ACA) [6] claim that in Catalonia there are more than 40 reservoirs of different capacities, types, and characteristics. Among them, the large reservoirs are the ones that have the capacity to supply the population. ACA defines as large reservoirs, by current regulations, such as those with a height of more than 5 m and a capacity exceeding 100,000 m^3. In Catalonia, there are 23 reservoirs that meet these characteristics that make us consider large reservoirs. Although these reservoirs may contain a large quantity of water, meteorological variations may influence its availability along the time. Hence, it is necessary not only the real-time water-level measurement also its prediction to manage the reservoir for optimal use. For instance, dry periods of weather may provoke the reduction of water levels and, in consequence, lack of supply. Then, a prediction of water levels may help to prepare a management plan for an optimal supply. Nevertheless, forecasting methods require data to predict future periods. Fortunately, thanks to governmental

© Springer Nature Switzerland AG 2019
V. Torra et al. (Eds.): MDAI 2019, LNAI 11676, pp. 164–176, 2019.
https://doi.org/10.1007/978-3-030-26773-5_15

strategies to collect environmental data and offer it openly, users can implement data mining techniques to provide worthy information for society. Üneş et al. [25] consider that the implementation of data science methods in the field of hydrology is important for the maintenance and control of infrastructure, pollution control, flood control, navigation, tourism, etc., as well as for the hydraulic structures that depend on it. Thus, for example, they emphasise the importance that the volume of water in reservoirs may have on economic activities, energy, policies, consumption, among others. Therefore, providing smart tools capable of predicting, displaying or deciding on data can be key in the efficient management of these infrastructures, both complex and economically expensive. As the researchers discussed, the prediction of the water level is equivalent to the prediction of a decisional variable in the management of the infrastructure. Jun-He Yang et al. [10] directly link the importance of this prediction to the economy of a country, tourism, crop irrigation, flood control, water supply, and hydroelectric power generation. Therefore, the forecast for the management of the reservoirs is key for the country in general, affecting many more areas than the hydrological one, and for the development of the area where they are located in particular. Nwobi-Okoye and Igboanugo [15] consider a serious problem for their country (Nigeria) to poorly plan electricity generation. Its objective is to predict the level of water in the Kainji dam, which is the one that provides water to the country's main hydroelectric power plant. Therefore, a good prediction of the water level in this reservoir can be good planning of the electricity produced and supplied, since the level of water is directly related to the capacity to produce energy, simply because it occurs with more force for the turbines and, therefore, it generates more electricity. This paper aims to implement state-of-the-art forecasting techniques in Catalan reservoirs. Our main contributions are, but not limited to:

- Analyse open data provided by governmental organisations.
- Compare state-of-the-art forecasting techniques.
- Predict reservoir water levels in longer periods of time series than literature.
- Predict reservoir water levels in higher accuracy than similar works in the state-of-the-art.

The remainder of this paper is organised as follows: Sect. 2 explores the literature for related works. We briefly describe the extraction of data procedure as the two approaches designed in Sect. 3. In Sect. 4 is presented the achieved product by highlighting the results, insights and future improvements. Finally, the paper is concluded in Sect. 5, summarising our contributions.

2   State of the Art

With respect to the use of techniques within data science for reservoir prediction, we find many works. A variant of the prediction of the water level in the marshes is the prediction of the flow of entry, which we can find in Valipour et al. [20], which compares between an autoregressive integrated moving average (ARIMA) model and a model based on neuronal networks. Kitsuchart and Siripen [17] analyse the prediction of the flow of rivers to elaborate better management of possible floods or alerts of danger. The prediction is made after the previous 72 h of the next 24 h of the level of the Chao Phraya River in Thailand. They have trained different models, where the best result is obtained with support vector machine (SVM) with radial base Kernel function. Mokhtar et al. [14] propose a neural network model with architecture (5-25-1) to try to model the time margin between rainfall and that this rain is reflected in the volume of water in the reservoir, at the Timah Tasoh of Malaysia. It uses data from the daily water level in the reservoir from 1999 to 2006. According to Valizadeh and El-Shafie [21], methods such as linear regression and the ARIMA model and its variants were the tools that were used until neuronal networks were imposed on these types of studies. Combinations of neural networks with other methods such as diffuse logic (Chaves and Kojiri [3]; El-Shafie et al. [5]) and SVM (Wieland et al. [23]; Kisi et al. [12]), for example, have tried to influence this area to improve predictions of models based on the predominant neural networks. There are many studies with variants in the use of neural networks, for instance, Moeeni et al. [13], draw up a model based on neuronal networks and genetic algorithms (ANN-GA) on the reservoir flows into the reservoir. And the integration of neuronal networks with fuzzy logic, the model adaptive neuro-fuzzy inference system (ANFIS) appears as an object of investigation in numerous articles on prediction in the hydrological field. The work done by Chang and Chang [2], Wang et al. [22] and Valizadeh and El-Shafie [21], can be an example. All of them carry out predictions of water volumes in reservoirs using the ANFIS technique. Hipni et al. [8] point out that ANFIS has been widely used in the predictive modeling of problems related to hydrology, emphasizing its ease of implementation, rapid and successful learning and a great generalization capacity, such as some of its causes popularity. Jain et al. [9] use the neuronal networks to predict the input flow into the reservoir and Ondimu and Murase [16] also for predictions of water level in reservoirs. In their case, the number of steps to predict reach up to 10 days. Our work aims to increase the prediction to fifteen days increasing management flexibility. In addition, we propose two approaches: using uniquely the water level data, uni-variant and, adding meteorological data called multi-variant.

3   Data Processing

This section aims to describe briefly the reservoirs to predict, data acquisition and processing and the features selection for both the uni-variant and the multi-variant approach.

3.1   Reservoirs

The reservoirs of the internal basins of Catalonia are of autonomous ownership, managed by the l'Agència Catalana de l'Aigua. In the basins, there are 9 large reservoirs: Darnius - Boadella, Sau, Siurana, Foix, Llosa del Cavall, Sant Ponç, la Baells, Susqueda, and Riudecanyes. The first 7 are owned by the ACA; On the other hand, the Susqueda reservoir is owned by Endesa and that of Riudecanyes is owned by the community of irrigators of Riudecanyes. These 9 reservoirs have a total capacity of 694 hm^3, designed to meet domestic, industrial and irrigation needs. We selected the Sau and La Baells reservoirs because of data quality. The election of these reservoirs is out of scope.

3.2   Data Extraction and Feature Selection

With the advancement of digital content, governmental institutions offer to society of environmental data to be used publicly. In our case, we acquire reservoirs status information from ACA [6] and meteorological data from the Servei Meteorológic de Catalunya (a.k.a. meteocat) [7]. In the first source, data can be downloaded directly while in meteocat, it is necessary to request meteorological data indicating specific locations. Note that reservoirs may not match in a location exactly with respect to the meteorological data. Table 1 contains an example of data generated by the Sau reservoir.

Table 1. Reservoir variables from ACA

Variable	Description	Unit
4165140	E06_Vilanova Sau_Sau_Cabal output	m^3/s
4159510	E06_Vilanova Sau_Sau_Volum reservoir	hm^3
4165141	E06_Vilanova Sau_Sau_Cabal input	m^3/s
4159509	E06_Vilanvova Sau_Sau_Nivell reservoir	m.s.n.m.
4159547	Vilanova Sau_Sau_Percentatge reservoir volume	%
3378678	E06_Emb Sau_Total group volumes 1+2	m^3/s

The above data can be collected daily from the reservoir construction and its generation of data until the present day. In our case, we established January 1st 1986 as initial collection data because of consistency among the rest of reservoirs. After preprocessing, we have the data in a temporary series format, where data on the volume in cubic hectometres of reservoirs is the most consistent data in all cases. Thus, we will consider this data by the prediction part from a single variable, we have called uni-variant analysis. Table 2 shows the variables of meteorological data requested to meteocat.

In the Sau reservoir, we have rain data from the reservoir itself. The rest of the meteorological data that they transmit to us are those of the population of Viladrau, the station closest to the marsh of the meteocat service, according to the data transmitted, that is to about 13 km in a straight line of the marsh. In the case of the La Baells reservoir, the data of both rain and other meteorological data are those of the small town of La Quar, the closest to the reservoir with a weather station according to the data transmitted to us. This town is at a

Table 2. Meteorological variables requested to meteocat

Variable	Description	Unit
TM	Average daily temperature	°C
TX	Maximum daily temperature	°C
TN	Minimum daily temperature	°C
HRM	Average daily relative humidity	%
PPT24h	Daily cumulative precipitation	mm
PM	Average daily atmospheric pressure	hPa
VVM2	Average daily wind speed at 2 m high	m/s
DVM2	Average daily wind direction at 2 m high	°
VVX2	Maximum daily wind gust at 2 m high	m/s
DVVX2	Direction of the maximum daily wind gust to 2 m high	°
VVM10	Average daily wind speed at 10 m high	m/s
DVM10	Average daily wind direction at 10 m high	°
VVX10	Maximum daily wind gust at 10 m high	m/s
DVVX10	Direction of the maximum daily wind gust to 10 m high	°

distance of about 7 km in a straight line from the swamp. We will dispense with the wind variables. We are not experts, but we make this decision not to overload the model with unnecessary variables, supporting us in the article by Jun-He Yang et al. [10], which eliminates the wind variables, which are the least relevant, while improving the model without considering them. Thus, we remain with the daily average temperature, relative humidity, atmospheric pressure and precipitation as attributes that we will consider to elaborate the models for the multi-variant approach.

4   Methodology and Results

This section describes the strategies and methods implemented on the selected reservoirs and, the results from the optimized simulations.

Since the goal of this research is the prediction on several days ahead, we follow a multi-step forecast where three different strategies can be distinguished [1]:

– **Iterative strategy:** A model which predicts the next one-step forecast and incorporates the predicted value to the model's entry to predict the following. In this way, the following time steps are predicted, iteratively, one by one, updating the model with the values of the predictions.
– **Direct strategy:** It generates a model for each desired output. There will be a model predicting the next value of the time series y_{t+1}, another one that predicts y_{t+2}, etc. But always based on real historical observations.
– **MIMO strategy (multi-input multi-output):** A single model that generates multiple outputs from multiple entries.

We explored the following state-of-the-art forecasting methods [18]:

- Neural Networks
  - Multilayer perceptron (MLP)
  - Convolutional neural network (CNN)
  - Long Short Term Memory (LSTM)
- Support Vector Machine (SVM)
- Random Forest (RF)

Combining the above list of strategies and techniques, we explored a large number of models. A model will be expressed with the technique acronym and the approach. For instance, using the support vector machine with the uni-variant approach would be named SVM-Uni. Due to space limitation, we will show uniquely the best six methods from each reservoir. Details about the chosen models and data specifications are displayed in Table 3. This table presents the number of samples used for training and test, the number of steps in and out and the strategy.

Table 3. Models and data specifications

Reservoir	Model	Train	Test	n_steps_in	n_steps_out	Strategy
Sau	SVM-Multi	3,412	854	11	15	MIMO
	MLP-Multi	3,416	854	7		
	SVM-Uni	2,823	706	17		
	LSTM-Uni	2,812	704	30		
	RF-Uni	5,463	1,366	5		
	RF-Multi	3,417	855	5		
La Baells	SVM-Multi	3,413	854	10		
	MLP-Multi	3,412	854	11		
	SVM-Uni	2,817	705	24		
	MLP-Uni	2,812	704	30		
	RF-Uni	5,463	1,366	5		
	RF-Multi	3,412	854	11		

The columns in Table are described as:

- Train: number of joint train records
- Test: number of test set records
- N_steps_in: previous days of observations from which we make the prediction.
- N_steps_out: number of future days predicted by the model. Strategy: input and output format.

Note that the implemented techniques have been optimised where parameter values are:

- LSTM
  - Neuron LSTM layer: 5
  - Epochs: 1000
  - Learning rate: 0.0045
  - Batch size: 128
  - Optimizer: Adam
  - Activation function: sigmoid
  - Input days: 30
- MLP La Baells-Multi / La Baells-Uni / Sau-Multi
  - Neuron Hidden layer: 5 / 30 / 48
  - Epochs: 773 / 600 / 337
  - Learning rate: 0.001267 / 0.0006 / 0.00079
  - Batch size: 47 / 16 / 126
  - Optimizer: Adam / Adam / Adam
  - Activation function: sigmoid / sigmoid / sigmoid
  - Input days: 11 / 30 / 7
- SVM La Baells-Multi / La Baells-Uni / Sau-Multi / Sau-Uni
  - C: 85 / 124 / 25 / 24
  - Epsilon: 0.0115 / 0.0144 / 0.015 / 0.0127
  - Gamma: 0.0101 / 0.0852 / 0.010 / 0.2788
  - Input days: 10 / 24 / 11 / 17
- RF La Baells-Uni / La Baells-Multi / Sau-Uni / Sau-Multi /
  - Number of estimators: 120 / 303 / 362 / 140
  - Maximum number of features: 5 / 84 / 5 / 5
  - Maximum tree depth: 15 / 11 / 83 / 91
  - Minimum number of samples to split: 26 / 29 / 21 / 11
  - Minimum number of samples per leaf: 14 / 3 / 44 / 18
  - Input days: 5 / 11 / 5 / 5

To optimise the parameters we have used different methodologies, such as loops for exhaustive, first with large jumps of values and then more closely related to the parameters that have yielded the best results; but the most efficient and best-performing method has been the use of 500 random combinations of parameters defined in ranks of values. Using this technique, we have seen that exhaustive search around the best parameters found was not necessary. The range of values for the optimisation of the neural networks is:

- No hidden cell neurons: integers in the range of values (5, 500)
- No of training periods: integers in the range of values (50, 1000)
- Learning rate: from 0.0001 to 0.01.
- Batch size: list values [8, 16, 32, 64, 128] or Batch size: from 8 to 128

By the SVM model:

- Parameter C: integers in the range of values (5, 125)

- Epsilon: Decimal numbers in the range of values (0.0001, 0.1)
- Gamma: decimals in the range of values (0.01, 0.9)

By the Random forest model:

- n_estimators: integers in the range of values (5, 500)
- max_features: integers in the range (1, no attributes)
- max_depth: integers in the range (1, 121)
- min_samples_split: integers in the range (2, 50)
- min_samples_leaf: integers in the range (1, 45)

Finally, the number of observations in the temporary series of models is optimised, maintaining the fixed output at 15.

Figures 1 and 2 compare the six best method and approach combinations with the lowest root mean square error (RMSE) values along the time for the Sau and La Baells reservoirs, respectively. Using the multi-variant SVM model, the RMSE in overall provides the lowest values along the time. Table 4 shows the RMSE, mean absolute error (MAE) and R-squared (R^2) results at the 15th day and global on Sau reservoir.

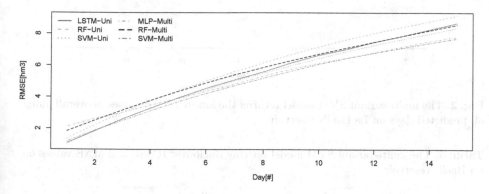

Fig. 1. The multi-variant SVM model returns the lowest RSME values from the 3rd predicted day on Sau reservoir

The best results are achieved with the multi-variant SVM model. Furthermore, the MAE error, which is expressed in the same units as the data and that RMSE, is lower in all cases, therefore, the most important errors are penalising the error of RMSE. We also note that, looking at the MAE, the classification varies: the uni-variant SVM would pass in front of multi-variant MLP and the uni-variant RF would be ahead of the multi-variant, to the extent global. This means that these models provoke fewer errors, but they make them bigger, which is why they penalise the RMSE. On the other hand, we can verify that we obtain a very acceptable R^2 determination coefficient, even in the worst model and day, which would be the 0.87 uni-variant RF on the fifteenth day. Table 5 shows the RMSE, MAE and R^2 results at the 15th day and global on La Baells reservoir.

Table 4. The multi-variant SVM model returns the lowest RMSE and MAE values on Sau reservoir

Model	15th day			Global		
	RMSE	MAE	R^2	RMSE	MAE	R^2
SVM-Multi	7.6907	5.7391	0.9130	5.2961	3.6088	0.9581
MLP-Multi	7.7908	5.9308	0.9107	5.4435	3.8811	0.9558
SVM-Uni	8.3699	6.3386	0.9065	5.6627	3.7640	0.9568
LSTM-Uni	8.6653	6.7872	0.8998	5.8064	3.9718	0.9546
RF-Multi	8.5605	6.7438	0.8921	5.9549	4.3192	0.9470
RF-Uni	9.1335	6.9229	0.8716	6.2832	4.2043	0.9662

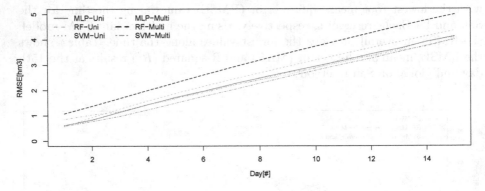

Fig. 2. The multi-variant SVM model returns the lowest RSME values in overall along all predicted days on La Baells reservoir

Table 5. The multi-variant SVM model returns the lowest RMSE and MAE values on La Baells reservoir

Model	15th day			Global		
	RMSE	MAE	R^2	RMSE	MAE	R^2
SVM-Multi	4.0029	2.4098	0.9515	2.5085	1.3506	0.9811
MLP-Multi	4.0330	2.7339	0.9508	2.6122	1.6148	0.9795
SVM-Uni	4.0985	2.6003	0.9537	2.6198	1.4503	0.9812
MLP-Uni	4.1279	2.8958	0.9531	2.6569	1.5905	0.9806
RF-Uni	4.3563	2.7258	0.9243	2.8514	1.4887	0.9677
RF-Multi	4.9869	3.5642	0.9248	3.3500	2.1372	0.9396

The worst model on the fifteenth day of La Baells has an R^2 higher than the best model of Sau, 0.9248 of multi-variant RF for the 0.9130 of the best model of Sau. We can also verify that global R^2s are really good enough. Analysing the MAE, we reaffirm that multi-variant SVM as the best model. If we consider the predictions of the Sau reservoir, then those in La Baells reservoir would be spectacular, since, although the RMSE is not directly comparable, dimension difference between both reservoirs do not justify the error gap from the models. However, the data from the Sau reservoir have been much more difficult to model than those of La Baells. Therefore, with worse results, the result of the best model of Sau can be more valuable than one of the best model of La Baells. Models studied in the literature predict the level of water in meters above sea level (MAMSL). This would explain their results with RMSE values close to zero. Table 6 compares the main articles that we have analysed with their predictions achieved.

Table 6. Our work outperforms related literature studies

Work	Unit	Intervals	Timesteps	RMSE	R^2	Others
Rani and Parekh [19]	MAMSL	135–148	10	0.82	0.95	
Üneş et al. [25]	MAMSL	23.5–24.6	1	0.057	0.893	
Onidmu and Murase [16]	MAMSL	1886–1888	4			0.12 (%MSE)
Kilinç and Cigizoglu [11]	Volum (hm^3)	47–153	1	7.63	0.86	
Dogan et al. [4]	MAMSL	1647–1650	1	0.035	0.93	
Çimen and Kisi [24]	MAMSL	1647–1650	1	0.073	0.985	

The Intervals columns corresponds to the minimum and maximum of the time series that each research is about. In the Timesteps column, we have the predicted future values of the corresponding time series of each proposed model. For instance, with a range of values ranging from 1647 to 1650 m, where predictions move at 3 m, the error will be smaller than predictions on wider ranges (i.e. 12 m). Therefore, this justifies these tiny RMSE values, which we can not compare to. Thus, when we predict the volume, as in the case of Kilinç and Cigizoglu [11], then we no longer have such ranges of values so short, as in the case of meters above sea level, and the error increases. In comparison with the literature, our research considers the prediction on volume and also through the determination coefficient, which evaluates the explanatory capacity of the model. In terms of R^2, our models are similar to the literature. The worst model is the uni-variant RF on Sau reservoir with a global value of 0.94, which is better than three out of the five results from Table 6. Our best model, the multi-variant SVM of La Baells returned an R^2 of 0.981, is better than all of them except Çimen and Kisi [24] with 0.985. Nevertheless, this last mentioned model makes a prediction at a time step, then we should compare it with the value of R^2 we get for the first day of our model, which is 0.99. Comparing our results with, far to the best of our knowledge, the only research article predicting the volume of water [11]. They obtained an RMSE of 7.63 hm^3 and, according to the range of volume of water from 47 to 153 hm^3, we may compare it with the study of the Sau reservoir (9.5 and 165 hm^3). In this case, the best and worst model returns an RMSE of 5.30 and 6.28 hm^3, respectively. Therefore, our models outperform state-of-the-art works. Note that models with the best overall result are those that have the best results on the fifteenth prediction day, specifically, and generally in the longest-term prediction. Since in the long term prediction it is where we accumulate more errors, therefore, an improvement in these farther days supposes an outstanding improvement in the global model. The models with the best overall result do not have to be those that further refine the prediction to one day, as in the case of multi-variant MLPs, which we have already seen are beginning to be the worst to end up practically in line with the best, both in the last days of prediction and in the overall result.

5   Conclusion and Future Work

The data of heat and weather coincided with good quality in temporarily large enough bands to be able to handle them. It is plausible to think that input and output flow data had helped multivariate prediction, especially input, that reflect the effect on the rain or snow dam. Thus, we might think that the effect of rainfall, the flow of entry, may be more relevant as a given to the rain itself. This would open us the most difficult debate of multivariate analysis, which is to know what attributes it is necessary to treat, a question that is not easy, since, for example, we have seen that the accumulation of sediments can vary the capacity of a reservoir. SVM, which in any case has been placed above the neuronal networks, both in the multi-variant and in the uni-variant analysis. Although results of the neural networks have been very close to the SMV, there is no point of comparison because of the computational cost. Results returned a longer prediction with higher precision than state-of-the-art related works. The models we have considered to make the comparison have been extracted from the reading of the state of the art [4,11,15,16,19,25], where neuronal networks are widely used for the prediction of time series in the hydrological field and we have seen that they are usually compared to the SVM model [8,14,24]. The review of an article used by Random Forest [10] made us also consider including this model in the study.

As future work, we plan but not limited to model a reservoir so that it has the possibility of being able to generalise, include different natural events such as snow and landslides or, study other sophisticated techniques such as deep learning.

References

1. Bao, Y., Xiong, T., Hu, Z.: Multi-step-ahead time series prediction usingmultiple-output support vector regression. Neurocomputing **129**, 482–493 (2014). https://doi.org/10.1016/j.neucom.2013.09.010. http://www.sciencedirect.com/science/article/pii/S092523121300917X

2. Chang, F.J., Chang, Y.T.: Adaptive neuro-fuzzy inference system for prediction of water level in reservoir. Adv. Water Resour. **29**(1), 1–10 (2006). https://doi.org/10.1016/j.advwatres.2005.04.015. http://www.sciencedirect.com/science/article/pii/S0309170805001338
3. Chaves, P., Kojiri, T.: Deriving reservoir operational strategies considering water quantity and quality objectives by stochastic fuzzy neural networks. Adv. Water Resour. **30**(5), 1329–1341 (2007). https://doi.org/10.1016/j.advwatres.2006.11.011. http://www.sciencedirect.com/science/article/pii/S0309170806002168
4. Doan, E., Kocamaz, U.E., Utkucu, M., Yldrm, E.: Modelling daily water level fluctuations of lake van (eastern turkey) using artificial neural networks. Fundam. Appl. Limnol. **187**(3), 177–189 (2016). https://doi.org/10.1127/fal/2015/0736. https://www.ingentaconnect.com/content/schweiz/fal/2016/00000187/00000003/art00001
5. El-Shafie, A., Jaafer, O., Seyed, A.: Adaptive neuro-fuzzy inference system based model for rainfall forecasting in klang river malaysia. Int. J. Phys. Sci. **6**(12), 2875–2888 (2011)
6. Generalitat de Catalunya: Agència Catalana de l'Aigua (2019). Accessed Mar 2019. http://aca.gencat.cat/ca/inici
7. Generalitat de Catalunya: Servei Meteorológic de Catalunya (2019). Accessed Mar 2019. http://www.meteo.cat/wpweb/serveis/formularis/peticio-dinformes-i-dades-meteorologiques/peticio-de-dades-meteorologiques
8. Hipni, A., El-shafie, A., Najah, A., Karim, O.A., Hussain, A., Mukhlisin, M.: Daily forecasting of dam water levels: comparing a support vector machine (svm) model with adaptive neuro fuzzy inference system (anfis). Water Resour. Manag. **27**(10), 3803–3823 (2013). https://doi.org/10.1007/s11269-013-0382-4
9. Jain, S.K., Das, A., Srivastava, D.K.: Application of ann for reservoir inflowprediction and operation. J. Water Resour. Plann. Manag. **125**(5), 263–271 (1999). https://doi.org/10.1061/(ASCE)0733-9496(1999)125:5(263). https://ascelibrary.org/doi/abs/10.1061/%28ASCE%290733-9496%281999%29125%3A5%28263%29
10. Jun-He, Y., Ching-Hsue, C., Chia-Pan, C.: A time-series water level forecasting model based on imputation and variable selection method. Comput. Intell. Neurosci. **2017**, 11 (2017)
11. Kilinç, I., Ciğizoğlu, K.: Reservoir management using artificial neural networks. In: 14th. Reg. Directorate of DSI (State Hydraulic Works) (2005)
12. Kisi, O., Nia, A.M., Gosheh, M.G., Tajabadi, M.R.J., Ahmadi, A.: Intermittentstreamflow forecasting by using several data driven techniques. Water Resour. Manag. **26**(2), 457–474 (2012). https://doi.org/10.1007/s11269-011-9926-7
13. Moeeni, H., Bonakdari, H., Fatemi, S.E., Zaji, A.H.: Assessment of stochastic models and a hybrid artificial neural network-genetic algorithm method in forecasting monthly reservoir inflow. INAE Lett. **2**(1), 13–23 (2017). https://doi.org/10.1007/s41403-017-0017-9
14. Mokhtar, S.A., Ishak, W.H.W., Norwawi, N.M.: Modelling of reservoir water release decision using neural network and temporal pattern of reservoir water level. In: 2014 5th International Conference on Intelligent Systems, Modelling and Simulation. pp. 127–130. January 2014. https://doi.org/10.1109/ISMS.2014.27
15. Nwobi-Okoye, C., Igboanugo, A.: Predicting water levels at kainji dam using artificial neural networks. Niger. J. Technol. (NIJOTECH) **32**(1), 129–136 (2013)

16. Ondimu, S., Murase, H.: Reservoir level forecasting using neural networks: lake naivasha. Biosyst. Eng. **96**(1), 135–138 (2007). https://doi.org/10.1016/j.biosystemseng.2006.09.003. http://www.sciencedirect.com/science/article/pii/S1537511006003059

17. Pasupa, K., Jungjareantrat, S.: Water levels forecast in thailand: a case study of chao phraya river. In: 2016 14th International Conference on Control, Automation, Robotics and Vision (ICARCV). pp. 1–6. November 2016. https://doi.org/10.1109/ICARCV.2016.7838716

18. Quiles, R.C., Roma, J.C., Roig, J.G., Alfonso, J.M.: Minería de datos: Modelos y algoritmos. UOC (2017)

19. Rani, S., Parekh, F.: Predicting reservoir water level using artificial neural network. Int. J. Innovative Res. Sci., Eng. Technol. 3(7) (2014)

20. Valipour, M., Banihabib, M.E., Behbahani, S.M.R.: Comparison of the arma, arima, and the autoregressive artificial neural network models in forecasting the monthly inflow of dez dam reservoir. J. Hydrol. **476**, 433–441 (2013). https://doi.org/10.1016/j.jhydrol.2012.11.017. http://www.sciencedirect.com/science/article/pii/S002216941200981X

21. Valizadeh, N., El-Shafie, A.: Forecasting the level of reservoirs using multiple input fuzzification in anfis. Water Resour. Manag. **27**(9), 3319–3331 (2013). https://doi.org/10.1007/s11269-013-0349-5

22. Wang, A., Liao, H., Chang, T.: Adaptive neuro-fuzzy inference system on downstream water level forecasting. In: 2008 Fifth International Conference on Fuzzy Systems and Knowledge Discovery. vol. 3, pp. 503–507. October 2008. https://doi.org/10.1109/FSKD.2008.671

23. Wieland, R., Mirschel, W., Zbell, B., Groth, K., Pechenick, A., Fukuda, K.: A new library to combine artificial neural networks and support vector machines with statistics and a database engine for application in environmental modeling. Environ. Modell. Softw. **25**(4), 412–420 (2010). https://doi.org/10.1016/j.envsoft.2009.11.006. http://www.sciencedirect.com/science/article/pii/S1364815209002953

24. Çimen, M., Kisi, O.: Comparison of two different data-driven techniques in modeling lake level fluctuations in turkey. J. Hydrol. **378**(3), 253–262 (2009). https://doi.org/10.1016/j.jhydrol.2009.09.029. http://www.sciencedirect.com/science/article/pii/S0022169409006015

25. Üneş, F., Demirci, M., Kişi, O.: Prediction of millers ferry dam reservoir level in usa using artificial neural network. Periodica Polytech. Civil Eng. **59**(3), 309–318 (2015). https://doi.org/10.3311/PPci.7379. https://pp.bme.hu/ci/article/view/7379

A Predictive Model for MicroRNA Expressions in Pediatric Multiple Sclerosis Detection

Gabriella Casalino[1,3], Giovanna Castellano[1,3(✉)], Arianna Consiglio[2,3], Maria Liguori[2], Nicoletta Nuzziello[2], and Davide Primiceri[1]

[1] Department of Computer Science, University of Bari Aldo Moro, Bari, Italy
giovanna.castellano@uniba.it
[2] Institute for Biomedical Technologies of Bari, Italian National Research Council, Bari, Italy
[3] Member of INDAM Research Group GNCS, Rome, Italy

Abstract. MicroRNAs (miRNAs) are a set of short non coding RNAs that play significant regulatory roles in cells. The study of miRNA data can be of valuable support for the early diagnosis of multifactorial diseases such as pediatric Multiple Sclerosis. However the analysis of miRNA expressions poses several challenges due to high dimensionality and imbalance of data. In this paper we present a data science workflow to develop a predictive model that is intended to support the clinicians in the diagnosis of Multiple Sclerosis starting from miRNA data produced by Next-Generation Sequencing. The goal is to create an effective model able to predict the pathological condition of a patient starting from his miRNA expression profile. Based on the proposed workflow, the miRNA dataset is firstly preprocessed in order to reduce its high dimensionality (from 1287 features to 40 features) and to mitigate class imbalance. Then a classification model is learnt from data via neural network training. Results show that the model defined by using the 40 data-driven selected features achieves an overall classification accuracy of 94% on test data and overcomes the model based on 42 features selected by the experts that achieves only 83% of overall accuracy.

Keywords: microRNA expression · Next-Generation Sequencing · Pediatric Multiple Sclerosis · Feature selection · Artificial neural networks · Classification.

1 Introduction

Physiological processes and cellular functions can be studied by quantifying the production (called *expression*) of RNAs in biological samples. Next-Generation Sequencing (NGS) techniques allow to quantify the expression of the whole set of RNAs that are active in the cells of a sample. MicroRNAs (miRNAs) are a class of small RNA that regulate the expression of other longer RNA and

© Springer Nature Switzerland AG 2019
V. Torra et al. (Eds.): MDAI 2019, LNAI 11676, pp. 177–188, 2019.
https://doi.org/10.1007/978-3-030-26773-5_16

the consequent production of proteins [3]. In the past few years, research on miRNA-related problems has become a hot field of bioinformatics because of miRNAs' essential biological function. Actually, the study of miRNA expression changes gives the possibility to identify biomarkers, i.e., molecules that are predictive of the clinical course or response to treatments, especially for complex and multifactorial diseases such as Multiple Sclerosis (MS).

Multiple Sclerosis (MS) is a demyelinating autoimmune disease of the CNS that usually affects young adults [16]. The onset during childhood and adolescence is being increasingly recognized [5], together with the demonstration of cognitive deficits in more than one-third of these patients [2]. The study of pediatric MS patients (PEMS) offers the unique opportunity of investigating the pathogenic mechanisms occurring at the earliest stages of the disease. To this aim, the analysis of miRNA expressions can be of valuable help. In a previous study [12,13] we investigated the transcriptome profile of peripheral blood samples in a cohort of PEMS patients, and we further validated (with specific laboratory assays and on a larger cohort of subjects) 12 miRNAs with statistically significant increase of expression and one miRNA with decreased expression in PEMS patients compared to pediatric healthy control (HCPE) subjects.

The bionformatic pipelines developed for miRNA expression analysis usually apply statistical tests to search for differentially expressed miRNAs in the comparison between healthy controls and diseased patients [14,15]. Such analysis allows the isolation of evident changes in expression, but it fails in extracting more complex interactions among different miRNAs that are correlated to the disease. Machine Learning (ML) algorithms, such as Artificial neural networks (ANNs) can be useful tools to capture complex interactions among miRNA expressions and their relation to the disease. However, to our knowledge, the use of ANNs to correlate miRNAs to autoimmune diseases has not been investigated so far. ANNs have been widely applied to microarray data [10], and some works have been presented on NGS data [11]. Few papers on ANNs applied to miRNA expression analysis have been published in the study of cancer [1,6].

The modeling of miRNA using ML methods such as ANNS poses several challenges. Redundant information is usually represented in datasets, moreover not all the features could be significant for class discrimination. This affects and sometimes invalidates the process of predictive modeling. For this reason feature selection techniques are commonly used to select a subset of relevant features [8]. Feature reduction has several advantages: model simplification, shorter training times, reduction of the over-fitting problem [19]. Moreover, while a large number of samples are required to create accurate predictive models using machine learning, in biological domains the number of features is very high, while the number of available observations is often rather low. In addition, biomedical datasets are often unbalanced since the number of positive samples (patients affected by the disease) is typically lower than the number of negative samples. The principal limitation of NGS is the high cost of sequencing, but when studying rare pediatric diseases it is both difficult to find an adequate number of patients and to recruit healthy children who accept to provide blood samples.

All these issues call for the use of data processing techniques that are necessary to reduce the dimension of the input space and to balance the dataset before applying ML algorithms. Moreover, since expert knowledge cannot be completely replaced by machines, intelligent techniques that combine human expertise and computational models for advanced data analysis are necessary to develop more reliable predictive models [4].

In this work, we propose a data science workflow that is intended to process miRNA expressions in order to extract a predictive model capable to detect the pathological condition of a patient. The dataset under analysis includes miRNA expressions of PEMS patients along with miRNA expressions of children affected by Attention deficit hyperactivity disorder (ADHD). Actually it has been observed that ADHD patients share some cognitive disorders with PEMS patients, hence deriving a predictive model capable to distinguish between these two diseases on the basis of miRNA expressions may be of valuable help for Biology experts. Our work is the first attempt to automatically analyze this specific dataset in order to derive a predictive model that is capable to discriminate between ADHD and PEMS on the basis of miRNA expressions.

The rest of the paper describes in detail all the phases of the workflow.

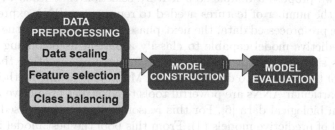

Fig. 1. The data science workflow.

2   The Workflow

The workflow designed to create a predictive model from miRNA expressions of PEMS patients includes the standard phases of a data science process (Fig. 1), namely data acquisition, data pre-processing, model construction, and model evaluation.

As concerns the data acquisition, the dataset was produced at the Institute for Biomedical Technologies of the Italian National Research Council (ITB-CNR) through small RNA sequencing of peripheral blood samples obtained from 47 children. The sequence data files produced were processed with a standard bionformatic pipeline. The sequences were compared with databases of known miRNAs, and sequence counts were computed to estimate miRNA expressions. The dataset includes expressions of 1287 miRNAs detected in 47 patients. The patients under study differ by pathological conditions. Besides healthy controls, we analyzed some patients affected by Multiple Sclerosis and some others affected by cognitive disorders. The number of patients for each class (pathological condition) is reported in Table 1.

Table 1. Data summary.

ClassID	Condition	Acronym	# Samples
0	Cognitive disorders	ADHD	8
1	Healthy controls	HCPE	20
2	Multiple Sclerosis	PEMS	19

The data pre-processing phase (Sect. 3) includes three steps that are dictated by the nature of the data under analysis. There are no missing values nor negative values in the data. However the dataset has a high dimensionality (1287 features) vs a low number of samples (only 60). Moreover the dataset is quite unbalanced. For these reasons, we apply a multi-stage data pre-processing in order to reduce drastically the dimensionality of the dataset and balance it. Specifically, we propose a double-step feature selection that leads to a drastic reduction of the number of features needed to represent miRNA expressions.

Given the pre-processed data, the next phase of the workflow is the construction of a predictive model capable to classify a patient as belonging to one of the 3 classes starting from a reduced set of features representing the miRNA expressions of the patient. To do this, several Machine Learning methods can be applied. In particular ANNs are powerful tools to construct predictive models by learning from biological data [6]. For this reason in this work we used ANNs to create a pool of predictive models (4). From this pool the best model is selected as final model to be evaluated on test data (Sect. 5). Moreover the final model has been evaluated by a comparison with a model created on the basis of 42 features selected by experts.

3 Data Preprocessing

To cope with high dimensionality and unbalancement, data are processed in order to obtain a more informative and easy-to-compute representation. Three pre-processing steps are performed, namely normalization, feature selection, and class balancing. Standard methods were applied for normalization and balancing, while a novel method was developed for a drastic and robust feature selection.

3.1 Normalization

The activity of miRNAs and their relative expression can have effect at different scales. This may hamper the quality of the predictive model derived from data.

Hence we apply a normalization so that all the features values are bounded in the interval $[0, 1]$. The *MinMax* scaling algorithm[1] is used.

From each value x_i in each feature x a scaled value $\hat{x}_i$ is computed as:

$$\hat{x}_i = \frac{x_i - \min(x)}{\max(x) - \min(x)}$$

This kind of normalization reduces the variance in data, as well as the effect of outliers [20].

3.2 Feature Selection

The number of features in the smallRNA dataset is very high. In order to significantly reduce the data dimensionality without loosing significant information, we propose to combine two feature selection methods: the *Variance Threshold* and the *Select K Best* (SKB) [20]. Both methods are implemented in the Scikit-learn Python Library[2] [17].

The Variance Threshold method was firstly used to remove the features with a variance lower than a fixed threshold. We used the mean variance computed over all the feature as threshold. In this way genes whose expression do not significantly vary over the samples are not selected, since they would not be useful to discriminate the different conditions. Using this method we selected 564 features out of the total 1287 features.

Then the SelectKBest method was applied to further reduce this subset of features. This method selects the most significant k features, based on the correlation information between each feature and the class information of each sample. The Chi-squared test is used to select the most discriminative features for the given classes.

3.3 Class Balancing

As previously discussed, the dataset presents class imbalance (see Table 1) since the number of patient affected by cognitive disorders is lower than the number of patients belonging to the other two classes. Class imbalance should be avoided since it could mislead the classification results. Common methods to balance the dataset are *undersampling* and *oversampling* [20]. In undersampling a subset of samples is removed from the class with more instances. In oversampling the minority class is over-sampled with replacement. Undersampling is not suitable for our dataset, since the total number of samples is small. For this reason, we applied oversampling[3] by adding some randomly selected samples from the under represented class so that each class is represented by 20 samples.

---

[1] https://scikit-learn.org/stable/modules/generated/sklearn.preprocessing. MinMaxScaler.html.

[2] https://scikit-learn.org/stable/modules/feature_selection.html.

[3] Oversampling algorithm is provided in scikit-learn module: https://imbalanced-learn.readthedocs.io/en/stable/.

4   Model Construction

To derive a predictive model from miRNA data we used multi-layered feed-forward neural networks [7]. The network architecture is composed of one input layer, two hidden layers and one output layer. The input layer contains k nodes, being k the number of features that were selected at the end of the pre-processing phase. The number of nodes for each hidden layer was empirical estimated, as discussed below. The *relu* activation function was used for all hidden nodes. The output layer is made of three nodes, one for each class. The *softmax* function has been used to encode the output as a probability, i.e, the output of the j-th node indicates the probability of the input sample to belong to class j. The class with the highest probability is associated to the given input sample.

The training of the neural network is based on optimization of the sparse categorical cross-entropy loss function:

$$loss = - \sum_{j=1}^{M} t_{p,j} \log(o_{p,j})$$

where M is the number of classes, $o_{p,j}$ is the j-th output of the network for the p-th input sample and $t_{p,j}$ is the j-th component of the target vector for the p-th sample (assuming *1-out-of-n* encoding for the target values). Optimization of the loss function can be performed in several ways. In our simulations we use the Adam (Adaptive moment estimation) method since it has been suggested to be the best overall choice among state-of-art gradient descent optimization algorithms [18]. The Adam method performs efficient stochastic optimization by computing only first-order gradients with little memory requirement [9]. Updates of network weights are computed iteratively on the basis of a running average of first and second moment of the gradient. The Python Keras library[4] running on Tensorflow[5] was used for the network training, using Jupyter Notebook[6] and taking advantage of the Colaboratory platform provided by Google[7].

To create a predictive model via neural network training, the dataset was divided into a training set (70% of the data, 42 examples) and test set (30% of the data, 18 examples). To perform model selection a stratified 5-fold cross validation was applied, taking into account class information to avoid constructing folds

Table 2. Experimental setup.

Parameter	
Number of features (k of SelectKBest)	10, 20, 30, 40
Number of hidden neurons	32x32, 64x32, 64x64

---

[4] https://keras.io/.
[5] https://www.tensorflow.org/.
[6] https://jupyter.org/.
[7] https://colab.research.google.com/notebooks/welcome.ipynb.

with unbalanced class distributions. In each run the training of the network was stopped at 100 epochs. Several training runs were performed to find the best configuration of the model that will be used to predict test data. The setup is reported in Table 2.

To evaluate the effectiveness of the model in correctly predicting the class of unknown samples we used the loss function as well as the classification accuracy:

$$Accuracy = \frac{Number\ of\ correct\ predictions}{Total\ number\ of\ predictions}$$

In Fig. 2 we show the trend of the loss function on both the training set and validation set during the training of the best model, for each subset of features. Low differences between the loss values on the training set and the validation set mean that the model is able to learn the relationships in data, without overfitting. It can be seen that the model trained with 40 features has the best loss trend. Table 3 reports the loss and accuracy values on the test set for models obtained with different configurations of hidden neurons and number of features. It can be observed that, as the number of the selected features increases, the loss decreases and the accuracy increases. The neural network with 40 features and 64 neurons in each hidden layer provides the best performance on test data (accuracy=0.94, loss=0.13) hence it was selected as the best model to perform the classification task.

The outcome of this experiment is that a good predictive model can be obtained with only 40 features. The subset of 40 features selected out of the initial 1287 features is reported in Table 4. It includes two out of the 13 differentially

Table 3. Loss and accuracy values on test set.

Hidden neurons	10 features		20 features		30 features		40 features	
	Loss	Acc	Loss	Acc	Loss	Acc	Loss	Acc
32 × 32	0.87	0.56	0.71	0.78	0.55	0.83	0.22	0.94
64 × 32	0.81	0.61	0.59	0.72	0.40	0.78	0.17	0.94
64 × 64	0.81	0.61	0.53	0.72	0.36	0.83	**0.13**	**0.94**

Table 4. List of 40 miRNA selected by our data-driven procedure.

miR-125a-3p	miR-1304-5p	miR-188-3p	miR-215-5p	miR-221-3p
miR-29a-3p	miR-29c-3p	miR-30d-3p	miR-3117-3p	miR-3120-3p
miR-3140-3p	miR-3179	miR-3682-3p	miR-3688-3p	miR-378h
miR-4440	miR-4443	miR-4489	miR-452-5p	miR-4665-5p
miR-4690-3p	miR-4731-3p	miR-4734	miR-4763-5p	miR-4804-5p
miR-483-3p	miR-502-5p	miR-582-3p	miR-589-3p	miR-619-5p
miR-6770-3p	miR-6820-3p	miR-6821-3p	miR-6883-3p	miR-6884-3p
miR-6894-5p	miR-7113-5p	miR-7-1-3p	miR-885-5p	miR-942-5p

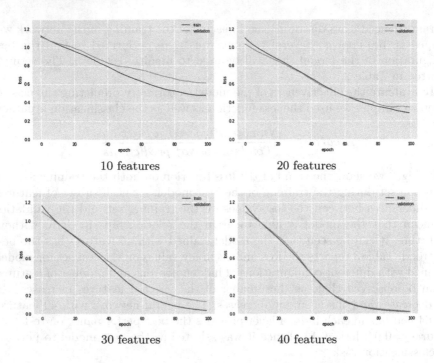

Fig. 2. Trend of the loss function during training with different subsets of features.

expressed miRNAs that were validated by further laboratory assays (miR-221-3p, miR-942-5p).

5   Model Evaluation

The selected best model was further evaluated on the test set composed of 18 examples (6 for each class). We evaluated True positive (TP), true negative (TN), false positive (FP), and false negative (FN) values, for each class c against the remaining two classes, and we computed the accuracy as well as the following classification measures:

- Positive Predictive Value (PPV): ratio of correctly classified samples w.r.t. those identified as pertaining to class c ($\text{PPV} = \frac{\text{TP}}{\text{TP+FP}}$)
- Negative Predictive Value (NPV): ratio of correctly classified samples w.r.t. those identified as not pertaining to class c ($\text{NPV} = \frac{\text{TN}}{\text{TN+FN}}$)
- True Positive Rate (TPR): ratio of samples correctly classified as belonging to class c w.r.t. those actually belonging to class c ($\text{TPR} = \frac{\text{TP}}{\text{TP+FN}}$)
- True Negative Rate (TNR): ratio of samples correctly classified as not belonging to class c w.r.t. those actually not belonging to class c ($\text{TNR} = \frac{\text{TN}}{\text{FP+TN}}$)

Table 6 reports the confusion matrix of the classification model, while the classification metrics are shown in Table 7. It can be seen that healthy patients

(HCPE) are correctly detected. For the class ADHD (patients affected by cognitive disorders) we observe high values for PPV and TPR measures. This means that the model is able to correctly detect the patients in this condition. As it can be observed in Table 6(a), only one patient has been wrongly classified as affected by Multiple Sclerosis (PEMS), this causes low values for TNR and NPV. The class of patients with Multiple Sclerosis is more difficult to be predicted than the other two classes. Indeed, one PEMS patient has been classified as ADHD. This result can be justified by the fact that ADHD patients share some cognitive disorders with PEMS patients, hence there may be a significant overlapping between the molecular pathways of these two classes (27% of PEMS patients show similar cognitive symptoms of ADHD patients [21]). In fact, the patient that is misclassified was diagnosed with a low grade of Multiple Sclerosis, but he showed some cognitive difficulties like ADHD patients.

Finally, to complete the evaluation of the model, we compared it with the best model obtained by training a neural network on a subset of 42 features proposed by the Biology experts (Table 5). This subset contains some results of the differential expression analysis previously performed on this dataset and a list of miRNAs that are known to be involved in neurodegeneration. We found that the model obtained by training the neural network on the features proposed by experts achieves an overall classification accuracy on the test of 83% which is lower than the overall accuracy of 94% achieved by the model obtained with our data-driven feature selection. Also the classification results shown in Table 6(b) indicate that the model based on the features selected by the experts has lower classification ability. Indeed, for all the three classes the classification measures present lower values than those obtained with the automatically selected features. It is interesting to observe that the model using the features selected by the experts can correctly identify all the patients affected by Multiple Sclerosis condition, but it misclassifies the patients belonging to the other two classes. In fact, the experts selected the features by an analysis of miRNAs that are differently expressed between HCPE and PEMS, while no experience had been gained on ADHD patients. Anyway, it misclassifies two healthy patients and one

Table 5. List of 42 miRNAs selected by domain experts.

let-7a-5p	let-7b-5p	let-7i-3p	let-7i-5p	miR-10a-5p
miR-125a-5p	miR-128-3p	miR-1304-3p	miR-1307-3p	miR-130b-3p
miR-140-3p	miR-144-5p	miR-148b-3p	miR-151a-3p	miR-151b
miR-15b-5p	miR-16-2-3p	miR-16-5p	miR-181a-2-3p	miR-181a-5p
miR-182-5p	miR-185-3p	miR-185-5p	miR-21-5p	miR-221-3p
miR-25-3p	miR-26a-5p	miR-26b-3p	miR-26b-5p	miR-27b-3p
miR-28-3p	miR-29a-3p	miR-30e-3p	miR-30e-5p	miR-320a
miR-3605-3p	miR-484	miR-501-3p	miR-652-3p	miR-6842-3p
miR-942-5p	miR-99b-5p			

affected by cognitive disorders as belonging to the PEMS class (see Table 6(b)). It is important to highlight how such false positives can be considered as serious mistakes because they represent erroneous Multiple Sclerosis diagnosis for healthy children.

Finally, it should be noted that the set of features selected by experts is only partially overlapped with the set of data-driven selected features. In particular, the two lists share three common miRNAs (miR-221-3p, miR-29a-3p, miR-942-5p), two complementary miRNAs (hsa-miR-125a-3p instead of 5p, hsa-miR-1304-5p instead of 3p), and a miRNA variant (hsa-miR-30d-3p instead of 30e). Moreover, the list of data-driven selected features has been submitted to Biology experts and they also found that other 25 miRNAs in the list are involved in genetic pathways that should be correlated to Multiple Sclerosis, like inflammation, immune response and neuronal functions. However, in order to fully assess the importance of the selected features, a deep biological investigation is required since most of the miRNA functions are still unknown.

Table 6. Confusion matrix of the model obtained by training the neural network on (a) the 40 data-driven selected features and (b) the 42 features selected by the experts.

(a)					(b)			
	Actual					Actual		
Predicted	ADHD	HCPE	PEMS		Predicted	ADHD	HCPE	PEMS
ADHD	6	0	1		ADHD	5	0	0
HCPE	0	6	0		HCPE	0	4	0
PEMS	0	0	5		PEMS	1	2	6

Table 7. Classification results of the model obtained by training the neural network on (a) the 40 data-driven selected features and (b) the 42 features selected by the experts.

(a)						(b)				
	ACC	TNR	TPR	PPV	NPV		ACC	TNR	TPR	PPV	NPV
ADHD	0.94	0.86	1.00	1.00	0.91	ADHD	0.94	1.00	0.92	0.83	1.00
HCPE	1.00	1.00	1.00	1.00	1.00	HCPE	0.89	1.00	0.85	0.66	1.00
PEMS	0.94	1	0.92	0.83	0.92	PEMS	0.83	0.67	1.00	1.00	0.75

6   Conclusions

In this paper we have presented a data science workflow for the classification of the pathological condition of a patient based on the miRNA expressions obtained with NGS technology. One significant contribution is the proposed data preprocessing phase that allows a drastic data-driven feature selection (from 1287 to only 40 features) useful for a good classification. Indeed, the selected features allow the creation of a predictive model via neural network training that achieves

94% of overall accuracy on test data. A comparison with a classification model built on the basis of 42 features proposed by the experts has confirmed the effectiveness of the proposed method in correctly identifying significant features that lead to good classification results. Of course the obtained results heavily depend on the available dataset which is quite small. Actually, the size of the current dataset was limited by the rarity of PEMS onset and includes all the patients that were recruited over a 3-years period. It would be useful to repeat the experiments proposed in this work with a bigger dataset. Moreover, as a future work, it would be interesting to consider other ML methods to accomplish the phase of model creation and compare their performance with ANNs.

As a final remark, we point out that this work is the first attempt to create a model useful to support the experts in the analysis of the miRNA dataset collected at the Institute for Biomedical Technologies of the Italian National Research Council (ITB-CNR). Such a model can be deployed as a tool to distinguish the three classes of the patients in a completely automatic way, discovering hidden relationships among the miRNAs that can not be derived by a classical differential expression analysis. Moreover the 40 miRNAs automatically selected by the proposed feature selection method can be further analyzed to derive other biological observations, such as an evaluation of the genes that are regulated by those miRNAs and an analysis of the molecular pathways involved in the activation of target genes, both for the study of pediatric Multiple Sclerosis and for novel investigations about ADHD. This work represents the first step toward the development of an intelligent system capable to support the expert in the analysis of miRNA expressions for early diagnosis of pediatric Multiple Sclerosis. To this aim, further work is in progress to combine data on miRNA expressions with clinical data about the patients, in order to derive more powerful diagnostic support models.

Acknowledgments. Molecular data for this investigation derive from a fully supported grant (cod 2014/R/10) by Fondazione Italiana Sclerosi Multipla (FISM).

References

1. Afshar, S., Afshar, S., Warden, E., Manochehri, H., Saidijam, M.: Application of artificial neural network in mirna biomarker selection and precise diagnosis of colorectal cancer. Iran. Biomed.J. **23**(3), 173–183 (2018)
2. Akbar, N., et al.: Aubert-Broche: altered resting-state functional connectivity in cognitively preserved pediatric-onset ms patients and relationship to structural damage and cognitive performance. Multiple Sclerosis J. **22**(6), 792–800 (2016)
3. Bartel, D.P.: Micrornas: genomics, biogenesis, mechanism, and function. Cell **116**(2), 281–297 (2004)
4. Casalino, G., Castiello, C., Del Buono, N., Mencar, C.: Intelligent Twitter data analysis based on nonnegative matrix factorizations. In: Gervasi, O., et al. (eds.) ICCSA 2017. LNCS, vol. 10404, pp. 188–202. Springer, Cham (2017). https://doi.org/10.1007/978-3-319-62392-4_14

5. Chitnis, T., Glanz, B., Jaffin, S., Healy, B.: Demographics of pediatric-onset multiple sclerosis in an ms center population from the northeastern united states. Multiple Sclerosis J. **15**(5), 627–631 (2009)
6. Elias, K.M., et al.: Diagnostic potential for a serum mirna neural network for detection of ovarian cancer. Elife **6**, e28932 (2017)
7. Haykin, S.: Neural Networks: A Comprehensive Foundation. Prentice Hall PTR, New Jersey (1994)
8. Inza, I., Larranaga, P., Saeys, Y.: A review of feature selection techniques in bioinformatics. Bioinformatics **23**(19), 2507–2517 (2007). https://doi.org/10.1093/bioinformatics/btm344
9. Kingma, D.P., Ba, J.: Adam: a method for stochastic optimization. arXiv preprint arXiv:1412.6980 (2014)
10. Lancashire, L.J., Lemetre, C., Ball, G.R.: An introduction to artificial neural networks in bioinformatics application to complex microarray and mass spectrometry datasets in cancer studies. Briefings Bioinform. **10**(3), 315–329 (2009)
11. Leung, M.K., Delong, A., Alipanahi, B., Frey, B.J.: Machine learning in genomic medicine: a review of computational problems and data sets. Proc. IEEE **104**(1), 176–197 (2016)
12. Liguori, M., et al.: Combined microrna and mrna expression analysis in pediatric multiple sclerosis: an integrated approach to uncover novel pathogenic mechanisms of the disease. Hum. Mol. Genet. **27**(1), 66–79 (2017)
13. Liguori, M., et al.: Association between mirnas expression and cognitive performances of pediatric multiple sclerosis patients: a pilot study. Brain Behav. **9**(2), e01199 (2019)
14. Love, M.I., Huber, W., Anders, S.: Moderated estimation of fold change and dispersion for rna-seq data with deseq2. Genome Biol. **15**(12), 550 (2014)
15. McCarthy, D.J., Chen, Y., Smyth, G.K.: Differential expression analysis of multifactor rna-seq experiments with respect to biological variation. Nucleic Acids Res. **40**(10), 4288–4297 (2012)
16. Olsson, T., Barcellos, L.F., Alfredsson, L.: Interactions between genetic, lifestyle and environmental risk factors for multiple sclerosis. Nat. Rev. Neurol. **13**(1), 25 (2017)
17. Pedregosa, F., et al.: Scikit-learn: machine learning in python. J. Mach. Learn. Res. **12**, 2825–2830 (2011)
18. Ruder, S.: An overview of gradient descent optimization algorithms. arXiv preprint arXiv:1609.04747 (2016)
19. Tang, J., Alelyani, S., Liu, H.: Feature selection for classification: a review. Data Classification: Algorithms and Applications p. 37 (2014)
20. Theodoridis, S., Koutroumbas, K., et al.: Pattern recognition. IEEE Trans. Neural Networks **19**(2), 376 (2008)
21. Weisbrot, D., et al.: Psychiatric diagnoses and cognitive impairment in pediatric multiple sclerosis. Multiple Sclerosis J. **20**(5), 588–593 (2014)

On Collaborative Filtering with Possibilistic Clustering for Spherical Data Based on Tsallis Entropy

Yuchi Kanzawa[(✉)]

Shibaura Institute of Technology, Tokyo, Japan
kanzawa@sic.shibaura-it.ac.jp

Abstract. This paper proposes a collaborative filtering (CF) method using possibilistic clustering for spherical data based on Tsallis entropy. This study was motivated by a previous work, which showed that adopting fuzzy clustering for spherical data in CF tasks provided better recommendation accuracy than fuzzy clustering for categorical-multivariate data. Moreover, possibilistic clustering algorithms are naturally more robust to noise than fuzzy clustering. The results of experiments conducted on an artificial dataset and one real dataset indicate that the proposed method is better than the conventional methods in terms of recommendation accuracy.

1 Introduction

Collaborative filtering (CF) is one of the most promising technique for recommender systems [1,2]. CF automatically predicts the interests of a target user by collecting preference-related information from other users who have preferences similar to those of the target user. There are various approaches to CF, including GroupLens [3], which is a representative CF method based on neighborhood models. Although neighborhood-based methods, such as GroupLens, are simple and time-efficient, in practice, the definition of a "neighborhood" is vague. However, it is natural to consider that users can be implicitly placed in groups based on their preferences. In case, if we are able to find such a user group, then we can define the neighborhood of a user as the group to which the user belongs. Thus, clustering involves finding implicit groups among users. Once clustered, the users in the same cluster are assumed to have similar preferences, whereas those in different clusters are assumed to have different preferences.

From among the many clustering methods, Kondo and Kanzawa [6] proposed using their clustering algorithm, referred to as q-divergence-based fuzzy clustering for spherical data (qFCS) [5], for CF tasks. They indicated that their proposed CF algorithm using qFCS outperforms GroupLens algorithm.

Fuzzy clustering methods are useful. However, their memberships do not always correspond well to the degree of belonging of the data. To address this weakness of fuzzy clustering methods, Krishnapuram and Keller [7] created a

© Springer Nature Switzerland AG 2019
V. Torra et al. (Eds.): MDAI 2019, LNAI 11676, pp. 189–200, 2019.
https://doi.org/10.1007/978-3-030-26773-5_17

possibilistic c-means (PCM) algorithm that uses a possibilistic membership function. Ménard et al. [8] proposed another possibilistic clustering technique that employs Tsallis entropy, referred to as Tsallis-entropy-regularized PCM (tPCM). Furthermore, Kanzawa devised a possibilistic clustering approach for spherical data, referred to as Tsallis-entropy-regularized possibilistic clustering for spherical data (tPCS) [9]. Note that q-divergence reduces to Tsallis entropy if the unit distribution serves as a reference for q-divergence. Then, we can say that tPCS is a possibilistic clustering corresponding to qFCS.

In this study, we propose a CF algorithm based on the tPCS method. First, to address the cases of missing data in the clustering phase of the CF task, all the unknown rating values are initialized with the lowest value from among all the known, rated values. Second, the tPCS method is applied to the matrix of rating values to obtain user clusters. The use of the tPCS method ensures that the users in the cluster to which the target user belongs are present in the appropriate neighborhood of the active user. Third, the GroupLens algorithm is applied to the cluster to which the active user belongs to predict the unknown rating values for this user. Finally, all items with predicted rating values higher than the preset threshold value are recommended to the active user. For all the numerical experiments, our proposed method is compared with other clustering-based CF algorithms on which the qFCS algorithm is based [5]. Our experiments are conducted on one artificial dataset and one real datasets. The results of our experiments show that the proposed method outperforms the other approaches in terms of accuracy.

The remainder of this paper is organized as follows. Section 2 discusses the GroupLens algorithm of the representative CF algorithm, namely the qFCS-based CF algorithm, which achieved the best recommendation accuracy in a previous study [6]. Section 2 also introduces a possibilistic clustering for spherical data based on Tsallis entropy, referred to as tPCS. Section 3 presents our proposed CF algorithm. The details of our numerical experiments are presented in Sect. 4. Finally, Sect. 5 concludes this paper.

2   Preliminaries

2.1   Conventional Collaborative Filtering Method: GroupLens

Neighborhood-based methods are the most frequently used algorithms for CF [3]. For these neighborhood-based methods, the subset of appropriate users is selected based on their similarity to an active user, and the weighted aggregate of their ratings is used to generate predictions for the active user. Let N and M be the number of users and items, respectively. Let $x \in \mathbb{R}_+^{N \times M}$ be a matrix whose (k, ℓ)-th element is the rating value of the k-th user for the ℓ-th item. It is important to note that some elements of x may be missing; the goal of CF is to predict such missing values. Then, a binary matrix $y \in \mathbb{R}^{N \times M}$ is defined by setting $y_{k,\ell}$ equal to one if the k-th user has a rating for the ℓ-th item; else, it is zero. $\hat{x}_{k,\ell}$ represents the prediction for the active user k for item

ℓ if $y_{k,\ell} = 0$. $\mathrm{sim}(k, k')$ is the similarity weight between the active user and a neighbor k', as defined by the following Pearson correlation coefficient:

$$\mathrm{sim}(k, k') = \frac{\displaystyle\sum_{\ell:y_{k,\ell}y_{k',\ell}=1} (x_{k,\ell} - \bar{x}_{k,\cdot})(x_{k',\ell} - \bar{x}_{k',\cdot})}{\sqrt{\displaystyle\sum_{\ell:y_{k,\ell}y_{k',\ell}=1} (x_{k,\ell} - \bar{x}_{k,\cdot})^2}\sqrt{\displaystyle\sum_{\ell:y_{k,\ell}y_{k',\ell}=1} (x_{k',\ell} - \bar{x}_{k',\cdot})^2}}, \tag{1}$$

where

$$\bar{x}_{k,\cdot} = \frac{\displaystyle\sum_{\ell:y_{k,\ell}y_{k',\ell}=1} x_{k,\ell}}{\displaystyle\sum_{\ell=1}^{M} y_{k,\ell}y_{k',\ell}}. \tag{2}$$

The GroupLens method [3] uses Pearson correlations to weigh the user similarity values that can be used by all available correlated neighbors, and then, it estimates the rating by computing the weighted average of deviations from the neighbors' mean. Further, the missing values for the active user, $\hat{x}_{k,\ell}$, are predicted from the evaluated values of the other users and their associated similarities using

$$\hat{x}_{k,\ell} = \bar{x}_{k,\cdot} + \frac{\sum_{k':\mathrm{sim}(k,k')\geq 0} \mathrm{sim}(k, k')(x_{k',\ell} - \bar{x}_{k',\cdot})}{\sum_{k':\mathrm{sim}(k,k')\geq 0} \mathrm{sim}(k, k')}, \tag{3}$$

where $x_{k',\ell}$ is replaced by $\bar{x}_{k',\cdot}$ if $x_{k',\ell}$ is missing.

The prediction method involved in the GroupLens algorithm is summarized by the following algorithm:

Algorithm 1 (GroupLens)

STEP 1. Calculate similarities using Eq. (1).
STEP 2. Calculate $\hat{x}$ using Eq. (3).   □

2.2 CF Using qFCS

In the GroupLens method, the neighborhood is defined heuristically as $\mathrm{sim}(k, k') \geq 0$. Note that this definition is not based on any theory. Neighborhoods for target users can be defined in many ways, one of them being to cluster users based on their preferences. From among the many clustering methods, Kondo and Kanzawa proposed the qFCS algorithm, as follows. Let $X = \{x_k \in \mathbb{R}^M \mid k \in \{1, \cdots, N\}, \|x_k\|_2 = 1\}$ be a dataset of $M - 1$-dimensional spherical points. The membership of x_k to the i-th cluster is denoted by $u_{i,k}$ ($i \in \{1, \cdots, C\}, k \in \{1, \cdots, N\}$), and the set of $u_{i,k}$ is denoted by u. The cluster center set is denoted by $v = \{v_i \mid v_i \in \mathbb{R}^M, i \in \{1, \cdots, C\}\}$. The variable controlling the i-th cluster size is denoted by π_i. The i-th element of vector π is denoted by π_i. The qFCS algorithm is obtained by solving the optimization problem

$$\underset{u,v,\pi}{\text{minimize}} \sum_{i=1}^{C} \sum_{k=1}^{N} (\pi_i)^{1-m} (u_{i,k})^m (1 - x_k^\mathsf{T} v_i) + \frac{\lambda^{-1}}{m-1} \sum_{i=1}^{C} \sum_{k=1}^{N} (u_{i,k})^m - u_{i,k} \quad (4)$$

$$\text{subject to} \sum_{i=1}^{C} u_{i,k} = 1, \sum_{i=1}^{C} \pi_i = 1, \|v_i\|_2 = 1, \quad (5)$$

and the algorithm is as bellow:

Algorithm 2

STEP 1. Set the number of clusters C, fuzzification parameters m and λ, initial cluster centers v, and initial variables controlling the cluster size π.

STEP 2. Calculate d using

$$d_{i,k} = 1 - x_k^\mathsf{T} v_i. \quad (6)$$

STEP 3. Calculate u as

$$u_{i,k} = \frac{\pi_i (1 + \lambda(1-m)x_k^\mathsf{T} v_i)^{\frac{1}{1-m}}}{\sum_{j=1}^{C} \pi_j (1 + \lambda(1-m)x_k^\mathsf{T} v_j)^{\frac{1}{1-m}}}. \quad (7)$$

STEP 4. Calculate v as

$$v_i = \frac{\sum_{k=1}^{N} (u_{i,k})^m x_k}{\left\| \sum_{k=1}^{N} (u_{i,k})^m x_k \right\|_2}. \quad (8)$$

STEP 5. Calculate π as

$$\pi_i = \frac{\left(\sum_{k=1}^{N} u_{i,k} (1 + \lambda(1-m)x_k^\mathsf{T} v_i) \right)^{\frac{1}{m-1}}}{\sum_{j=1}^{C} \left(\sum_{k=1}^{N} u_{j,k} (1 + \lambda(1-m)x_k^\mathsf{T} v_j) \right)^{\frac{1}{m-1}}}. \quad (9)$$

STEP 6. Check the limiting criterion for (u, v, π). If the criterion is not satisfied, go to Step 2.

Furthermore, Kondo and Kanzawa proposed using the above qFCS algorithm for CF tasks as follows:

Algorithm 3

STEP 1. Set a threshold value, $\check{x}$.

STEP 2. Replace each missing value with the lowest value among all the ratings values.

STEP 3. Process Algorithm 2.

STEP 4. Calculate $\hat{x}$ using

$$\hat{x}_{k,\ell} = \bar{x}_{k,\cdot} + \frac{\sum_{f(x_{k'})\equiv f(x_k)} \text{sim}(k,k')(x_{k',\ell} - \bar{x}_{k',\cdot})}{\sum_{f(x_{k'})\equiv f(x_k)} \text{sim}(k,k')}. \tag{10}$$

STEP 5. Recommend all items to the active user $\#k$ with $\hat{x}_{k,\ell} \geq \check{x}$ and $y_{k,\ell} = 0$.
□

Kondo and Kanzawa indicated through some numerical experiments that their proposed CF algorithm using qFCS outperforms not only the GroupLens algorithm but also CF using FCCMM [4].

2.3 Possibilistic Clustering for Spherical Data Based on Tsallis Entropy

Kanzawa [9] proposed a possibilistic clustering methods for spherical data using Tsallis entropy, defined as

$$\text{minimize}_{u,v} \sum_{i=1}^{C}\sum_{k=1}^{N} u_{i,k}^m (1 - x_k^\mathsf{T} v_i) + \frac{\lambda^{-1}}{m-1} \sum_{i=1}^{C}\sum_{k=1}^{N}(u_{i,k}^m - u_{i,k}) - \lambda^{-1}\sum_{i=1}^{C}\sum_{k=1}^{N} u_{i,k},$$
$$\tag{11}$$

subject to $\|v_i\|_2 = 1$, $\qquad\qquad\qquad\qquad\qquad\qquad\qquad\qquad$ (12)

where all the symbols correspond to those for qFCS. The tPCS algorithm is described as follows:

Algorithm 4

STEP 1. Select a subset of objects as initial cluster centers. It is possible consider all objects: $C = N$; $v_i = x_i$ $i \in \{1, \cdots, C\}$. Set fuzzification parameters m and λ.

STEP 2. Perform possibilistic clustering as follows:

STEP (A). Calculate u as

$$u_{i,k} = (1 - \lambda(1-m)(1 - x_k^\mathsf{T} v_i))^{\frac{1}{1-m}}. \tag{13}$$

STEP (B). Calculate v as

$$v_i = \frac{\sum_{k=1}^{N}(x_k)^m}{\|\sum_{k=1}^{N}(x_k)^m\|_2}. \tag{14}$$

STEP (C). If (u,v) converges, terminate the possibilistic clustering. Otherwise, go to STEP. (A).

Then, we obtain C cluster centers.

STEP 3. Merge the cluster centers among which the distances are negligible.

3   Proposed Method

As mentioned previously, fuzzy clustering methods are useful; however, their memberships do not always correspond well to the degree of belonging of the data. To address this weakness of fuzzy clustering methods, Krishnapuram and Keller [7] proposed a possibilistic c-means (PCM) algorithm that uses a possibilistic membership function. In particular, possibilistic clustering algorithms are supposedly naturally immune to noise relative to fuzzy clustering.

While some users in the recommendation system faithfully express their true opinion, many provide noisy ratings, which can be detrimental to the quality of the generated recommendations. The presence of noise can violate modeling assumptions and may thus lead to instabilities in estimation and prediction. Even worse, malicious users can deliberately insert attack profiles in an attempt to bias the recommender system to their benefit.

Then, adopting possibilistic clustering in CF tasks is expected to produce more accurate recommendation results than fuzzy clustering. A previous study [6] showed that CF based on qFCS outperforms CF based on qFCCMM in terms of recommendation accuracy. The tPCS algorithm is a possibilistic counterpart to the qFCS algorithm. Then, adopting tPCS instead of qFCS in CF tasks is expected to produce more accurate recommendation results than CF based on qFCS.

If Algorithm 4 is implemented, the missing values are obtained using the proposed methods.

Algorithm 5

STEP 1. Set a threshold value, $\check{x}$.
STEP 2. Replace each missing value with the lowest value among all the ratings' values.
STEP 3. Process Algorithm 4.
STEP 4. Calculate $\hat{x}$ using Eq. (10).
STEP 5. Recommend all items to the active user $\#k$ with $\hat{x}_{k,\ell} \geq \check{x}$ and $y_{k,\ell} = 0$.
$\qquad\qquad\qquad\qquad\qquad\qquad\qquad\qquad\qquad\qquad\qquad\qquad\qquad$ □

Though this algorithm is time consuming, in practice, however, the first four steps of the algorithm can be preformed before an active user tries the recommendation system.

Tables 1, 2, 3, 4 and 5 show the flow of Algorithm 5. In particular, Table 1 shows an initial rating matrix, where four users evaluate five items, and users $\#2$ and $\#4$ do not evaluate items $\#4$ and $\#2$, respectively, denoted by "N/A." Table 2 shows the rating matrix after STEP 2 of Algorithm 5 was applied to Table 1. Thus, "N/A" is replaced with $\min\limits_{\substack{1 \leq k \leq 4 \\ 1 \leq \ell \leq 5 \\ (k,\ell) \notin \{(2,4),(4,2)\}}} x_{k,\ell}$. Further, Table 3 shows the rating matrix after STEP 3 of Algorithm 5 was applied to Table 2, where users $\#1$ and $\#3$ are placed in cluster $\#1$, and users $\#2$ and $\#4$ are placed in cluster $\#2$. Then, Table 4 shows the rating matrix immediately before

STEP 4 of Algorithm 5 is applied to cluster #2 in Table 2, where the value $\min\limits_{\substack{1\le k\le 4 \\ 1\le \ell\le 5 \\ (k,\ell)\notin\{(2,4),(4,2)\}}} x_{k,\ell}$ is restored to "N/A," to be predicted. Table 5 shows the rating matrix immediately after STEP 4 of Algorithm 5 was applied to cluster #2 in Table 2, where the restored "N/A" is replaced with the predicted rating value. If these predicted values are higher than a given threshold value $\tilde{x}$, the corresponding items are recommended to the relevant users.

Table 1. Example of initial rating matrix: $N = 4$, $M = 5$, and $\{x_{k,\ell}\}_{(k,\ell)=(1,1)}^{(4,5)}$ are actual rating values from the users, and $x_{2,4}$ and $x_{4,2}$ need to be predicted

Item / User	#1	#2	#3	#4	#5
#1	$x_{1,1}$	$x_{1,2}$	$x_{1,3}$	$x_{1,4}$	$x_{1,5}$
#2	$x_{2,1}$	$x_{2,2}$	$x_{2,3}$	N/A	$x_{2,5}$
#3	$x_{3,1}$	$x_{3,2}$	$x_{3,3}$	$x_{3,4}$	$x_{3,5}$
#4	$x_{4,1}$	N/A	$x_{4,3}$	$x_{4,4}$	$x_{4,5}$

Table 2. Example of rating matrix to be clustered: $N = 4$ and $M = 5$. $x_{2,4}$ and $x_{4,2}$ are set as the minimal value of $\{x_{k,\ell}\}_{(k,\ell)=(1,1)}^{(4,5)}$

Item / User	#1	#2	#3	#4	#5
#1	$x_{1,1}$	$x_{1,2}$	$x_{1,3}$	$x_{1,4}$	$x_{1,5}$
#2	$x_{2,1}$	$x_{2,2}$	$x_{2,3}$	$\min\limits_{\substack{1\le k\le 4 \\ 1\le \ell\le 5 \\ (k,\ell)\notin\{(2,4),(4,2)\}}} x_{k,\ell}$	$x_{2,5}$
#3	$x_{3,1}$	$x_{3,2}$	$x_{3,3}$	$x_{3,4}$	$x_{3,5}$
#4	$x_{4,1}$	$\min\limits_{\substack{1\le k\le 4 \\ 1\le \ell\le 5 \\ (k,\ell)\notin\{(2,4),(4,2)\}}} x_{k,\ell}$	$x_{4,3}$	$x_{4,4}$	$x_{4,5}$

4   Numerical Experiments

This section describes the four example datasets used to evaluate the proposed algorithm, comprising one artificial dataset and one real datasets. We compare the CF accuracy of the following three algorithms: Algorithms 1, 3, and 5.

An artificial 100×100 rating matrix composed of 100 users and 100 items is shown in Table 6, which includes exactly 5 user clusters. In the dataset, users and items #1–#20 have identical ratings for all values. Correspondingly, users and items #21–#40, #61–#80, and #81–#100 also have the same ratings

Table 3. Example of rating matrix clustered using qFCCMM: $N = 4$, $M = 5$, and $C = 2$.

Cluster \ User \ Item		#1	#2	#3	#4	#5
#1	#1	$x_{1,1}$	$x_{1,2}$	$x_{1,3}$	$x_{1,4}$	$x_{1,5}$
	#3	$x_{3,1}$	$x_{3,2}$	$x_{3,3}$	$x_{3,4}$	$x_{3,5}$
#2	#2	$x_{2,1}$	$x_{2,2}$	$x_{2,3}$	$\min\limits_{\substack{1\leq k\leq 4\\ 1\leq \ell\leq 5\\ (k,\ell)\notin\{(2,4),(4,2)\}}} x_{k,\ell}$	$x_{2,5}$
	#4	$x_{4,1}$	$\min\limits_{\substack{1\leq k\leq 4\\ 1\leq \ell\leq 5\\ (k,\ell)\notin\{(2,4),(4,2)\}}} x_{k,\ell}$	$x_{4,3}$	$x_{4,4}$	$x_{4,5}$

Table 4. Example of the rating matrix to be applied to GroupLens: $N = 2$ and $M = 5$. $x_{2,4}$ and $x_{4,2}$ are predicted from the user ratings in cluster #2.

Cluster \ User \ Item		#1	#2	#3	#4	#5
#2	#2	$x_{2,1}$	$x_{2,2}$	$x_{2,3}$	N/A	$x_{2,5}$
	#4	$x_{4,1}$	N/A	$x_{4,3}$	$x_{4,4}$	$x_{4,5}$

Table 5. Example of the rating matrix applied to GroupLens: $N = 2$ and $M = 5$. $x_{2,4}$ and $x_{4,2}$ are replaced with the predicted values, $\hat{x}_{2,4}$ and $\hat{x}_{4,2}$, respectively. If these predicted values are higher than a given threshold value $\check{x}$, then the corresponding items are recommended to the corresponding users.

Cluster \ User \ Item		#1	#2	#3	#4	#5
#2	#2	$x_{2,1}$	$x_{2,2}$	$x_{2,3}$	$\hat{x}_{2,4}$	$x_{2,5}$
	#4	$x_{4,1}$	$\hat{x}_{4,2}$	$x_{4,3}$	$x_{4,4}$	$x_{4,5}$

among their groups. The ideal memberships of five user clusters are depicted in Fig. 1, in which each row shows the 100-dimensional membership vector $u_i = (u_{i,1}, \cdots, u_{i,100})^\mathsf{T}$ in gray scale (white and black are for $u_{\max}$ and 0, respectively). The aim of the experiment is to extract a similar structure from the dataset.

The experiment was conducted as follows. Algorithm 1 contained no parameter settings. In Algorithm 3, the cluster number was set as $C = 5$, which is the actual value. In addition, the fuzzification parameters were set as $m \in \{1.0001, 1.0004, 1.0007, 1.0010\}$ and $\lambda \in \{2^0/10, 2^1/10, \ldots, 2^5/10\}$ for Algorithm 3, and the initial membership was set based on actual information. In Algorithm 5, the cluster number and initial membership did not need to be

Fig. 1. Ideal membership of the artificial rating matrix

set, and the fuzzification parameters were set similar to those in Algorithm 3. Then, we applied three algorithms to this dataset with 7500 missing values; the rating values were selected randomly and were then considered missing in this dataset for purposes of the experiment. For each missing value, the outcome of the proposed method is probabilistic and depends on which entries were randomly deleted. In our experiment, five trials were performed using five different sets of incomplete data in order to produce more significant, reproducible results. The three algorithms specified above were applied to the abovementioned setting. In our study, we used the mean absolute error (MAE) metric to evaluate the prediction accuracy of the algorithms. MAE measures the average error between the predicted rating and true rating. Let $x_{k,\ell}^*$ be the true ratings, and $\hat{x}_{k,\ell}$ be the ratings predicted by CF. Further, let W be the number of user-item pairs for which CF suggested predictions. Then, MAE is defined as follows:

$$\text{MAE} = \frac{\sum_{k=1}^{N} \sum_{\ell=1}^{M} |\hat{x}_{k,\ell} - x_{k,\ell}^*|}{W}. \tag{15}$$

It is obvious that lower MAE values are preferred.

The fuzzification parameter values were selected as the lowest sum of the MAE values for all five missing patterns. The lowest sum of MAE values and the corresponding fuzzification parameter value for each cluster number are shown in Table 7. This table shows that the MAE value obtained from Algorithm 5 is the lowest (best) among those obtained from the three algorithms. This result indicates that the possibilistic clustering algorithm (Algorithm 4) is able to identify the group to which active users belong based on their preferences.

Table 6. Artificial dataset

User \ Item	1	⋯	20	21	⋯	40	41	⋯	60	61	⋯	80	81	⋯	100
1	1	⋯	1	2	⋯	2	3	⋯	3	4	⋯	4	5	⋯	5
⋮															
20	1	⋯	1	2	⋯	2	3	⋯	3	4	⋯	4	5	⋯	5
21	5	⋯	5	1	⋯	1	2	⋯	2	3	⋯	3	4	⋯	4
⋮															
40	5	⋯	5	1	⋯	1	2	⋯	2	3	⋯	3	4	⋯	4
41	4	⋯	4	5	⋯	5	1	⋯	1	2	⋯	2	3	⋯	3
⋮															
60	4	⋯	4	5	⋯	5	1	⋯	1	2	⋯	2	3	⋯	3
61	3	⋯	3	4	⋯	4	5	⋯	5	1	⋯	1	2	⋯	2
⋮															
80	3	⋯	3	4	⋯	4	5	⋯	5	1	⋯	1	2	⋯	2
81	2	⋯	2	3	⋯	3	4	⋯	4	5	⋯	5	1	⋯	1
⋮															
100	2	⋯	2	3	⋯	3	4	⋯	4	5	⋯	5	1	⋯	1

Table 7. Lowest sum of MAE for the artificial dataset and the corresponding fuzzification parameter value for Algorithms 1, 3, and 5

Algorithm	Lowest sum of MAE	Fuzzification parameter (λ, t)
1	0.628045	—
3	0.628045	$(1.2, 32)$
5	0.536231	$(2, 32)$

The remaining dataset is "MovieLens" [10] released by the GroupLens Research Project at the University of Minnesota. This dataset was compiled through the "MovieLens" web-site [10], and contains the reactions of users asked to evaluate movies they watched. In "MovieLens," 6,040 users recorded 1,000,000 ratings for 3,900 movies, where, the ratings are scaled from 1 to 5, with 5 being the best score. We used 277,546 ratings from 905 users for 684 movies in our experiments. It is because the original dataset is too sparse to define neighbors of active users for all the methods compared in this experiment. Therefore, each movie was evaluated by more than 240 people, and each user rated over 200 movies. Restricting subsets both users and movies makes clear the difference among results obtained from methods.

We applied the three algorithms, namely Algorithms 1, 3, and 5, to the real dataset, and compared the obtained prediction accuracy using the area underneath the receiver operating characteristic (ROC) curve (AUC), defined as follows. AUC is the area under the ROC curve [11,12], and the ROC curves are constructed as follows. First, all the recommendations were ranked as per the rating score. Recall and Fallout were calculated for each rating cut-off pair, as seen below.

$$\text{Recall} = \frac{\text{TP}}{\text{TP} + \text{FN}}, \tag{16}$$

$$\text{Fallout} = \frac{\text{FP}}{\text{FP} + \text{TN}} \tag{17}$$

where True Positive (TP) is the number of correctly recommended items from the selected items, False Positive (FP) is the number of incorrectly recommended items from the selected items, False Negative (FN) is the number of correctly recommended items from the non-selected items, and True Negative (TN) is the number of incorrectly recommended items from the non-selected items. In this experiment, the AUC was calculated using the discrete threshold values from 0.1 to the maximal rating value in increments of 0.1.

Algorithm 1 did not contain parameter settings. In Algorithm 3, the cluster numbers were set as $C \in \{2, 3, \ldots, 7\}$. In addition, the fuzzification parameters were set as $m \in \{1.0001, 1.0004, 1.0007, 1.001, 1.01, \ldots, 1.1\}$ and $\lambda \in \{10^1, \ldots, 10^5\}$ for Algorithm 3. In STEP 1 of Algorithm 2 (this algorithm was used as STEP 3 of Algorithm 3), the initial item typicality values were provided in a manner similar to that of the k-means++ method [13]. In particular, the first item membership was the normalizing value selected uniformly at random from the data points being clustered. Thereafter, each subsequent item membership was the normalizing value selected from the remaining data points with a probability inversely proportional to its pseudo similarity from the point's closest existing initial membership. For the 10 initial settings, the clustering result with the maximal objective function value was selected for STEP 3 in Algorithm 3. In Algorithm 5, the cluster number and initial membership did not need to be set, and the fuzzification parameters were set as $m \in \{1.01, \ldots, 1.1\}$ and $\lambda \in \{10^1, \ldots, 10^5\}$.

The experiment was performed as follows. First, 100,000 rating values in the "MovieLens" dataset, was randomly selected to be missing. Next, Algorithm 1, Algorithm 3, and Algorithm 5 were applied to the dataset for five settings of missing values. Finally, the average of the five AUC values were calculated for each dataset.

The highest AUC value for each method and the parameter value at which the highest AUC value was achieved are shown in Table 8, where the highest AUC value among the three methods is underlined. Table 8 shows that the AUC value obtained from Algorithm 5 is higher than those obtained from the other methods. Thus, we can say that the proposed algorithm outperforms other methods in terms of CF accuracy.

Table 8. The highest AUC value for each method and the corresponding parameter values for the "MovieLens" dataset

Method	AUC	Parameter value		
		m	λ	C
Algorithm 1	0.787015			
Algorithm 3	0.788122	1.01	100000	6
Algorithm 5	0.790486	1.3	100	

5 Conclusion

In this study, we proposed a CF algorithm for a recommendation system based on possibilistic clustering for spherical data. The results of experiments conducted on four datasets indicate that the CF method based on the possibilistic clustering algorithm for spherical data outperforms the conventional methods in terms of recommendation accuracy.

In future research, the proposed method will be compared with other methods further detailed setting, e.g., applied to a large number of real datasets, along with measuring trends of accuracy as increasing the number of missing values.

References

1. Paul, R., Neophytos, I., Mitesh, S., Peter, S., Jhon, R.: GroupLens: an open architecture for collaborative filtering of netnews. In: Proceedings of Computer Supported Cooperative Work of the ACM, pp. 175–186 (1994)
2. Sarwar, B., Karypis, G., Riedl, J.: Item-based collaborative filtering recommendation algorithms. In: Proceedings of the 10th International Conference on World Wide Web, pp. 285–295 (2001)
3. Herlocker, J.L., Konstan, J.A., Borchers, A., Riedl, J.: An algorithmic framework for performing collaborative filtering. In: Proceedings of the 22nd Annual International ACM SIGIR Conference on Research and Development in Information Retrieval, pp. 230–237 (1999)
4. Kondo, T., Kanzawa, Y.: Collaborative filtering using fuzzy clustering for categorical multivariate data based on q-divergence. JACIII **23**(3), 493–501 (2019)
5. Higashi, M., Kondo, T., Kanzawa, Y.: Fuzzy clustering method for spherical data based on q-divergence. JACIII **23**(3), 561–570 (2019)
6. Kondo, T., Kanzawa, Y.: Performance comparison of collaborative filtering using fuzzy clustering for spherical data. In: Proceedings of SCIS&ISIS 2018, pp. 644–647 (2018)
7. Krishnapuram, R., Keller, J.M.: A possibilistic approach to clustering. IEEE Trans. Fuzzy Syst. **1**, 98–110 (1993)
8. Menard, M., Courboulay, V., Dardignac, P.: Possibilistic and probabilistic fuzzy clustering: unification within the framework of the non-extensive thermostatistics. Pattern Recogn. **36**, 1325–1342 (2003)
9. Kanzawa, Y.: On possibilistic clustering methods based on Shannon/Tsallis-entropy for spherical data and categorical multivariate data. In: Torra, V., Narukawa, Y. (eds.) MDAI 2015. LNCS (LNAI), vol. 9321, pp. 115–128. Springer, Cham (2015). https://doi.org/10.1007/978-3-319-23240-9_10
10. GroupLens: MovieLens. http://grouplens.org/datasets/movielens/
11. Swets, J.A.: ROC analysis applied to the evaluation of medical imaging techniques. Proc. Investig. Radiol. **14**, 109–121 (1979)
12. Hanley, J.A., McNeil, B.J.: The meaning and use of the area under a receiver operating characteristic (ROC) curve. Proc. Radiol. **143**, 29–36 (1982)
13. Arthur, D., Vassilvitskii, S.: k-means++: the advantages of careful seeding. In: Proceedings of the 8th Annual ACM-SIAM Symposium on Discrete Algorithms, pp. 1027–1035 (2007)

Programmed Inefficiencies in DSS-Supported Human Decision Making

Federico Cabitza(✉) (iD), Andrea Campagner, Davide Ciucci (iD),
and Andrea Seveso

University of Milano–Bicocca, Viale Sarca 336, Milan, Italy
federico.cabitza@unimib.it

Abstract. Machine learning–based decision support systems (DSS) are attracting the interest of the medical community. Their usage, however, could have deep consequences in terms of biasing the doctor's interpretation of a case through *automation bias* and *deskilling*. In this work we address the design of DSS with the goal of minimizing these biases through the design and implementation of *programmed inefficiencies* (PIs), that is, features with the stated purpose of making the reliance of the human doctor on the DSS less obvious (or more difficult). We illustrate this concept by presenting a real-life medical DSS, called DataWise, embedding different PIs and currently undergoing iterative prototyping and testing with the medical users in two clinical settings. We describe the main features of DataWise, and show how different PIs have been conveyed by prompting doctors for multiple input and using qualitative visualizations instead of precise, but possibly misleading, indications. Finally, we discuss the implications of this design approach to naturalistic decision making, especially in life-saving domains like medicine is.

Keywords: Human decision making · Uncertainty · DSS

1 Motivation and Background

In the decision support systems (DSS) community, researchers are generally aware that decision making is a human activity that technology can at best support, but never replace. However, DSS are different from any other technology that could support decision makers do their job, e.g. different from systems that can allow decision makers to get access to the relevant information, whose availability can either suggest or confirm the right decision. DSS are *directly* involved in human decision making, for their role in presenting viable options (if not the "one best option"), among many potential alternatives, and possibly present these already ranked for importance, confidence or plausibility. In other words, their output can influence decisions more than any other technology used in decision making, and this is their explicit and primary purpose: to influence decisions so that the rate of decision error could be lower than without their

V. Torra et al. (Eds.): MDAI 2019, LNAI 11676, pp. 201–212, 2019.
https://doi.org/10.1007/978-3-030-26773-5_18

support, as observed in [2], for Deep Learning–based techniques, or in [17], for Swarm Intelligence–based techniques.

DSS based on machine learning and large amounts (big) of annotated data (that is, past accurate decisions) are exhibiting lower error rates than ever before and hence come with a strong promise to improve decision making in many fields where this is a delicate and life-saving process, like medicine [21]. These systems are often called *black-box algorithms*, as neither users, nor their developers, can easily get a full understanding of *how* such a DSS comes to suggest one best option against the others and, most notably, *why*. In medicine this can have deep consequences, both in the short term, as the DSS can bias and prime the doctor's interpretation of a case, leading to *automation bias* [19] (that is the use of DSS as replacements for vigilant information seeking and processing, leading to overreliance on the DSS' output); and in the long term, as it has been speculated that this kind of DSS can undermine the doctors' judgment capability and *deskill* them [3].

In our research, we address this twofold potential unintended consequence of embedding highly accurate, black-box DSSs into medical human decision making: we try to assess the impact of these systems on doctors' performance, and design real-life DSS aimed at minimizing automation bias and at preserving the decision autonomy of humans. We pursue this latter aim, which is the focus of this paper, by means of the concept of *programmed inefficiency* (PI).

A PI is a *DSS feature purposely developed to make DSS-assisted human decision making less efficient*, that is possibly longer, more difficult, less immediate than if performed without this feature[1]. A PI is aimed at making human reliance on the DSS advice less obvious, e.g., by purposely and slightly *undermining* the DSS accuracy and making the user aware of this, or just letting users be aware of the intrinsic and unavoidable uncertainty of any automatic classification and prediction. In what follows, we will illustrate how we embedded a number of PIs in a real-life DSS, called DataWise. This system is under iterative prototyping and testing to support more than 100 surgeons working in an Italian large teaching hospital, where they perform more than 5,000 joint replacement surgeries yearly. In particular, DataWise is currently used for two main tasks: first, the pre-operative prognostic evaluation of the odds different treatments proposed for a given patient will be effective. As such, DataWise is both a tool to support shared decision making in doctor-patient communication, and a tool to assess the appropriateness of different surgical alternatives, among which the one of not undertaking surgery at all. This latter one is an important purpose to pursue, for its role in enabling and positively informing policies of so called *value-based health care*, that is health care that transitions from a fee-for-service model (an orientation that can bring to opportunistic overuse), to reimbursement models that are directly informed by the observed outcome of any treatment option in large groups of patients, as this is assessed on the basis of the available records [22]. The second task regards follow-up management, that is DataWise

---

[1] As such a PI is something in between what Tenner calls *inspired inefficiency* [20] and Ohm and Frankle desirable [13].

as a tool for the post-operative phase, where its prognostic capabilities, and in particular its feature to estimate the risk of complication (at any follow-up step, starting from the discharge of the patient from the hospital), can help doctors adapt the intensity of care and of the monitoring tasks on-the-fly, so as to focus on the most problematic (to be) patients, save resources, and make health care a more sustainable activity.

2   Programmed Inefficiencies in DataWise

We designed some *programmed inefficiencies* in DataWise to test its accept-ability and utility in naturalistic decision making [10] in the ambit mentioned above: the main PI is that DataWise can abstain from producing a prediction. This occurs whenever its confidence on any potential target class is below a certain threshold [4]. Furthermore, we purposely avoided giving doctors quan-titative estimates of confidence, performance rates, probabilities and prediction scores, unless explicitly requested. Rather, we exploited Gestalt elements instead (i.e., shape- or color-related elements) to convey the same information but endow it with a more concrete idea of the uncertainty that affects any DSS prediction and statistical estimation. For the interpretation of the prediction, we largely employ intuitive visualizations that nudge the decision makers to stick to the available evidence (i.e., the data about the single case, or other similar cases) instead of thinking in purely abstract and model-oriented terms. In so doing, we aim to promote a naturalistic approach to the design of human decision making support, that is solutions emphasizing the process nature of decision making, instead of reducing it to the consultation of an oracle, however accurate, and encouraging the doctors to take fully into account the situated, contingent, and interpretative elements available to them [11]. Also in so doing, we address the requirement, raised by Raiffa in 1968, for "a methodology that brings informa-tion, however vague and imprecise, into the analysis, rather than a method that suppresses information in the name of scientific objectivity" [1]. In what follows, we will first outline the main use case of DataWise; then we will focus on specific subtasks (prediction visualization, prediction interpretation) and illustrate how we introduced gentle PIs to have humans engage with the system.

3   The DataWise Use Case

DataWise is the DSS module of a larger application, called DataReg: this is an electronic health registry that has been in use for 2 years to date and that allows to collect and manage both medical data (as those reported by doctors in surgery diaries, discharge letters, followup encounters forms) and Patient-Reported Outcome (PRO) data, which are data reported by the patients about their own health conditions (and how they perceive the outcome of specific treat-ments they had undergone) by means of standard and validated questionnaires at regular times, mainly 3, 6 and 12 months after a specific treatment (in our case, surgery). The data collected by DataReg are periodically used as training data for the development of the predictive models employed in DataWise. Basically, DataWise provides doctors with predictions about the prospective PRO that a specific patient will report at some future steps in the surgery followup.

The main use case of the human decision making process supported by DataWise unfolds along a 5-step sequential workflow, described in the following sections.

3.1   Patient and Model Selection

First, the doctor has to select a patient from the DataReg registry. DataWise then imports the necessary predictor data, and prompts for the input of any other data that is not yet available (e.g. the surgical procedure if the system is used at the pre-operative step). The doctor also has the option of asking a prediction for a patient not currently in the DataReg list; in this case DataWise will present the doctor a form where to impute the necessary predictor values.

Then, the doctor has to select a prediction model among the available ones (e.g. if DataWise is used at pre–operative time, it allows to predict either the 3– or 6– months PRO, that is, how the selected patient should feel after those time interval), at a specific "time step" (in the followup path). DataWise will also ask the doctor to express their preference (if any) among specificity (i.e. false positives avoidance) and sensitivity (i.e. false negatives avoidance), as well as to provide what they deem is the most likely outcome (in terms of the prospective target score) and tell how much confident they feel about their prediction.

As any PI, also these requests may seem to require an additional, useless, effort by the doctor. However, asking their for specific or sensitive predictions allows the doctor to provide a weight to make the advice potentially less accurate but more aligned with their values and attitudes, which can depend on contextual factors: e.g., the need to minimize overuse [18] or to adapt to the patient's preferences and attitudes towards interventional treatments.

On the other hand, the latter PI allows to compare the doctor's prognosis with DataWise's prediction, and assess the performance of both when the patient will eventually give a feedback after the treatment. This is done not to foster a competitive attitude of the doctors towards the AI support (although comparing the diagnostic/prognostic performance between machine and humans is a recurrent theme in the specialist literature, e.g. [8,9]); but rather to have doctors keep exercising the difficult art of interpretation and forecasting, and prevent them from over-relying on the newly available computational support: moreover, incentives can be associated to the highest accuracy rates and proposed for the doctors exhibiting them, beyond direct gratification and visibility.

3.2   Prognostic Support

In the Result page (see Fig. 1), DataWise displays four visualizations:

1. On the top-left side of this page, a bidimensional diagram depicts the score predicted by the regression model, where this is put in relation with the

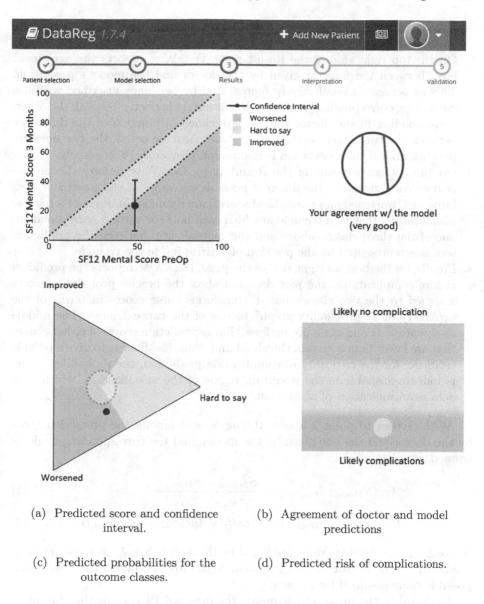

(a) Predicted score and confidence interval.

(b) Agreement of doctor and model predictions

(c) Predicted probabilities for the outcome classes.

(d) Predicted risk of complications.

Fig. 1. The results page.

present score. The score estimate is rendered with its 95% confidence interval and placed in a plane that is partitioned in three colored regions; these represent different prognostic categories, i.e., *improved, worsened,* and *hard to say.* The latter partition corresponds to a buffer zone whose width is estimated according to the *Minimal clinically important difference* (MCID) associated to the predicted score (if this is available in the specialist literature [15]), or to some safety margin suggested by the doctor in the analysis phase (usually around 10%).

2. On the top right side of the Result page, DataWise reports the agreement level between the forecast given by the doctor and the model's prediction. This agreement is qualitatively represented by two lines inscribed within a circle: one, corresponding to the model output, is perfectly vertical; the other, corresponding to the doctor's guess, is inclined in function of the difference between the two scores, so that, if the two scores are equal, the two lines are parallel, and, if their deviation is maximum, they cross orthogonally.

3. On the bottom left side of the Result page, DataWise reports the prediction scores assigned to the different possible outcomes of the treatment (i.e., improved, worsened and normalized scores) are rendered in terms of a *tripolar visualization*, that is, a triangle in which each pole (or vertex) corresponds to one of the three classes above and the mutual ratio between the prediction scores is represented by the position of a circle inside the triangle.

4. Finally, on the bottom right side of the page, DataWise reports the predicted risk of complications: the plot does not show the precise probability scores assigned to the two classes but it transforms these scores in terms of the proximity of a "probability glyph" to one of the two extremes. The middle area represents uncertain predictions, that is prediction scores of either classes that are lower than a certain threshold and, thus, insufficient to give a reliable estimate: for the criticality of complication prediction, the probability glyph is indistinguishable in the uncertain region of the visualization, which then acts as an indication of abstention.

With respect to point 3 above, the agreement among the prediction given by the doctor and the one given by the model, and the corresponding angle, is defined as:

$$Agr(S_{Model}, S_{Doctor}) = \frac{|S_{Model} - S_{Doctor}|}{|S_{Max} - S_{Min}|} \tag{1}$$

$$Deg(S_{Model}, S_{Doctor}) = arcsin(\sqrt{Agr(S_{Model}, S_{Doctor})}) \tag{2}$$

where S_{Model} is the score value predicted by the model, S_{Doctor} is the score value predicted by the doctor and S_{Max} (resp. S_{Min}) are the maximum (resp. min) possible value assumed for the score.

In regard to the prognostic support, the designed PI regards the abstention and the qualitative and number-less visualization of the predictions. In regard to this latter PI, we designed DataWise to purposely avoid to convey any result in terms of quantities (e.g., numbers, percentages). These, although of immediate comprehension, could induce a misplaced sense of accuracy and precision in regard to prediction that are nevertheless affected by some degree of uncertainty. Rather, the indications are given in topological terms (i.e., how close a circle is to some pole or extreme), position (i.e., in which region a circle is located) or shades of colors, and the doctor is free to get an informed idea of how reliable the provided indication is.

With respect to the abstention PI, it is noteworthy that, in the classification model described in point 3, the *hard to say* is not to be considered (in the training and predictions of the model) as a third class to predict, but rather as the neutral response of the model when it abstains from giving a binary prediction, as illustrated in more detail in [4]. The tripolar diagram then represents the output of a three–way–out classification task (and not a multi-class one). In order to implement this approach and compute three *probability scores*, the system performs transformations inspired by [5]: in short, given three decision costs $\lambda(Improved|Worsened)$ (the cost of saying that a patient improved when, in reality, she is worsened), $\lambda(Worsened|Improved)$ (the cost of saying that a patient worsened when, in reality, she is improved), and α (the cost of abstaining) and the basic probability scores given by the prediction algorithm $P(Improved), P(Worsened)$ we compute the respective *prediction risks*:

$$r(\text{Improved}) = \lambda(Improved|Worsened) * P(Worsened) \qquad (3)$$
$$r(\text{Worsened}) = \lambda(Worsened|Improved) * P(Improved) \qquad (4)$$
$$r(\text{Hard to Say}) = \alpha \qquad (5)$$

Thus, let $m = max\{r(Improved), r(Worsened), r(\text{Hard to Say})\}$, we then compute

$$s(\cdot) = 1 - \frac{r(\cdot)}{m} \qquad (6)$$

for each of *Improved*, *Worsened* and *Hard to Say* and normalize the three values in order to make them sum to 1. In doing so, we obtain the probability scores which are plotted on the tripolar diagram. The central zone of the triangle (the one inside the dashed circle) represents an area of *second–order uncertainty*, i.e. it represents the area in which the probability scores of the three alternatives are so close to each other that the predictive model is uncertain on its own prediction (which, in its turn, encompasses the uncertain outcome, when the model abstains from giving a binary outcome, as explained above).

3.3   Interpretation Support

In the following Interpretation page, the doctor can explore the model and the given predictions by leveraging explainable AI techniques. More specifically, DataWise allows doctors to:

1. See the local feature *relevance* (computed using Local Surrogate Models [16]), in order to understand the impact of each feature on the model prediction.
2. Conduct a *counterfactual* analysis [23] of the given indications: in so doing, the doctor can try to understand what prediction would the model give if any patient values *were different* and, the other way round, see how small changes in the predictive features could alter the model's prediction;
3. Perform an *analogical analysis*, on the basis of the most similar patients that the system has found with respect to the patient under examination;

(a) Counterfactual exploration (background color gradient of the value cells) and feature importances (background color of the heading cells).

(b) Similarity betweem the selected patient and its nearest neighbors, the dashed circle represents the diversity w.r.t. the reference population. The table shows the anagraphics and clinical data for the depicted patients

Fig. 2. The Interpretation page.

The Interpretation page interface is divided in three distinct parts (see Fig. 2): The topmost part of the page displays the patient values, with respect to the predictor features. The header of the patient record supports the interpretation of the feature relevance in a qualitative manner: the darker the background of a column name, the more important the corresponding feature for the given prediction. This gradient is displayed according to the feature relevance scores (obtained with methods from the LIME library [16]): each feature is associated with weights of a corresponding linear regression model, and these are converted into relevance scores by taking their absolute values and normalizing by the maximum value.

On the other hand, the background of the value fields below addresses the counterfactual analysis in the same qualitative manner: according to a gradient scale, the darker the color, the lower the absolute distance (on either ordinal or scalar fields) between the current value and a different value that would make the prediction change. In regard to nominal scales (like, e.g., gender), the darkest color indicates that any change in that field would have the effect to change the target value. We are currently working on a prototype that would allow the doctor to manually alter the values of the selected patient's features in order to observe, in real-time, the causal effect of these interventions in the indications given by the underlying ML model.

The middle and bottom parts of the screen show information about the most similar patients to the one currently under examination (with respect to a PCA analysis). This is done to allow the doctor to reason *analogically* and explore the features and records of those patients to look for similarities or cues that could be relevant for the patient at hand. This latter one is compared with all of the other patients whose record is stored in DataReg in order to find the 5 most similar patients. These patients are then represented as data points, whose color represents the degree of similarity with respect to the patient at hand, and whose position on a 2D scatter plot represents their values in regard to the 2 most relevant features (computed considering the global feature importance mentioned above). The degree of diversity of the patient with respect to the training set is represented in terms of the offset between the center of a dashed circle and the patient circle (the higher the offset, the less representative the training set of the patient under consideration). Below this twofold visualization, the system reports a summary of the records of these patients (basic identity information and the values of the predictor features), with a link to the corresponding patients records.

With respect to point 1, in order to compute the color gradient indicating how large a change would be required to make the selected patient change prediction outcome, the following formula is used:

$$G(a) = B(a) = \begin{cases} \frac{|v_a^p - v_a^s|}{|v_a^{max} - v_a^{min}|} & a \text{ is continuous} \\ \frac{|r(v_a^p) - r(v_a^s)|}{|V_a| - 1} & a \text{ is ordinal} \\ \mathbb{1}_{v_a^p \neq v_a^s} & a \text{ is categorical} \end{cases} \tag{7}$$

where a is a specific predictor feature, $R(a) = 1$ (resp. $B(a), G(a)$) is the red (resp. blue, green) color channel value for feature a, v_a^p is the value for feature a for the selected patient p, v_a^s is the value for feature a for the patient s most similar to p in the other outcome class, $r(v_a^p)$ (resp. $r(v_a^s)$) is the rank of value v_a^p (resp. $r(v_a^s)$), V_a is the number of values feature a can assume and $\mathbb{1}$ is the indicator function for a given predicate.

With respect to point 2, in order to compute the patients most similar to the selected one, first the categorical variables are transformed into dummy values using one–hot encoding then the Euclidean distance is computed (normalizing the distances of each feature, in order to ignore distortions due to magnitude effects), from which the 5 patients with minimal distance value are selected.

In order to compute the offset of the patient circle with respect to the center of the dashed circle, first the centroid c of the training set D is computed, and the distance between the patient's record and the centroid is computed as $dist(p, c) = \sqrt{\sum_{a \in A} (\frac{v_a^p - v_a^c}{v_a^{max} - v_a^{min}})^2}$.

Denoting as $dist(D) = \{dist(p', c) | p' \in D\}$, we compute the offset as[2]

$$\frac{dist(p, c)}{max\{dist(p', c) | \text{p' is not an outlier}\}}$$

After the interpretation step, doctors can (and actually are invited to) validate the process and evaluate the extent the system did actually support their decision making task. This is done through a short questionnaire regarding both the *usability* and the *usefulness* of the system. In particular, usefulness is evaluated in terms of the pragmatic value of the tool to have doctors reflect on the case at hand, and induce any change in their initial opinion. This information is useful for the improvement of the user experience and for future retraining of the model.

4   Conclusion

In this paper we have presented a set of features of a real-world machine learning-based DSS that we are testing in the health care domain, in particular in the prognostic and therapeutic management of musculoskeletal disorders. These features are illustrated both from the point of view of the interaction design, and of the underlying computational methods we developed to deploy them. In addition, we have provided a more general design framework, whose main idea is to endow an accurate DSS with purposely *programmed inefficiencies*.

Other different explanatory tools have been developed, the most recent one being Google's What-If Tool [7]. This tool shares some features with DataWise: visualizations of the most similar patients and conterfactual analysis. What-If also provides the possibility of editing a data-point to check if prediction changes and allows to compare multiple models. DataWise, on the other hand, provides qualitative-based information nudging the doctor towards a deeper analysis of the evidence at hand (see the concept of PI), for example making more explicit the model's uncertainty (through the use of abstention).

The features presented in this paper are intended to be a preliminary proof of the concept and a driver to enable further analyses of the doctors' performance and satisfaction. In particular, the PIs that we have developed in our medical DSS, called DataWise, and that we have described in this paper, are currently under evaluation both in terms of user acceptance and perceived informativeness

---

[2] The value offset(p) could be greater than 1 (thus, resulting in the patient's circle being outside of the dashed circle); this means that the training set is unrepresentative of the patient under consideration.

and utility. As stated above, programmed inefficiencies are a kind of purposely designed "computational friction" [6], that is any feature, contrivance and device that calls for specific actions to turn data, in this case the DSS predictions, into useful information. In this mould, we intend PI as way to make DSS-assisted (and in particular AI-assisted) human decision making a more "viscous" process, and therefore as a way to counteract known phenomena emerging from human decision making when highly accurate DSSs offer their predictions and interpretation in a more straightforward way: *automation bias* and *decision deskilling*. Designs with Programmed Inefficiencies (PI) require decision makers to consider *more* information than those that do not encompass them, and therefore to devote more time in making sense of the data provided (hence the concept of inefficiency). However, PI designs do not undermine efficiency by means of a low-usability tool or by information overload. On the contrary, they require the doctors to build a comprehension from less quantitative, numeric and easy-to-interpret (as well as to-misinterpret) indications, in favour of analog, visual (even pictorial) and possibly more uncertain and vague visualizations: doing so aims to avoid the quantitative fallacy [12,14], for which AI systems can look even more accurate than they actually are. Medical AI is recognized as a great opportunity for medical work to increase its effectiveness, efficiency and safety; however, it can also pose issues if deployed recklessly [3], like the disruption of clinical workflow. PI-endowed DSSs are proposed as a solution to "enable AI assistance without encouraging clinical decision making passivity" [24] by trading off a small amount of efficiency to minimize the odds of automation bias and deskilling, by requiring doctors to keep exercising their "decision muscle". In particular, PIs are conceived and developed to keep humans in the loop and, even more than this, "force" them to exert both perceptual and cognitive skills to make sense of the DSS output. Their introduction ground on the awareness that AI-based DSS will inevitably change the process of the traditional human decision making (hopefully making it more effective, more accurate and more safe [21]), but will also require humans to develop new skills to make sense of its output. Our future work will be devoted in tuning the number and kind of PIs to optimize performance (in terms of decision accuracy) and user satisfaction; in particular, we are planning to test the acceptance of two features: one by which doctors can interact with the prediction visualizations to change the sensitivity/specificity trade-off and see how predicted outcome change accordingly; the other one by which the system renders complication risk by means of pictures according to Gestalt principles (like closure and good form) instead of in terms of linear scales and length (like in the current version, see Fig. 1). Lastly, we aim to compare the performance of medical teams that are supported by our tool with the performance of similar teams that do not use it in real-world conditions, to assess its actual impact on human decision making.

References

1. Berg, M.: Rationalizing Medical Work: Decision-Support Techniques and Medical Practices. MIT Press, Cambridge (1997)

2. Bien, N., Rajpurkar, P., Ball, R.: Deep-learning-assisted diagnosis for knee magnetic resonance imaging: development and retrospective validation of MRNet. PLoS Med. **15**(11), e1002699 (2018)
3. Cabitza, F., Rasoini, R., Gensini, G.F.: Unintended consequences of machine learning in medicine. JAMA **318**(6), 517–518 (2017)
4. Campagner, A., Cabitza, F., Ciucci, D.: Exploring medical data classification with three-way decision tree. In: HEALTHINF 2019. SCITEPRESS (2019)
5. Campagner, A., Ciucci, D.: Three-way and semi-supervised decision tree learning based on orthopartitions. In: Medina, J., et al. (eds.) IPMU 2018. CCIS, vol. 854, pp. 748–759. Springer, Cham (2018). https://doi.org/10.1007/978-3-319-91476-3_61
6. Frischmann, B., Selinger, E.: Re-engineering Humanity. Cambridge University Press, New York (2018)
7. Google What-If Tool. https://pair-code.github.io/what-if-tool/index.html
8. Haenssle, H., et al.: Man against machine: diagnostic performance of a deep learning convolutional neural network for dermoscopic melanoma recognition in comparison to 58 dermatologists. Ann. Oncol. **29**(8), 1836–1842 (2018)
9. Han, S.S., Park, G.H., Lim, W., et al.: Deep neural networks show an equivalent and often superior performance to dermatologists in onychomycosis diagnosis. PloS One **13**(1), e0191493 (2018)
10. Klein, G.: Naturalistic decision making. Hum. Factors **50**(3), 456–460 (2008)
11. Mongtomery, K.: How Doctors Think: Clinical Judgment and the Practice of Medicine. Oxford University Press, New York (2005)
12. Muller, J.Z.: The Tyranny of Metrics. Princeton University Press, Princeton (2018)
13. Ohm, P., Frankle, J.: Desirable inefficiency. Florida Law Review (777) (2017)
14. O'Mahony, S.: Medicine and the mcnamara fallacy. JRCPE **47**(3), 281–287 (2017)
15. Revicki, D., et al.: Recommended methods for determining responsiveness and minimally important differences for patient-reported outcomes. J. Clin. Epidemiol. **61**(2), 102–109 (2008)
16. Ribeiro, M.T., Singh, S., Guestrin, C.: "Why should I trust you?": explaining the predictions of any classifier. CoRR abs/1602.04938 (2016)
17. Rosenberg, L., Lungren, M., Halabi, S., et al.: Artificial swarm intelligence employed to amplify diagnostic accuracy in radiology. In: IEMCON 2018, pp. 1186–1191 (2018)
18. Rumball-Smith, J., Shekelle, P.G., Bates, D.W.: Using the electronic health record to understand and minimize overuse. JAMA **317**(3), 257–258 (2017)
19. Skitka, L.J., Mosier, K.L., Burdick, M.: Does automation bias decision-making? IJHCS **51**(5), 991–1006 (1999)
20. Tenner, E.: The Efficiency Paradox: What Big Data Can't Do. Knopf, New York (2018)
21. Topol, E.J.: High-performance medicine: the convergence of human and artificial intelligence. Nat. Med. **25**(1), 44 (2019)
22. Vetter, T.R., Uhler, L.M., Bozic, K.J.: Value-based healthcare: a novel transitional care service strives to improve patient experience and outcomes. Clin. Orthop. Relat. Res. **475**(11), 2638–2642 (2017)
23. Wachter, S., Mittelstadt, B., Russell, C.: Counterfactual explanations without opening the black box: automated decisions and the GDPR. Harv. J. Law Technol. **31**(2), 2018 (2017)
24. Yu, K.H., Kohane, I.S.: Framing the challenges of artificial intelligence in medicine. BMJ Qual. Saf. **28**(3), 238–241 (2019)

Multilayer Identification: Combining N-Grams, TF-IDF and Monge-Elkan in Massive Real Time Processing

Ignacio González and Alfonso Mateos[(✉)]

Decision Analysis and Statistics Group, Departamento de Inteligencia Artificial,
Universidad Politécnica de Madrid, Madrid, Spain
igmigonzalezgarcia@gmail.com, alfonso.mateos@upm.es

Abstract. In modern societies control is based on information. Nowadays, in many countries, companies are obligated to provide to tax administrations all their invoices and withholders and financial entities to provide information that is used to offer prefilled tax declaration. In the case of Spain, the Tax Agency (AEAT) receives 180 million invoices by month and must process in a few days at the end of January more than 500 millions of registers to prefill Income Tax forms. Hundreds of thousands of these data are not correctly identified by the provider and must be returned to the sender or stored as not identified and analyzed afterwards. Traditionally this process consumed many technical and human resources. AEAT has been able to provide for first time a solution for identification in real time with enormous throughput that fulfil its needs. It is based in a combination of six algorithms, based in three different ideas, n-gram, TI-ILF, and Monge-Elkan that has surpassed any previous expectative.

Keywords: Identification · Monge-Elkan · Edit distance · Hybrid metrics · NLP

1 Introduction

Public administrations need, in many cases, to identify citizens knowing only some aspects of their activity, and, in others to attribute them wealth or income after crossing their data with those imputed by third parties as withholders or financial entities. The objective may be to detect discrepancies, which are used in risk analysis or to help the taxpayer, as it happens when a prefilled declaration [1] is offered.

Prepopulated forms have become a key element of the quality of the service provided by tax administrations and evaluated by organisms as International Monetary Fund, that includes this indicator in the guide of TADAT (Tax Administration Diagnostic Assessment Tool), in Performance Outcome Area 4 (Timely Filing of Tax Declarations) [2].

The quality of the process of identification is critical. IMF guide defines as best practice: "At time of filing, *automatically checks the taxpayer's identity against the registration database*, records the date of filing, performs arithmetic checks, records the tax liability, and stores declaration data" If this it is not done correctly, the estimated

© Springer Nature Switzerland AG 2019
V. Torra et al. (Eds.): MDAI 2019, LNAI 11676, pp. 213–223, 2019.
https://doi.org/10.1007/978-3-030-26773-5_19

risks are erroneous, and taxpayers are selected for inspection due to wrongly imputed information or the prefilled forms are delivered with errors. Bad identification can cause annoyance to millions of people and to thousands of merchants. The information that is wrongly rejected originates expenses and complaints and false positives move to undue inspections. The information stored as not - identified is almost useless.

Nowadays the problem has reached a new dimension. In many countries tax administrations require e-invoices[1] (SII system in Spain) and prefilled declarations are extended from Income Tax to VAT. Authorities access with crawlers and scrappers to Internet to download data necessary for the control of the collaborative economy. While a few years ago AEAT had to identify hundreds of millions of data by year, now has to identify billions of data by month.

Spain has the highest rate in the process of these forms, (returns/minute), followed by Sweden (7.1) and Denmark (4.7) [OECD, p. 13].

In January 2019, during the period established to receive from companies the data that must be used to prefill Income Tax declaration, AEAT (Tax Agency), has received 464,178,746 records from 12,479 declarants, corresponding to 15 forms (withholdings, yields of financial products, subsidies, etc.) containing the largest of them 35,543,350 records. The number of invoices received in the SII system, a simultaneous process, designed to enforce VAT declaration, has been 180 million by month.

In the past information was presented in magnetic support, using Value Added Networks and afterwards using Internet. The format of the files was validated during the reception and identification was made later. The result was communicated to the interested parties including the detail of errors, that should be corrected.

By Order HFP/231/2018 the possibility of providing information in magnetic support was eliminated and in 2019, for the first time, real-time identification was done in real time. The description of these forms, of the processes and a text portal is public[2].

It has been necessary to create an innovative platform to process data in memory, reaching a speed of 140,000 fully validated and processed records per minute. The type of process has let to increase the quality of the process of identification dramatically.

We distinguish three moments in the process of identification:

Previous. The data provided by a taxpayer about itself are clearly associated with their NIF (Tax Identification Number). It is not the case for data attributed to them by third parties. It happens that the person obligated to inform, merchants or Internet platforms, neither must nor can they verify some data of their clients. Internet platforms request data to connect their users, but often they provide only a name or nickname and a contact telephone and there are no rules to structure these data.

During the process. Inconsistencies appear as a consequence of the existence of erroneous data. Sometimes the error is intentional, as it happens when the interested parties that know that their data will be delivered to the Treasury, hide part of their names, use abbreviations, misspell surnames or invent identification codes since the

---

[1] https://www.ciat.org/la-factura-electronica-en-america-latina/

[2] https://www.agenciatributaria.es/AEAT.internet/en_gb/informativas.shtml

algorithm is public. With these restrictions it must be decided in real time if the information received is reliable and attributable or not.

After the presentation. Inspectors need to consult the data base with incomplete data. It happens when it is necessary to locate taxpayers with common surnames and names, and their identification code is not known, but some additional data are known such as the city of residence, or age. Duplication must be minimized. Sometimes the information available, although associated with different keys, belongs to the same person. For instance, information related to an account opened by a non-resident who afterwards is resident in the country with a new identification code.

The algorithm for creating the control digit is public and it is possible to create a fictitious code using a string of numbers (e.g.: 11111111), to search for its control digit, obtaining H, and to use the combination 1111111H, Juan Español Español in some invoices. The system should detect falsehood. In this case, it is easy due to the lack of connection between the code and the name and surnames. In others is not so easy. There are several types of problems to be solved: (i) *invented codes* as NIF 11111111H that should be rejected; (ii) *correct NIF's with misspelled names*, Juan M. Gonzales Gonzales; (iii) *Shortened names-* Instead of Juan Manuel Gonzalez Martinez the sender of the invoice refers to Juan Gonzalez or Juan Gonzalez González.

The purpose of this communication is to show the combination of algorithms that have led the quality of the identification to a new frontier, its components and its results.

2 Multilayer Identification

The new identification process combines six algorithms based on three different philosophies: (i) a layer of modified Monge-Elkan algorithms; (ii) a layer of traditional cleaning and comparison algorithms that take advantage of the features of the installation, such as identifying anagrams and one other based on the concept of n-gram, (iii) a layer based in an abstract *engine of recovery* using TF-ILF. The structure is:

Previous step: Cleaning of data, translation and pseudo-transcription.

Layer 1: MONGE-ELKAN_LEVENSHTEIN, MONGE-ELKAN_TRIGRAM and MONGE-ELKAN_BIGRAM.
Layer 2: TRIGRAM and COMPARA.
Layer 3: TF-IDF.

2.1 The Nucleus of the System

The AEAT developed in the last century, when the identification process was carried out in a mainframe, an algorithm (COMPARA), developed in PLI, that was executed in batch. It used as input either "NumberOfFiscalIdentification + Surname/Name" or "NumberOfFiscal_Identification + 3CharactersOfAnagram". This option allows to use information that is received when the taxpayers, besides names and surnames, send an anagram (code of four letters that is a combination of characters of names and surnames) provided by AEAT.

In a previous step data are processed with a character-based tool that eliminates points, commas, capitals, accents, whites, double blanks, restore standardized abbreviations, eliminate irrelevant words, such as "de", "el", "la" (articles), and abbreviations such as Exmo, Snas, Mr., Epouse, etc. After cleaning, abbreviated surnames are replaced, Basque and Catalan names are translated to their equivalents in Spanish and a phonetic pseudo-transcription is made to some Asian names. COMPARA considers the strings as character sequences and detects misspellings and typographical errors. This was very useful with manual data entry. This algorithm is maintained, although the situation has evolved, because it is useful although must be complemented.

Using a new technology and programming in Java, three layers containing new algorithms have been added. The first one includes uses of TRIGRAM, a Java implementation of 3-gram. An n-gram is a sequence of words and a bigram is a 2-gram [3]. The underlying idea is that when two chains are very similar, they have many n-grams in common. Their implementation is now standard[3] [4]. Gomma and Fahmy [5] explained the three main categories of *text similarity approach*, but did not discuss about algorithms performance. Gelbuckh [6] edited a text with bibliography about the many *similarity metrics* [7]. It is known that the best string distance to use depends on the situation and that character-based measures are useful for recognizing typographical errors, but not recognizing rearranged terms, and [8] that in Name-Matching tasks the combination of TF-IDF is useful because TF-IDF performed best among several string-edit based metrics. The second one TRIGRAM and COMPARA.

2.2   Upgrading the System. TF-IDF

We begin with the description of a non-standard component of the solution that was initially used to solve a general problem of data mining and that has evolved. If we want to identify, without knowledge of his NIF, a taxpayer, who is male, with annual income close to 24,000 euros, that lives in the town of Valencia, that has children and that has a current account in BBVA, the search should use some specific values of variables Y_j (e.g.: Y_1 = Genre; Y_2 = Income; Y_3 = Children, etc.)

$$y_1 = male, y_2 \sim 24,000, y_3 = Valencia, y_4 = Yes, y_5 = BBVA$$

The system should return the best candidates providing their NIFs and data, and a score between 0 and 1 for the similarity. The problem is that data are dispersed in many datamarts with many columns (>20,000) in tables with hundreds of millions of registers. A *recovery engine*, as explained in Fig. 1 is needed.

Retrieval Theory uses *measures of similarity*. Some of them use an *edit distances*, that is the minimum number of single-character edit operations needed to transform one string to another [9]. Many types of edit distances and many elaborations of underlaying statistical models have been proposed; Hamming [10], Levenshtein [11],

---

[3] http://sourceforge.net/projects/secondstring/;   http://sourceforce.net/projects/simmetrics;   http://geo graphiclib.sourceforge.net.

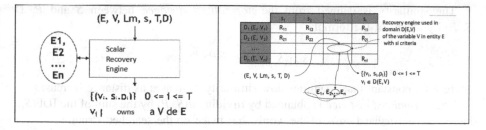

Fig. 1. Scheme of the recovery engine

Damerau-Levenshtein [12, 13], Needleman-Wunsch [14], Jaro [15, 16], Jaro-Winkler [17], N-gram [18]. One alternative is the use of *resemblance coefficients. Static token - based measures* that compare tokens. Monge and Elkan [19, 20] proposed one hybrid similarity measure that combines the benefits of sequence-based and token based methods.

Combining these ideas with TF_IDF [21, 22] a very abstract recovery system was developed. TF-IDF (Term frequency. Inverse document frequency) are numeric values used to measure how important is a word in a document. TF refers to how many times a given term appears in a document. IDF measures if the word is rare o common in the document. The intuition is that frequent terms are less important. Each word has both scores. For a term t in a document d, the weight $W_{t,d}$ of term t in document d is given by: $W_{t,d} = \text{TF}_{t,d} \log (N/\text{IDF}_t)$. Google has been using TF*IDF as a *ranking factor*.

In Fig. 1, the variables represented are: E: datamart where the search is defined; V: variable of Entity E on which the search is defined; Lm: value that is sought; s: similarity criterion; T, dimension of the maximum number of values that are expected as a result; D: dimension for maximum distance. The result of the query is: $\{(v_i, s_i, p_i)\}$, $0 \leq i \leq T$. The set is formed by the T values; $v_i \in D(E, V)$ with a greater similarity with the pattern Lm, given s_i; in this case the *popularity* of the value, p_i is the number of rows with this value.

The general approach is: Let considerer n variables $Y_1, Y_2, \ldots Y_n$ and look forward the case in which their values are $(Y_1 = y_1, Y_2 = y_2, \ldots)$. We will find records of a certain database whose values in these magnitudes *are as close as possible* to the targeted n-tuple. In the case that we are describing, the database, SYSBASE, was accessed through a dictionary. When the n-vector space formed by the n (multidimensional) variables Y_j is built and the database is projected, some records will appear as a conglomerate because they are close to the target n-tuple. The underlying logic was to decide the best candidates using a logistic curve. It was implemented also the control of conjunctive variables, if one value is not given can not be given the other, using β factors that graded the AND nature of each variable.

Similarity, s, after redundancy correction is:

$$s = \left(\sum_{i=1}^{n} \alpha_i x_i \right) \left(\prod_{i=1}^{n} (1 - \beta_i (1 - x_i))^{1/\sum_{j=1}^{n} \gamma_{ij}} \right),$$

where $\sum_{i=1}^{n} \alpha_i = 1; \beta_i \in [0, 1]$.

The similarity is defined from the *normalized distance* between S and P, D (S, P) by means of the formula:

$$sim(S, P) = e^{-K_2 D(S,P)},$$

where K_2 a constant that quantifies how similarity decreases as distance increases.

The *normalized distance* is obtained by dividing $d(S, P)$ by the sum of the IDF(S_i), that is, the normalized costs of the words that make up the searched string.

$$idf(S_i) = log(N/n_i),$$

where N is the total number of occurrences of words and n_i the number of occurrences of the word S_i in the collection. Normalization consists in dividing the value of $idf(S_i)$ by the maximum value of $idf(S_i)$ in the collection.

In the implemented model, the operations that can be performed are: (i) Delete (S_i), consume the term S_i with a cost equal to $idf(S_i)$; (ii) Substitute (S_i, P_j), consume the terms S_i and P_j when their similarity measure is greater than the threshold U. The cost of the operation is:

$$\left(Idf(S_i) + Idf(P_j)\right)\left(1 - K(S_i, P_j)sim(S_i, P_j)\right),$$

where $K(S_i, P_j)$ is 1 if no change is necessary.

A new version in Java of TF_IDF was included in layer 3. The combination of these algorithms in three layers was used until 2018 using thresholds (COMPARA = 0.8; TRIGAM = 0.8; and TF-IDF = 0.8). But the solution did not let to process data in real time with more than a 99% of identified registers that was our objective.

3   A Variation of Monge-Elkan Algorithm

It was necessary a deeper understanding of the problem. While in other cases as the detection of plagiarism, complete sentences are searched in long texts, in this case subtle differences in short texts must be evaluated. We knew [8] that Jaro and Jaro-Wrinkler are primarily intended for short strings (e.g.: personal first and last names), but the real problem was different. It has three dimensions: (a) Similarity between *names or surnames* is high and the risk of false positives increases. For example, between two sisters *Ana* González Martínez and *Eva* González Martínez the difference is very small and these are very common surnames (Martínez, González, etc.); (b) In massive information provided by merchant data are not false but incomplete. Each person has, in the general case, two names and two surnames and in many cases uses *only two or three of the four,* that is only two or three of the four components arrive in the input because there those known by the merchants. Are included two names and a surname, or not? What is the right order? The information of the NIF is correct? If the merchant only knows that his client is *Juan Martinez,* instead of Juan *Manuel* Martinez or *Yan Hue* and provides this information with the correct code (NIF) it is an unnecessary inconvenience to reject the information.

Given to texts A, B with $|A|$ and $|B|$ being the respective number of tokens (e.g.: name), and an external intertoken similarity measure, the Monge-Elkan measure is computed as follows:

$$sim_{MongeElkan}(A, B) = \frac{1}{|A|} \sum_{i=1}^{|A|} max_{j=1}^{|B|} \{ sim'(a_i, b_j) \}.$$

Informally, this measure is the average of the similarity values between the more similar token pairs in both strings and it is a good approach to the problem of SII declarants.

It was considered, after the study of previous investigation in the field [25] the possibility of combining Monge-Elkan (1966) with the previous solution but introducing some changes. The solution built was: A matrix W_{ij} is created with the comparison of all the words of the name and of the pattern, those that exist. Next, for each word of the name, the word with maximum similarity is searched, starting with the maximum. Once the maximum tuple is located, the algorithm eliminates (resets the level of similarity) in this said word of the pattern (if we have already used that word from the pattern we eliminated so that it is not considered in the following words in which to look for similarity), since nobody has the name Juan Juan, and continues with the rest of the words.

4 Fine Tuning and Results

4.1 Final Solution

The solution, combination of three layers, was initially evaluated with a file with 68,999 records built by experts with real data, selecting all possible cases, very common surnames, names of men and women, from different areas of the country, etc., omitting one of the names of pile etc.

Each layer delivered the lowest percentage of errors with a different threshold. Although all the similarity functions give values between 0 and 1, unless they are normalized, the singular points (minimum values) are in different zones of the values of the thresholds. With these data a first solution was designed.

In Fig. 2 are presented the false positives and negatives for each type of metrics obtained in the analysis. In x-axis is presented the relative distances to the point where both curves (false positives and false negatives) intersect. The curve ME_LEV_FP (Monge-Elkan_Levensthien_FalsePositives) begins with the point (1) with false positives = 6,098, false negatives = 10 at threshold (in its own scale) 0.40 The intersection occurs in a point with threshold 0.5575. The graph is rescaled so this first point has an abscise of 0.1575. This process is repeated with all metrics. Clearly using Lovenshtein distance inside Monge-Elkan metrics smoother curves are obtained.

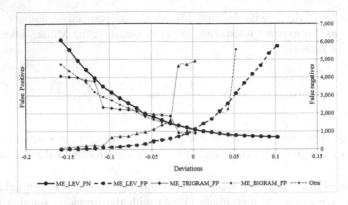

Fig. 2. Errors by type, threshold and metrics

4.2 Fine Tuning

Information received of electronic invoices (180 M registers by month) has been used fine tune the thresholds. The platform is based in x86 servers. There are 225 blades with 900 cores of 2,3 GHz and 2163 of 2 GHz with 45 Tb of memory. The description of the technology of the TGVI (Transmission of Big Volumes of Information) is public [25]. The resources dedicated to the reception of the information and to identification 14 blades with 24 cores (4,376 Tb). The size of the census loaded in memory was 0.4 Tb. From 1/2/2019 to 7/03/2019 information was processed with initial thresholds. It was tested against the file of control and samples of the data received that the rate of false positives was less than 1%.

The idea used in fine-tuning is explained in Fig. 3 with a diagram based on Signal Detection Theory [24] and the rationale used in a ROC (Receiver Operating Characteristic) curve construction. We consider two metrics α, β. A register with a name has $sim_\alpha = 0.7$ and $sim_\beta = 0.8$. Both values are greater than thresholds T_i and this register would be accepted and imputed but with a risk of "false positive" CBD. Conceptually is similar to the erroneous decision adopted when accepting noise as signal. If threshold α is increased this register it is not identified in the first layer, is considered as noise, and it is processed in the second one, probably more appropriate (other type of distance) to solve the problem placed by the string. If the order and thresholds are optimized the best possible solution is adopted.

The decision about the order of the layers was adopted during first evaluation. The aim in this phase of the investigation was to optimize the thresholds. After a few days of testing, from 08/03/ to 13/03/2019, the threshold of COMPARA was adjusted to 0.7.

From 14/03/2019 to 18/04/2019 the thresholds of the first layer was reduced letting more registers pass. This was a clear expression of the quality of the decision realized in the front end using the combination of three hybrids algorithms, included Monge-Elkan. In Fig. 4 we see that percentage of unidentified continuously diminish from more than 3.5% to less than 0.8% when the total structure of layers is tuned with throughput of more than 4 millions of registers processed in real time by day.

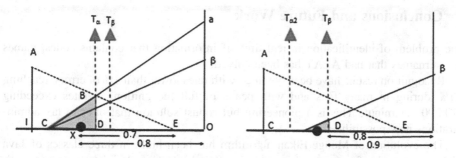

Fig. 3. Tuning of the thresholds

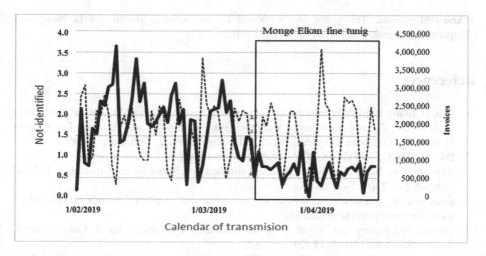

Fig. 4. Process of fine tuning

Nowadays, layer 1 implements Monge-Elkan_Levenshtein (75%), Monge-Elkan_Trigram (66%) and Monge-Elkan-Bigram (75%). Second layer Trigram (50%) and COMPARA (75%); third layer TF-IDF (70%). Call to algorithms is made sequentially and in this specific order. In each case if threshold is surpassed identification is accepted. If not, at the end registers considered unidentified. During the evolution some peaks with more than 99.9% of records identified has been reached. The quality of the service provided to taxpayers has increased in dramatic form. During the monthly process of 180 millions of invoices the number of undue errors communicated to taxpayers has been reduced in 5,370,000. Companies have avoided the administrative burden needed to solve this impressive amount of incidents only in one month.

5   Conclusions and Future Work

The problem of identification in real time of information that contains codes, names and surnames that had AEAT has been solved.

Recognition results have been obtained with rates lower than 1% of errors, reaching 0.1% during in many files and with peaks of full recognition, at rates exceeding 140,000 per minute This is a pioneering but robust solution that offer to tax administrations new possibilities.

The evolution of Monge-Elkan algorithm has been built, in three classes of Java without external references and its code is offered to other administrations and institutions in the webpage of the AEAT [26].

Acknowledgements. This paper was supported by the Spanish Ministry of Economy and Competitiveness project MTM2017-86875-C3-3-R.

References

1. OECD: Using Third Party Information Reports to Assist Taxpayers Meet their Return Filing Obligations—Country Experiences with the Use of Pre-populated Personal Tax Returns Prepared by Forum on Tax Administration Taxpayer Services Sub-group (2006)
2. IMF: TADAT. Field Guide (2016). http://www.tadat.org
3. Akinwale, A., Niewiadomski, A.: Efficient similarity measures for texts matching. J. Appl. Comput. Sci. **23**(1), 7–28 (2015)
4. Chapman, S.: SimMetrics: a java & c# .net library of similarity metrics (2006). http://sourceforge.net/projects/simmetrics/
5. Gomaa, W., Fahmy, A.: A survey of text similarity approaches. Int. J. Comput. Appl. (0975 – 8887) **68**(13), 13–18 (2013)
6. Gelbukh, A. (ed.): CICLing 2009. LNCS, vol. 5449, pp. 17–23. Springer, Heidelberg (2009). https://doi.org/10.1007/978-3-642-00382-0
7. Naumann, F.: Similarity measures. Hasso Plattner Institut Universität Postdam (2013). https://hpi.de
8. William, W.C., Ravikumar, P., Fienberg, S.E.: Comparison for string distance metrics for name-matching Tasks. American Association for Artificial Intelligence (2003). https://cs.cmu.edu
9. Cormode, G., Mutukhrishnan, S.: The string edit distance matching problem with move. ACM Trans. Algorithms (TALG) **3**(1), 2 (2006)
10. Hamming, R.W.: Error detecting and error correcting codes. Bell Syst. Tech. J. **29**(2), 147–160 (1950)
11. Levenshtein, V.I.: Binary codes capable of correcting spurious insertions and deletions of ones. Probl. Inf. Transm. **1**(1), 8–17 (1965)
12. Levenshtein, V.I.: Binary codes capable of correcting deletions, insertions and reversals. Dokl. Akad. Nauk SSSR **163**(4), 845–848 (1965)
13. Damerau, F.J.: A technique for computer detection and correction of spelling errors. Commun. ACM **7**(3), 171–176 (1964)
14. Needleman, S.B., Wunsch, C.D.: A general method applicable to the search for similarities in the amino acid sequence of two proteins. J. Mol. Biol. **48**(3), 443–453 (1970)

15. Jaro, M.A.: Advances in record-linkage methodology as applied to matching the 1985 census of Tampa, Florida. J. Am. Stat. Assoc. **84**(406), 414–420 (1989)
16. Jaro, M.: Probabililistic linkeage of large public health data files. Stat. Med. **14**, 491–498 (1995)
17. Winkler, W.E.: String comparator metrics and enhanced decision rules in the Fellegi-Sunter model of record linkage, pp. 354–359 (1990)
18. Kondrak, G.: N-gram similarity and distance. In: Consens, M., Navarro, G. (eds.) SPIRE 2005. LNCS, vol. 3772, pp. 115–126. Springer, Heidelberg (2005). https://doi.org/10.1007/11575832_13
19. Monge, A., Elkan, C.: The field matching problem: algorithms and applications. In: Proceedings of the Second International Conference on Knowledge Discovery and Data Mining (1996)
20. Monge, A., Elkan, C.: An efficient domain-independent algorithm for detecting approximately duplicate data records. In: The Proceedings of the SIGMOD 1997 Workshop on Data Mining and Knowledge Discovery (1997)
21. Bafna, P., Pramod, D., Vaidya, A.: Document clustering: TF-IDF approach. In: International Conference on Electrical, Electronics, and Optimization Techniques (ICEEOT), Chennai, pp. 61–66 (2016)
22. Prasetya, D., Wibawa, A., Hirashima, T.: The performance of text similarity algorithms. Int. J. Adv. Intell. Inform. **4**(1), 63–69 (2018)
23. Jimenez, S., Becerra, C., Gelbukh, A., Gonzalez, F.: Generalized Mongue-Elkan method for approximate text string comparison. In: Gelbukh, A. (ed.) CICLing 2009. LNCS, vol. 5449, pp. 559–570. Springer, Heidelberg (2009). https://doi.org/10.1007/978-3-642-00382-0_45
24. Spackman, K.: Signal detection theory: valuable tools for evaluating inductive learning. In: Proceedings of the Sixth International Workshop on Machine Learning, pp. 160–163 (1989)
25. AEAT.   https://www.agenciatributaria.es/AEAT.internet/Inicio/La_Agencia_Tributaria/Campanas/_Campanas_/Declaraciones_informativas/_SERVICIOS_DE_AYUDA/Identificacion_fiscal/Identificacion_fiscal.shtml
26. AEAT: Implementation of a Monge-Elkan based system of identification (2019). https://www.agenciatributaria.es/AEAT.desarrolladores/Desarrolladores/Desarrolladores.html

An Evidential Clustering
for Collaborative Filtering Based
on Users' Preferences

Raoua Abdelkhalek[✉], Imen Boukhris[✉], and Zied Elouedi[✉]

LARODEC, Institut Supérieur de Gestion de Tunis,
Université de Tunis, Tunis, Tunisia
abdelkhalek_raoua@live.fr, imen.boukhris@hotmail.com, zied.elouedi@gmx.fr

Abstract. Users are often surrounded by a large variety of items. For this purpose, Recommender Systems (RSs) have emerged aiming to help and to guide users towards items of interest. Collaborative Filtering (CF) is among the most popular recommendation approaches, which seeks to pick out the most similar users to the active one in order to provide recommendations. In CF, clustering techniques can be used for grouping the most similar users into some clusters. Nonetheless, the impact of uncertainty involved throughout the clusters' assignments as well as the final predictions should also be considered. Therefore, in this paper, we propose a clustering approach for user-based CF based on the belief function theory. This theory, also referred to as evidence theory, is known for its strength and flexibility when dealing with uncertainty. In our approach, an evidential clustering process is performed to cluster users based on their preferences and predictions are then generated accordingly.

Keywords: Recommender Systems ·
User-based collaborative filtering · Uncertain reasoning ·
Belief function theory · Evidential clustering

1 Introduction

Recommender Systems (RSs) [1] have sprung up as a convenient solution to deal with the information overload problem. They are considered as helpful tools to guide users in their decision making process in a very personalized way. More specifically, they filter data, predict users' preferences and provide them with the appropriate predictions. Collaborative Filtering (CF) [2] is a very popular approach in RSs field, and has received a great deal of attention. The recommendation process in a CF system consists on predicting the users' preferences based on the users or the items having similar ratings. In this kind of RS, a typical matrix of user-item ratings is exploited to compute similarities between users (user-based) or items (item-based). Predictions are then performed based on the computed similarities. Although CF strategies are very simple and intuitive, they reveal some limitations such as the scalability problem [3]. Indeed, the

© Springer Nature Switzerland AG 2019
V. Torra et al. (Eds.): MDAI 2019, LNAI 11676, pp. 224–235, 2019.
https://doi.org/10.1007/978-3-030-26773-5_20

whole user-item matrix needs to be browsed in order to compute similarities and generate predictions accordingly. In this situation, a lot of heavy computations is required, which leads to a poor scalability performance. Therefore, clustering techniques can be adopted to partition the set of items or users based on the historical rating data. Hence, predictions can be generated independently within each partition. Ideally, clustering would improve the quality of the provided predictions and increase the scalability of CF systems [4]. In this paper, we are particularly interested in the user-based CF category. Actually, when performing the clustering process of users in the system, they are likely to bear upon more than only one cluster, which is referred to as soft clustering. This imperfection may affect the correlation between the users and consequently the quality of the produced predictions. Therefore, the uncertainty spreading around the clusters' assignment should be considered. To tackle this issue, we opt for the belief function theory (BFT), also referred to as evidence theory [5–7]. This latter is known for its strength and flexibility when dealing with uncertainty. It allows a rich representation of knowledge uncertainty in different levels ranging from the complete ignorance to the total certainty. For these reasons, we propose, in this work an evidential clustering approach for user-based CF where a clustering process is performed under the BFT to group the users based on their preferences and generate predictions accordingly. To do so, we involve the Evidential C-Means (ECM) [8] which allows us to handle uncertainty for objects' assignment. Indeed, a preliminary work [9] has been addressed in this context based only on the item-based CF category. Contrariwise, the proposed approach covers an evidential clustering for user-based CF where users are assigned to soft clusters rather than items. Moreover, the uncertainty of the final predictions is also quantified and represented to the active user based on the Evidential K-Nearest Neighbors [10] formalism. We assume that such representation may increase the intelligibility and the transparency of the provided predictions.

This paper is organized as follows: Sect. 2 recalls the basic concepts of the belief function theory and introduces both the Evidential C-Means and the Evidential K-Nearest Neighbors techniques. Section 3 presents briefly some related works on clustering CF as well as CF under the belief function framework. Our proposed recommendation approach is presented in Sect. 4. Section 5 exposes the experimental results conducted on a real world data set. Finally, the paper is concluded in Sect. 6.

2   Background Related to Evidence Theory

In this section, we recall the basic concepts and notations related to the Evidence theory or belief function theory [5–7]. Besides, two machine learning techniques in an evidential framework will be presented namely ECM [8] and EKNN [10].

2.1   Basic Concepts and Notations

Belief function theory, also called Evidence theory or Dempster-Shafer theory [5,6], is a rich and flexible framework for modeling and quantifying imperfect knowledge. Let Ω be the frame of discernment defined as a finite set of variables w. Such set refers to n elementary events to a given problem such that: $\Omega = \{w_1, w_2, \cdots, w_n\}$. The key point of the belief function theory is the basic belief assignment (bba) which represents the belief committed to each element of Ω such that $m : 2^\Omega \rightarrow [0,1]$ and $\sum_{E \subseteq \Omega} m(E) = 1$. The mass $m(E)$ quantifies the degree of belief exactly assigned to an element E of Ω. The plausibility function denoted pl, quantifies the maximum amount of belief that could be given to a subset E of Ω. It is defined as: $pl(E) = \sum_{E \cap F \neq \varnothing} m(F)$.

Evidence may be ambiguous or incomplete. Consequently, it may not be equally trustworthy. That is why, a discounting operation can be applied to get the discounted bba denoted by m^δ such that:

$$m^\delta(E) = (1 - \delta) \cdot m(E), \forall E \subset \Omega \quad and \quad m^\delta(\Omega) = \delta + (1 - \delta) \cdot m(\Omega) \qquad (1)$$

where $\delta \in [0,1]$ is the discounting factor.

Two bba's m_1 and m_2 derived from two reliable and independent information sources can be fused using Dempster's rule of combination defined as:

$$(m_1 \oplus m_2)(E) = k. \sum_{F,G \subseteq \Theta : F \cap G = E} m_1(F) \cdot m_2(G) \qquad (2)$$

$$where \quad (m_1 \oplus m_2)(\varnothing) = 0 \quad and \quad k^{-1} = 1 - \sum_{F,G \subseteq \Omega : F \cap G = \varnothing} m_1(F) \cdot m_2(G)$$

Ultimately, in order to making decisions, several solutions have been proposed in the frame of belief function theory such as *the maximum of plausibility pl* and the *pignistic probability* denoted $BetP$. $BetP$ is defined as follows:

$$BetP(E) = \sum_{F \subseteq \Omega} \frac{|E \cap F|}{|F|} \frac{m(F)}{(1 - m(\varnothing))} \ for \ all \ E \in \Omega \qquad (3)$$

Within the belief function theory, the decision can be made by choosing the hypothesis having the highest value of the plausibility function pl or the pignistic probability $BetP$.

2.2 Evidential C-Means (ECM)

The Evidential C-Means (ECM) [8] is an evidential clustering technique which generalizes both the hard k-means and Fuzzy C-Means (FCM) methods [11]. The core idea of this technique is to assign each object to the different subsets of clusters with degrees of belief. In fact, each cluster w_k is presented by its center v_k which is also referred to as prototype. However, unlike FCM, one object can belong not only to a singleton cluster but also to a partition of clusters ($E_j \subseteq \Omega$)

that can be called a *meta-cluster*. Correspondingly, the meta-cluster E_j is also represented by a prototype denoted $\overline{v_j}$ and defined as follows:

$$\overline{v_j} = \frac{1}{|E_j|} \sum_{k=1}^{c} s_{kj} v_k \tag{4}$$

where c is the number of clusters, $s_{kj} = 1$ if $w_k \in E_j$ and $s_{kj} = 0$ otherwise.

The credal partition of each object is obtained by minimizing the following objective function J_{ECM} for n objects and c clusters:

$$J_{ECM} = \sum_{i=1}^{n} \sum_{j/E_j \neq \emptyset, E_j \subseteq \Omega} |E_j|^\alpha m_{ij}^\beta d_{ij}^2 + \sum_{i=1}^{n} \delta^2 m_{i\emptyset}^\beta \tag{5}$$

where m_{ij} denotes $m_i(E_j)$ and d_{ij} is the euclidean distance between the ith object and the jth partition's prototype. The parameter α consists of controlling the degree of penalization for subsets with high cardinality. For β and δ, they present two parameters for treating noisy objects.

2.3   Evidential K-Nearest Neighbors

The Evidential K-Nearest Neighbors (EKNN) classifier [10] extends the classical K-Nearest Neighbors (KNN) methods by incorporating classifier output uncertainties. Let $X = \{(x^i = x_1^i, \cdots, x_P^i) \mid i = 1, \cdots, C\}$ be the set of training samples. Each object x^i is characterized by a class label $L^i \in \{1, \cdots, C\}$ indicating its membership to a specific class in $\Omega = \{w_1, \cdots, w_C\}$. Considering that y is a new pattern to be classified using the information of the training set. We denote Γ^K the set of the K-nearest neighbors of y in X. Each object $x_i \in \Gamma^K$ represents an independent piece of evidence regarding the class membership of y. Hence, each neighbor induces a basic belief assignment defined as follows:

$$m(\{w_q\}|x^i) = \alpha_0 \phi_q(d^i)$$
$$m(\Omega|x^i) = 1 - \alpha_0 \phi_q(d^i) \tag{6}$$
$$m(E|x^i) = 0, \ \forall E \in 2^\Omega \setminus \{\Omega, \{w_q\}\}$$

where w_q refers to the class label of x^i, α_0 is a parameter where $0 < \alpha_0 < 1$.

The constant α_0 has been heuristically fixed to a value of 0.95. d^i corresponds to the euclidean distance between the object to be classified and the other objects in the training set while ϕ_q is a decreasing function verifying: $\phi_q(d) = exp(-\gamma_q d^2)$. Note that γ_q has been defined as a positive parameter assigned to each class w_q. It is considered as the inverse of the mean distance between all the training patterns belonging to the class w_q. Since each neighbor represents a particular source of evidence, we obtain K bba's that can be aggregated into a unique one m_y using Dempster's rule of combination such as:

$$m_y = m(.|x^1) \oplus \cdots \oplus m(.|x^K) \tag{7}$$

where $\{1, \cdots, K\}$ is the set containing the indexes of the K-nearest neighbors.

3   Related Works on Collaborative Filtering

A panoply of clustering-based CF approaches that rely on a cluster model to provide predictions have been proposed in RSs. They consist on identifying users or items sharing similar ratings and group them in clusters. Thereafter, only users or items associated to the same partition are considered in the prediction. For instance, authors in [3] have proposed to partition the users of a CF system using a clustering algorithm and used the obtained partitions as neighborhoods. They first applied a clustering technique (i.e. k-means) on the user-item ratings matrix to obtain different clusters. In the next step, the neighborhood for the active user is selected based on the cluster where he belongs. Finally, predictions scores are computed based on aggregating ratings of active user's neighbors using traditional CF technique. Similarly, a CF approach has been introduced in [12] based on users' preferences clustering. They supposed that users in the ratings matrix can be grouped into three different user's clusters namely, optimistic user cluster, in which users prefer to use high ratings, pessimistic user cluster, in which users prefer to use low ratings and neutral user cluster, in which users tend to give reasonable ratings for items. In [13], authors have proposed a smoothing-based CF approach where clusters are generated from the training data and predictions are performed on this basis. The authors in [14] presented a different CF approach in which, instead of users, they group items into several clusters using the Pearson Correlation similarity measure and the kMetis graph partitioning algorithm. The author in [4] proposed a personalized CF approach that joins both the user clustering and item clustering strategies. In fact, users are grouped according to their past preferences towards items, and each group of users has a cluster center. The nearest neighbors of the active are selected based on the similarity between the active user and cluster centers and a smoothing of the given predictions is performed where necessary. In the final step, an item clustering process for CF is applied to produce the recommendations.

Clustering allows to alleviate the scalability problem and improve the recommendation performance. On the other hand, handling uncertainty that reigns throughout the whole prediction process is considered also as an important challenge in real-world RSs problems. Thus, the belief function theory [5–7] is regarded as a convenient and a rich framework for dealing with uncertainty. Recent studies have emphasized the favors of using such theory for uncertain reasoning in RSs area. Authors in [15–17] have proposed to model the user's preferences within the evidential framework and incorporate context information and social network to provide recommendations. A new method for combining information about users' preferences has been proposed in [18] based on the belief function theory to deal with highly conflicting mass functions. In [19,20], the authors have proposed an evidential extension of the standard item-based CF where the nearest items have been treated as the pieces of evidence conducting the resulted predictions. In the same context, they adopted in [9] an evidential clustering for item-based CF using the Evidential C-Means technique to cluster items based on their ratings. When it comes to user-based CF, which is the framework of this work, a preliminary work has been performed in [21] where the user-based CF has been represented under the belief function framework. Their recommendation process consisted in computing the similarities between the active user and the other users by exploiting the current user-item matrix. An aggregation of the nearest users' ratings has been performed to provide predictions. However, a lot of heavy computations are required in such situation. This problem is referred to as the scalability problem which we tackle in our proposed recommendation approach.

4   EC-UBCF: Evidential Clustering User-Based for CF

Fundamentally, our purpose is to improve the existing user-based CF under the belief function theory. Especially, we aim to cope with the scalability problem occurring in user-based CF by performing a clustering process where uncertainty is also handled. We assume that considering the uncertainty emerged during the clusters' assignment would retain or rather improve as well as possible the scalability and predictions performance. Besides, the uncertainty pervaded in the final predictions is also taken into account to reflect more reliable and credible recommendations. The proposed approach goes through five main steps namely evidential users' clustering, identifying clusters, users' neighborhood selection, modeling users' neighborhood ratings and generating users' neighborhood predictions.

4.1   Evidential Users' Clustering

During this first step, we aim to carry out an evidential clustering process for the given users in the system. The process consists on embracing the belief function theory in order to handle uncertainty about users' assignments to clusters. We define $\Omega_1 = \{w_1, w_2, \ldots, w_c\}$ where c is the number of clusters. In the proposed recommendation approach, each user can belong to all clusters with a degree of belief. Moreover, the evidential clustering of the users bestows a credal partition that enables a given user to be assigned to multiple clusters, or rather multiple partitions of clusters. For this purpose, we opt for the Evidential C-Means (ECM) [8], an efficient soft clustering technique which allows us to generate the credal partition of users. In other words, each user in the system will be allocated a mass of belief not only to single clusters, but also to any subsets of the frame of discernment Ω_1. At the start, we make use of the user-item ratings matrix to randomly initialize the cluster centers commonly referred to as prototypes. The euclidean distance is then computed between the users and the non empty sets of Ω_1. Finally, the convergence and the minimization of the objective function (Eq. 5) generates the final credal partition. Table 1 illustrates an example of a credal partition generated for five users where the number of clusters is $c = 3$.

4.2   Identifying Clusters

Until now, we have obtained a credal partition corresponding to each user in the system. In fact, each *bba* can be transformed into a pignistic probability

Table 1. Example of credal partition corresponding to five users

Users	$\emptyset$	$\{w_1\}$	$\{w_2\}$	$\{w_1, w_2\}$	$\{w_3\}$	$\{w_1, w_3\}$	$\{w_2, w_3\}$	Ω_1
Eric	0.0025	0.9682	0.009	0.0078	0.0046	0.0043	0.0018	0.0017
Alice	0.0468	0.2946	0.2715	0.1106	0.1135	0.0731	0.0516	0.0382
John	0.0005	0.0010	0.0018	0.0004	0.9934	0.0009	0.0017	0.0004
Maria	0.0062	0.0212	0.8856	0.0174	0.0247	0.0107	0.0246	0.0097
Peter	0.0366	0.1484	0.4931	0.0909	0.0947	0.0479	0.0556	0.0327

$BetP(w_k)$ (Eq. 3). Thereupon, we apply throughout this second step, these pignistic probabilities for the purpose of making a final decision about the cluster
corresponding to the current user. These values are interpreted as the degree of
membership of the user u to cluster w_k. Finally, a hard partition can be easily
obtained by assigning each user to the cluster with the highest $BetP$ value as
illustrated in Table 2. Note that in some cases where the pignistic probability
values are equal or quite close, the plausibility function pl can be computed and
each user can be assigned to the cluster having the highest plausibility value.

Table 2. Example of pignistic probabilities corresponding to five users

Users	$BetP(\{w_1\})$	$BetP(\{w_2\})$	$BetP(\{w_3\})$	Selected cluster
Eric	0.9773	0.0144	0.0083	w_1
Alice	0.4188	0.3833	0.1979	w_1
John	0.0017	0.0029	0.9953	w_3
Maria	0.0387	0.9155	0.0458	w_2
Peter	0.2374	0.5992	0.1633	w_2

4.3 Users' Neighborhood Selection

Once the users' preferences have been analyzed and the cluster model has been
built, we should now consider only the users belonging to the same cluster as
the active user for the neighborhood selection. Hence, based on the obtained
clusters, this step consists on identifying the set of the K-nearest neighbors of
the active user U_a. To this end, the distances between U_a and the whole users
in this cluster are computed as follows:

$$dist(U_a, U_i) = \frac{1}{|I(a,i)|} \sqrt{\sum_{I \in I(a,i)} (r_{a,j} - r_{i,j})^2} \tag{8}$$

$|I(a,i)|$ corresponds to the number of items rated by the active user U_a and the
user U_i, $r_{a,j}$ and $r_{i,j}$ denote respectively the ratings of the user U_a and U_i for
the item I_j. According to the obtained distances, we pick up the Top-K most
similar users leading to the users' neighborhood formation.

4.4   Modeling Users' Neighborhood Ratings

Let Γ_k be the set of the K-nearest neighbors of the active user. We define the frame of discernment $\Omega_{pref} = \{r_1, r_2, \cdots, r_L\}$ as a rank-order set of L preference labels. In this phase, the rating of each user U_i belonging to Γ_k is transformed into a mass function spanning over the frame of discernment Ω_{pref} in order to model uncertainty. This representation is defined as follows:

$$m_{U_a,U_i}(\{r_p\}) = \alpha_0 \exp^{-(\gamma_{r_p}^2 \times (dis(U_a,U_i))^2} \tag{9}$$

$$m_{U_a,U_i}(\Omega_{pref}) = 1 - \alpha_0 \exp^{-(\gamma_{r_p}^2 \times (dis(U_a,U_i))^2}$$

Following [10], α_0 is set to the value of 0.95 and γ_{r_p} is the inverse of the average distance between each pair of users who gave the same ratings r_p. To evaluate the reliability of each neighbor, these *bba*'s are then discounted as follows:

$$m_{U_a,U_i}^\delta(\{r_p\}) = (1 - \delta) \cdot m_{U_a,U_i}(\{r_p\}) \tag{10}$$
$$m_{U_a,U_i}^\delta(\Omega_{pref}) = \delta + (1 - \delta) \cdot m_{U_a,U_i}(\Omega_{pref})$$

Where δ corresponds to the discounting factor. It is defined on the basis of the users' distances such that: $\delta = \frac{dis(U_a,U_i)}{max(dist)}$ where $max(dist)$ is the maximum value of the computed distances.

4.5   Generating Users' Neighborhood Predictions

After the representation and the discounting of the evidence corresponding to each similar user, the obtained *bba's* are combined follows:

$$m^\delta(\{r_p\}) = \frac{1}{N}(1 - \prod_{U \in \Gamma_k}(1 - \alpha_{r_p})) \cdot \prod_{r_p \neq r_q} \prod_{U \in \Gamma_k}(1 - \alpha_{r_q}) \qquad \forall r_p \in \{r_1, \cdots, r_{Nb}\} \tag{11}$$

$$m^\delta(\Omega_{pref}) = \frac{1}{N} \prod_{p=1}^{Nb}(1 - \prod_{U \in \Gamma_k}(1 - \alpha_{r_p}))$$

Nb denotes the number of the ratings provided by the similar users, Γ_k is the set of the K-nearest neighbors of the active user, α_{r_p} is the belief committed to the rating r_p, α_{r_q} is the belief committed to the rating $r_q \neq r_p$ and N is a normalized factor defined by [10]:

$$N = \sum_{p=1}^{Nb}(1 - \prod_{U \in \Gamma_k}(1 - \alpha_{r_p}) \prod_{r_q \neq r_p} \prod_{U \in \Gamma_k}(1 - \alpha_{r_q}) + \prod_{p=1}^{Nb}(\prod_{U \in \Gamma_k}(1 - \alpha_{r_q}))) \tag{12}$$

The main advantage behind this representation is that the final prediction must be a basic belief assignment which reflects more credible results. The decision about the final rating that should be provided to the active user is made with the pignistic probability. We assume that the rating within the greatest pignistic probability is more likely to be the potential future one. On the other hand, all the other possible preferences generated by this approach are also displayed to the active user as illustrated in Fig. 1.

	$\star$	$\star\star$	$\star\star\star$	$\star\star\star\star$	$\star\star\star\star\star$	Ω_{pref}
 Active User	0.1539	-	0.0938	-	0.2872	0.465

Fig. 1. Example of evidential ratings predictions

5   Experiments and Discussions

MovieLens[1], one of the widely used real world data set in the CF area has been used in our experiments. In this data set, a collection of 100.000 ratings obtained from 943 users on 1682 movies has been performed. In our experiments, we perform a comparative evaluation over our proposed method as well as the traditional evidential user-based CF proposed in [21]. We followed the strategy conducted in [22] where movies are first ranked based on the number of their corresponding ratings. Different subsets containing the ratings given by the users for 20 movies are then extracted by progressively increasing the number of the missing rates. Hence, each subset will contain a specific number of ratings leading to different degrees of sparsity. For all our experiments, we used 5 different subsets where, for each one, we randomly extract 20% of the available ratings as a testing data and the remaining 80% were considered as a training data.

5.1   Evaluation Metrics

Two evaluation metrics typically used in RSs have been considered to evaluate our proposal namely the *Mean Absolute Error* (MAE) and the *Root Mean Squared Error* (RMSE) defined by:

$$MAE = \frac{\sum_{u,i}|p_{u,i} - r_{u,i}|}{\|p_{u,i}\|} \quad and \quad RMSE = \sqrt{\frac{\sum_{u,i}(p_{u,i} - r_{u,i})^2}{\|p_{u,i}\|}} \tag{13}$$

Where $r_{u,i}$ is the real rating for the user u on the item i, $p_{u,i}$ is the predicted rating, $\|p_{u,i}\|$ is the total number of the predicted ratings over all the users.

Lowest values of these measures imply a better prediction accuracy and a higher performance.

---

[1] http://movielens.org.

5.2 Results

We carried out experiments over the selected subsets by switching each time the number of clusters c. We used $c = 2$, $c = 3$, $c = 4$ and $c = 5$. For each selected cluster, we used 10 different neighborhood sizes and the average results of 10 repetitions were represented for that cluster. Finally, the results corresponding to the different number of clusters used in the experiments are also averaged. That is to say, we compute the MAE and the RMSE measure for each value of c and we note the overall results. For all our experiments, we used $\alpha = 2$, $\beta = 2$ and $\delta^2 = 10$ as invoked in [8]. Considering different sparsity degrees, the experimental results are summarize in Table 3.

Table 3. The comparison results in terms of MAE and RMSE

Evaluation metrics	Sparsity degrees	Evidential UB-CF	Evidential clustering UB-CF
MAE	60.95%	0.740	0.691
RMSE		1.089	1.069
MAE	65%	0.766	0.678
RMSE		0.965	0.888
MAE	70%	0.737	0.742
RMSE		0.821	0.807
MAE	75%	0.685	0.716
RMSE		0.850	0.891
MAE	80.8%	0.650	0.707
RMSE		0.854	0.902
Overall MAE		0.716	**0.707**
Overall RMSE		0.916	**0.911**

The performance of the proposed evidential clustering user-based CF is compared to that of the traditional evidential user-based CF [21]. It can be seen that the clustering based approach provides a slightly better performance than the other evidential approach. In fact, EC-UBCF acquires the lowest average values in terms of MAE and RMSE (0.707 compared to 0.716 for MAE) and (0.911 compared to 0.916 for RMSE). These results show that the evidential clustering process would maintain a good prediction quality while improving the scalability performance which is an important challenge in RSs.

5.3 Scalability Performance

We perform the scalability of our approach by varying the sparsity degree. We compare the results to the standard evidential UB-CF as depicted in Fig. 2.

According to Fig. 2, the elapsed time corresponding to the Evidential Clustering UB-CF is substantially lower than the basic Evidential UB-CF.

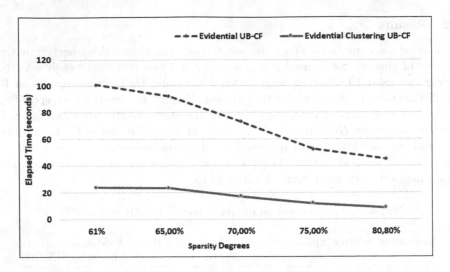

Fig. 2. Elapsed time of evidential clustering UB-CF vs. evidential UB-CF

These results are explained by the fact that standard user-based CF method needs to search the closest neighbors to the active user in the whole users' space, which leads to huge computing amount. However, in our approach, predictions are made based on the neighbors belonging to the active user's cluster, which consequently reduces the time needed for the neighborhood formation.

6 Conclusion

In this paper, we have proposed a new evidential clustering user-based CF approach. We first build a clustering model according to the users' past preferences. Based on the obtained clusters, the K-nearest users are selected and the predictions are then performed and provided to the active user. Compared to a previous user-based CF method under the belief function theory, elapsed time has been significantly improved while maintaining a good recommendation performance.

References

1. Bobadilla, J., Ortega, F., Hernando, A., Gutiérrez, A.: Recommender systems survey. Knowl. Based Syst. **46**, 109–132 (2013)
2. Su, X., Khoshgoftaar, T.M.: A survey of collaborative filtering techniques. Adv. Artif. Intell. **2009**, 1–19 (2009)
3. Sarwar, B.M., Karypis, G., Konstan, J., Riedl, J.: Recommender systems for large-scale e-commerce: scalable neighborhood formation using clustering. In: International Conference on Computer and Information Technology. IEEE, Dhaka (2002)
4. Gong, S.: A collaborative filtering recommendation algorithm based on user clustering and item clustering. JSW **5**(7), 745–752 (2010)

5. Dempster, A.P.: A generalization of bayesian inference. J. R. Stat. Soc. Ser. B (Methodol.) **30**, 205–247 (1968)
6. Shafer, G.: A Mathematical Theory of Evidence, 1. Princeton University Press, Princeton (1976)
7. Smets, P.: The transferable belief model for quantified belief representation. In: Smets, P. (ed.) Quantified Representation of Uncertainty and Imprecision. HDRUMS, vol. 1, pp. 267–301. Springer, Dordrecht (1998). https://doi.org/10.1007/978-94-017-1735-9_9
8. Masson, M.H., Denoeux, T.: ECM: an evidential version of the fuzzy c-means algorithm. Pattern Recognit. **41**(4), 1384–1397 (2008)
9. Abdelkhalek, R., Boukhris, I., Elouedi, Z.: A clustering approach for collaborative filtering under the belief function framework. In: Antonucci, A., Cholvy, L., Papini, O. (eds.) ECSQARU 2017. LNCS (LNAI), vol. 10369, pp. 169–178. Springer, Cham (2017). https://doi.org/10.1007/978-3-319-61581-3_16
10. Denoeux, T.: A K-nearest neighbor classification rule based on Dempster-Shafer theory. IEEE Trans. Syst. Man Cybern. **25**, 804–813 (1995)
11. Bezdek, J.C., Ehrlich, R., Full, W.: FCM: the fuzzy c-means clustering algorithm. Comput. Geosci. **10**, 191–203 (1984)
12. Zhang, J., Lin, Y., Lin, M.: An effective collaborative filtering algorithm based on user preference clustering. Appl. Intell. **45**(2), 230–240 (2016)
13. Xue, G.R., et al.: Scalable collaborative filtering using cluster-based smoothing. In: ACM SIGIR Conference on Research and Development in Information Retrieval, pp. 114–121. ACM (2005)
14. O'Connor, M., Herlocker, J.: Clustering items for collaborative filtering. In: ACM SIGIR Workshop on Recommender Systems, vol. 128, UC Berkeley (1999)
15. Nguyen, V.-D., Huynh, V.-N.: A reliably weighted collaborative filtering system. In: Destercke, S., Denoeux, T. (eds.) ECSQARU 2015. LNCS (LNAI), vol. 9161, pp. 429–439. Springer, Cham (2015). https://doi.org/10.1007/978-3-319-20807-7_39
16. Nguyen, V.D., Huynh, V.N.: Using community preference for overcoming sparsity and cold-start problems in collaborative filtering system offering soft ratings. Electron. Commer. Res. Appl. **26**, 101–108 (2017)
17. Nguyen, V.D., Huynh, V.N.: Integrating community context information into a reliably weighted collaborative filtering system using soft ratings. IEEE Trans. Syst. Man Cybern. Syst. **99**, 1–13 (2017)
18. Nguyen, V.D., Huynh, V.N.: Two-probabilities focused combination in recommender systems. Int. J. Approx. Reason. **80**, 225–238 (2017)
19. Abdelkhalek, R., Boukhris, I., Elouedi, Z.: Evidential item-based collaborative filtering. In: Lehner, F., Fteimi, N. (eds.) KSEM 2016. LNCS (LNAI), vol. 9983, pp. 628–639. Springer, Cham (2016). https://doi.org/10.1007/978-3-319-47650-6_49
20. Abdelkhalek, R., Boukhris, I., Elouedi, Z.: Assessing items reliability for collaborative filtering within the belief function framework. In: Jallouli, R., Zaïane, O.R., Bach Tobji, M.A., Srarfi Tabbane, R., Nijholt, A. (eds.) ICDEc 2017. LNBIP, vol. 290, pp. 208–217. Springer, Cham (2017). https://doi.org/10.1007/978-3-319-62737-3_18
21. Abdelkhalek, R., Boukhris, I., Elouedi, Z.: A new user-based collaborative filtering under the belief function theory. In: Benferhat, S., Tabia, K., Ali, M. (eds.) IEA/AIE 2017. LNCS (LNAI), vol. 10350, pp. 315–324. Springer, Cham (2017). https://doi.org/10.1007/978-3-319-60042-0_37
22. Su, X., Khoshgoftaar, T.M.: Collaborative filtering for multi-class data using Bayesian networks. Int. J. Artif. Intell. Tools **17**, 71–85 (2008)

Analysing the Impact of Rationality on the Italian Electricity Market

Sara Bevilacqua[1], Célia da Costa Pereira[2(✉)], Eric Guerci[3], Frédéric Precioso[2], and Claudio Sartori[1]

[1] Università di Bologna, Bologna, Italy
{sara.bevilacqua,claudio.sartori}@unibo.it
[2] Université Côte d'Azur, CNRS, I3S Lab, Sophia Antipolis, France
{celia.pereira,frederic.precioso}@unice.fr
[3] Université Côte d'Azur, CNRS - GREDEG, Sophia Antipolis, France
eric.guerci@unice.fr

Abstract. We analyze the behavior of the Italian electricity market with an agent-based model. In particular, we are interested in testing the assumption that the market participants are fully rational in the economical sense. To this aim, we extend a previous model by considering a wider class of cases. After checking that the new model is a correct generalization of the existing model, we compare three optimization methods to implement the agents rationality and we verify that the model exhibits a very good fit to the real data. This leads us to conclude that our model can be used to predict the behavior of this market.

1 Introduction and Related Work

The need for understanding the evolution of the prices in the electrical power markets has increased with the new trends (the emergence of the liberalized market) of the electrical market in many countries [13,14]. The use of Artificial Intelligence techniques has already proven to be effective in modeling the electricity market. Faia *et al.* proposed in [3] a Genetic Algorithm (GA) based approach to solve the portfolio optimization problem for simulating the Iberian electricity market. The results show that their GA based method is able to reach better results than previous implementations of Particle Swarm Optimization (PS) and Simulated Annealing (SA) methods. Santos *et al.* proposed in [12] a new version of the Multi-Agent System for Competitive Electricity Markets (MASCEM, [11]) with the aim of optimizing it with repect to the results as well as to the execution time.

Other models have been proposed, like the one presented by Urielli *et al.* [17], in which the authors study the impact of the Time-Of-Use (TOU) tariffs in a competitive electricity market place. A very interesting and recent survey

Célia da Costa Pereira—Acknowledges support of the project PEPS AIRINFO funded by the CNRS.

V. Torra et al. (Eds.): MDAI 2019, LNAI 11676, pp. 236–247, 2019.
https://doi.org/10.1007/978-3-030-26773-5_21

of potential design changes in the electricity market and their consequences, has been proposed by Ela *et al.* in [2].

In this paper, we propose a framework which helps analysing the behavior of the participants in the Italian electrical power market [15]. We would like to stress that our interest is in understanding how the market behaves as a consequence of the actions of its participants to make profit, and also in analysing the behavior of the market in order to maximize the social welfare from an economical rational point of view [16,18], i.e., with respect to the electricity producers as well as with respect to the electricity consumers. To this aim, we first reproduce and then extend an existing economical-based model for the Italian electricity market [7].

The paper is organized as follows. Section 2 briefly presents the three optimization methods used in the paper. Section 3 presents the mechanism of exchanges in the Italian market proposed in the literature. Section 4 presents the extended model as well as the obtained results. Section 5 concludes the paper.

2 Some Background: A Brief Description of the Used Methods

In this section, we will briefly present the three methods used in our work to model the rationality of the market participants.

A *Genetic Algorithm (GA)* [6,8] is a computational technique inspired by biology. The basic idea of a GA is to mimic the Darwinian principle of survival according to which species with a high capacity of adaptation have an higher probability to survive and then to reproduce. The algorithm considers a population of individuals represented by their genes. Three operators can be used to mimic the evolution of these individuals: *mutation* which randomly changes some bits of a gene, *crossover* which mimics the sexual reproduction of the living beings, and *selection* which consists of deciding which among the individuals in the population will survive in the next generation. This choice is made thanks to a *fitness function* which is an objective function allowing to compute the extent to which an individual of the population is adapted to solve the considered problem.

In *Monte Carlo Optimization* [1], an approximation to the optimum of an objective function is obtained by drawing random points from a probability distribution, evaluating them, and keeping the one for which the value of the objective function is the greatest (if a maximum is sought for) or the least (if a minimum is sought for). As the number of points increases, the approximation converges to the optimum.

Particle Swarm Optimization (PS) [9,10] is a meta-heuristic method inspired by the behavior or rules that guide the group of animals, for example bird flocks. According to these rules, the members of the swarm need to balance two opposite behaviors in order to reach the goal: individualistic behavior, in which each element searches for an optimal solution, and social behavior, which allows the swarm to be compact. Therefore, individuals take advantage from other searches moving toward a promising region. In this algorithm, the evolution of the population is re-created by the changing of the velocity of the particles. The idea is to tweak the values of a group of variables in order to make them become closer to the member of the group whose value is closest to the considered target. PS is similar to genetic algorithms (GAs). It is also a population-based method with the particularity that the elements of the population are iteratively modified until a termination criterion is satisfied.

3   The Italian Electricity Market

3.1   The Market Configuration

The reality of the Italian Electricity Market which takes place in the Italian Power Exchange (IPEX), considers a two-settlement market configuration with a generic forward market and the Day-Ahead Market (DAM). The DAM price value is commonly adopted as underlying for forward contracts; therefore, as in Guerci et *al.* [7], we will refer to DAM as the spot (i.e. immediate, instant) market session for simplicity. The forward market session is modeled by assuming a common, zone-independent, and unique forward market price P^f for all market participants and by determining the exact historical quantity commitments for each generating unit.

Definition 1 (Generating Company). *A generating company (GenCo) is an agent g, (with $g = 1, 2, ..., G$, and G is the number of GenCos) which owns N_g generators[1]. The ith generator (where $i = 1, 2, ..., N_g$) has lower $\underline{Q}_{i,g}$ and upper $\overline{Q}_{i,g}$ production limits, which define the feasible production interval for its hourly real-power production level $\hat{Q}_{i,g,h} = \hat{Q}^f_{i,g,h} + \hat{Q}^s_{i,g,h}$ ([MW]), with $\underline{Q}_{i,g} \leq \hat{Q}_{i,g,h} \leq \overline{Q}_{i,g}$ where $\hat{Q}^f_{i,g,h}$ and $\hat{Q}^s_{i,g,h}$ are respectively the quantity sold in the forward market and the quantity accepted in the DAM.*

It is assumed that the company g takes a long position in the forward market (it means that the company makes agreement with the market operator with large advance) for each owned generator i, corresponding to a fraction $f_{i,g,h}$ (where h indicates the hour of the day) of its hourly production capacity, that is $\hat{Q}^f_{i,g,h} = f_{i,g,h} \cdot \overline{Q}_{i,g}$. The value of such fraction varies throughout the day, indeed forward contracts are commonly sold according to standard daily profiles. The value of $f_{i,g,h}$ has been estimated by looking at historical data and thus corresponds to a realistic daily profile for each generator.

Definition 2 (Revenues for the forward and spot markets). *The revenue, $R^f_{g,h}$ ([€h]), from forward contracts for company g is:*

$$R^f_{g,h} = \sum_{i=1}^{N_g} \hat{Q}^f_{i,g,h} \cdot P^f \tag{1}$$

---

[1] In the following we will use the terms *generator* and *power plant* interchangeably.

The spot revenue, $R^s_{g,h}$, per hour for GenCo g is obtained as follows:

$$R^s_{g,h} = \sum_{z=1}^{Z} \hat{Q}^s_{z,g,h} \cdot P^s_{z,h} \tag{2}$$

where $P^s_{z,h}$ is the price in the spot market in zone z at hour h, and Z is the total number of zones.

Let $C_{i,g,h}$ ($[\text{€}/\text{h}])^2$ be the total cost (of production) function of the ith generator of GenCo g. The total profit per hour, $\pi_{g,h}$, $[\text{€}/\text{h}]$ for GenCo g is computed as follows:

$$\pi_{g,h} = R^s_{g,h} + R^f_{g,h} - \sum_{i=1}^{N_g} C_{i,g,h}(\hat{Q}_{i,g,h}) \tag{3}$$

The considered set of thermal power plants, independently owned by GenCos, consists of up to 224 generating units, using 5 different technologies. The number of generation companies and generating units offering in the DAM varies throughout the day. Based on historical data, it has been determined for each period (day and hour) the thermal power plants that offered in DAM.[3] For each power plant in the dataset, information on the maximum and minimum capacity limits is available, as well as on the parameters needed to compute the cost.

3.2   Market Exchanges

A GenCo g submits to the DAM a bid consisting of a pair of values corresponding to the limit price P^s_i ($[\text{€}/\text{MW}]$) and the maximum quantity of power $Q^s_i \leq \overline{Q}_{i,g} - \hat{Q}^f_{i,g}$($[\text{MW}]$) that it is willing to be paid and to produce, respectively. After receiving all generators' bids, the market operator clears the DAM by performing a social welfare maximization, subject to the following constraints:

- the zonal energy balance (Kirchhoff's laws),
- the maximum and minimum capacity of each power plant,
- the inter-zonal transmission limits.

It is worth noting that the Italian demand curve in the DAM is price-inelastic, i.e., it is unaffected when the price changes. Therefore, the social welfare maximization can be transformed into a minimization of the total reported production costs, i.e., of the bid prices (see Eq. 4). This mechanism determines both the unit commitments for each generator and the Locational Marginal Price (LMP) for each connection bus. However, the Italian market introduces two slight modifications. Firstly, sellers are paid the zonal prices (LMP), therefore, this fact has to

---

[2] The details about the function can be found in [7].

[3] Notice that bid data are publicly available on the power exchange website with a one-week delay, therefore, information about what plants were actually present and the like is supposed to be common knowledge.

be explicitly considered in the model, whereas buyers pay a unique national price (PUN, Prezzo Unico Nazionale) common for the whole market and computed as a weighted average of the zonal prices with respect to the zonal loads. Secondly, transmission power-flow constraints differ according to the flow direction.

The factor which has to be minimized by solving the linear program is the following:

$$\min \sum_{g=1}^{G} \sum_{i=1}^{N_g} P_{i,g,h}^s \hat{Q}_{i,g,h}^s \tag{4}$$

It is subjected to the following constraints:

- Active power generation limits: $\underline{Q}_{i,g} \leq \hat{Q}_{i,g,h} = \hat{Q}_{i,g,h}^s + \hat{Q}_{i,g,h}^f \leq \overline{Q}_{i,g}$ [MW]
- Active power balance equations for each zone z:
$\sum_{g=1}^{G} \sum_{j \in z} \hat{Q}_{j,g,h}^s - Q_{z,load,h} = Q_{z,inject,h}$ [MW]
being $\sum_{g=1}^{G} \sum_{j \in z} \hat{Q}_{j,g,h}^s$ the sum of all the productions over all generators located in zone z, $Q_{z,load,h}$, the load demand at zone z in hour h and $Q_{z,inject,h}$, the net oriented power injection in the network at zone z in hour h.
- Real power flow limits of line, l: $Q_{l,st} \leq \overline{Q}_{l,st}$ [MW] and $Q_{l,ts} \leq \overline{Q}_{l,ts}$ [MW] being $Q_{l,st}$ the power flowing from zone s to zone t of line l and $\overline{Q}_{l,st}$ the maximum transmission capacity of line l in the same direction. $Q_{l,st}$ are calculated with the standard DC power flow model [4].

The solution consists of the set of the active powers $\hat{Q}_{i,g,h}^s$ generated by each plant i and the set of zonal prices P_z^s (LMPs) for each zone $z \in [1, 2, \ldots, Z]$, where Z is the number of zones.

4   Relaxing the Zonal Constraint

In this section we will present the new "relaxed" framework as well as the results we have obtained from our experiments.

4.1   Model Description

Each GenCo g must submit to the DAM a bid, i.e., a set of prices for each of its own power plants. Therefore, each GenCo has an action space for each power plant, which is a set of possible prices that the GenCo can choose. This set is represented by Vector $AS_{i,g}$ which is obtained with the following product:

$$AS_{i,g} = MC_{i,g} \cdot MKset, \tag{5}$$

where, $AS_{i,g}$ represents the action space of power plant i of Genco g, $MC_{i,g}$ is the marginal cost of the same power plant, and $MKset = [1.00, 1.04, \ldots, 5.00]$ is the vector with the mark-up levels. In this way, GenCos are sure not to propose a price lower than the costs.

The Multi-agent System. The multi-agent system is depicted in Fig. 1. The G GenCos are reported on the top of the figure. These GenCos repeatedly interact with each other at the end of each period $r \in \{1,\dots,R\}$, that is they all submit bids to the DAM according to their current beliefs on opponents' strategies.

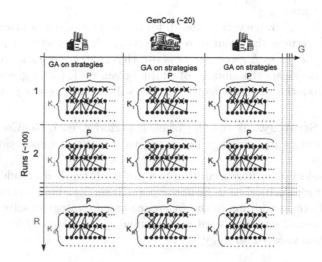

Fig. 1. A schematic representation of a simulation.

At the beginning of each period r, GenCos need to study the current market situation in order to identify a better reply to the opponents, to be played at period $r+1$.

In order to choose the most competitive strategy, GenCos need to repeatedly solve the market for different private strategies. This corresponds to an optimization problem.

4.2 The Optimization Process

In this context, the goal of the optimisation algorithm is to keep a large population of candidate strategies and to improve at the same time their fitness/performance in the market. Thus, a population of size P, (see Fig. 1), of strategies is defined, which will evolve throughout the K_r generations. Unlike Guerci et al. [7], *we consider that each power plant can have its own strategy independently of the zone.*

A Strategy. In this new framework, a strategy for a GenCo g, depicted as a black dot in Fig. 1, is a vector of prices in the action space, *one for each of the N_g power plants of GenCo g.*

The Two Used Fitness Functions. In [7], for a given GenCo g, a fitness function, $f_g : \mathbb{R}^{A_g} \to \mathbb{R}$ returns the profit of a strategy which is a vector of prices in the action space, *one for each collection of power plants*, (A_g), *situated in the same zone and using the same technology*. As it has been defined in [7], such a fitness reinforces the weight of the power plants with low prices. We will name it *type 1 fitness*.

In our "relaxed" framework, instead, for a given GenCo g, a fitness function, $f'_g : \mathbb{R}^{N_g} \to \mathbb{R}$ returns the profit of a strategy which is a vector of prices in the action space, *one for each single power plant*. Here, we propose a more "realistic" fitness which considers the amount of profit (given by Eq. 3) a given individual (strategy) allows the power plans to get. We will name it *type 2 fitness*.

Selecting a Strategy. At the end of each period r, each GenCo bids to the market by selecting one strategy belonging to its current population of candidates.

In [7] the selection is done according to a probabilistic choice model in order to favor the most represented strategy in the population (i.e., based on the *frequency probability*). Here, in addition to the *frequency based* strategy of selection used in [7], we use a second strategy *based on the value of the fitness of the individuals*. We name it *fitness-based* strategy.

4.3 Evaluation of the Proposed Approach

We have considered and tested two different market models:

- *ApproxGenco*, which replicates the market model proposed in [7] — the GenCos adopt a common strategy for all the power plants with a common technology, situated in the same zone.
- *RealGenco*, which relaxes the above constraint and makes it possible to model the fact that each power plant may adopt a different strategy — this is the way it happens in the reality.

The Goal of the Experiments. We have evaluated the effectiveness of the proposed framework. To this aim we have:

- shown that our model really extends the model proposed in [7],
- verified if the fact of relaxing the assumptions according to which all the power plants in the same zone should be associated to the same strategy compromises the results,
- compared the results obtained with a simplest statistical method relying on repeated random sampling as the Monte Carlo method with the results obtained with the two studied models,
- verified if another population based-method like the *Particle Swarm Optimisation (PS)* method, lying on the hypothesis of cooperation between the individuals in the population, instead of *Genetic Algorithms (GA)* in which the individuals of the population can be perceived as being in competition, may improve the results.

The Different Scenarios. In order to test the ability and the robustness of each optimization method to reproduce the daily PUN time series, we have considered the three methods (Genetic Algorithm, Particle Swarm, Montecarlo), the two fitness functions, (type 1 and type 2 fitness) and the two types of strategy choice (probability based and fitness based), obtaining thus the different scenarios listed in Table 1.

Table 1. Experimental scenarios

Acronym	Fitness function of the genetic algorithm	Best strategy choice based on
GaFreq1	Type 1	Frequency
GaFreq2	Type 2	Frequency
GaFitness1	Type 1	Fitness
GaFitness2	Type 2	Fitness
PsFreq1	Type 1	Frequency
PsFreq2	Type 2	Frequency
PsFitness1	Type 1	Fitness
PsFitness2	Type 2	Fitness
Montecarlo	Type 2	Fitness

Data. The demand of energy for each zone is provided in a *load* matrix with the following information: a first column which contains the zones, the second which contains the maximum limit prices and the third column which contains the demand quantities of electricity.

All the characteristics of the power plants are collected in a structure with the following features:

- the names of the GenCos (for example ATEL, EDISON, ...),
- the names of the used technologies (for example coal, combined cycle gas turbine,...),
- the prices of the fuels,
- information related to the Italian power plants: the columns indicates respectively the zone, maximum production quantity, minimum production quantity, coefficient a, coefficient b, coefficient c (see Sect. 3), Genco's id, technology index, and fuel index and power plant's id.
- the production quantity data from other power plants (i.e. not produced by the GenCo).

The PUN historical values used in the experiments are public data which can be found in [5].

Implementation and Results. The framework described has been implemented in MATLAB R2017a with Optimization and Global Optimization toolboxes. Experiments were performed on a computer running Windows 7 and based on an Intel©CoreTMi7-3610QM @2.30 GHz microprocessor with 8 GB RAM.

In all the simulations, the number of GenCos participating in the market varies between 15 and 19, while the number of power plants for each GenCo varies between 1 and 90. The three optimisation methods used the matlab default parameters and they are allocated the same number of objective function evaluations.

The results of our experiments lead us to conclude that the *ApproxGenco* model can be considered as a reliable replication of the model proposed in [7]. Indeed, under the same conditions (the ones supported by the old model), it reproduces exactly the same market mechanism — we obtain the same result, i.e., the same PUN with the two models.

In Fig. 2, we can see the historical values (red line), as well as the values obtained with [7] (we named it *old*) which are represented with dashed (purple) lines and the values obtained with our *relaxed* model under the same assumptions that the ones made in [7] (we named it *GAfreq1*, they are represented by the green line with squares).

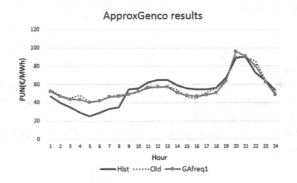

Fig. 2. Reproducibility – Real and simulated PUNs provided by ApproxGenco. (Color figure online)

We can now proceed with its comparison with the alternative scenarios.

Two interesting points emerge. We can observe, thanks to the results illustrated in Fig. 3, that despite having "relaxed" the constraint reducing the action space, and under the same assumptions, (i) *the results are similar for both ApproxGenco and RealGenco models*, i.e., the simulated PUN is similar, (ii) *the time needed for running the ApproxGenco and RealGenco models is similar*. More precisely, the running time of each iteration for ApproxGenco with a population of 50 elements is 11.62 s for GA, 15.76 s for PS, and 11.34 s for Monte Carlo. Concerning the RealGenco the values are: 14.56 s for GA, 18.68 s for PS and 10,5 s for Monte Carlo.

We can then conclude that we can *take into account the differences between the power plants as for example their different efficiency levels without worsening the quality of the results.*

Evaluation of the RMSD. These considerations at macro-level are supported by the evaluation of the root-mean-square deviation (RMSD) which is a frequently used measure of the difference between values predicted by a model and the values actually observed. The RMSD represents the sample standard deviation of the differences between predicted values and observed values. The formula we have used is the following:

$$RMSD = \sqrt{\frac{\sum_{h=1}^{24}(\hat{y}_h - y_h)^2}{24}} \tag{6}$$

where h represents the hour of the day (therefore it varies between 1 and 24), $\hat{y}_h$ and y_h are respectively the predicted value and the observed value of the PUN at hour h.

Figure 3 shows the RMSD of all the scenarios we have considered for Approx-Genco's and RealGenco's.

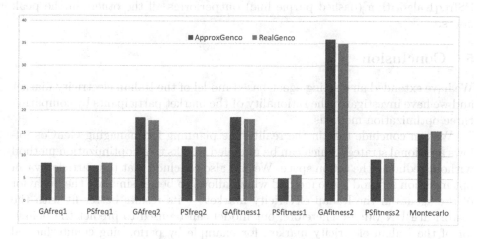

Fig. 3. RMSD for ApproxGenco and RealGenco methods

Evaluation of the Absolute Error. These considerations at macro-level are also supported by the results concerning the absolute error which, in our work, represents the distance between the curve with the historical PUNs values and the one obtained from the different scenarios. The formula used is the following:

$$AE_h = |pr_h - hist_h| \tag{7}$$

where pr_h is the prediction at hour h and $hist_h$ is the historical value of the PUN at the same hour.

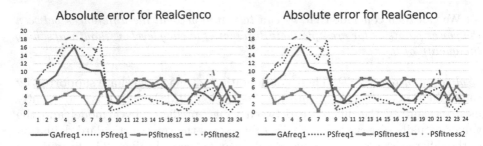

Fig. 4. Absolute errors for the best ApproxGenco model and for the RealGenco model.

Figure 4 illustrates the absolute errors for the three best optimization methods for both ApproxGenco and RealGenco models. We can notice that in the ApproxGenco model the results obtained by the three PS algorithms are better than the ones obtained by the best GA algorithm in the peak hours. Instead, during the off-peak hours the GA algorithm outperforms two out of three PS algorithms. Things are slightly different with the RealGenco model, where PSfitness1 (cyan line with squares) outperforms all the other algorithms in the off-peak hours while it produces the worse results in the peak hours. However, the PSfreq1 algorithm (dashed purple line) outperforms all the others in the peak hours.

5   Conclusion

We have extended an existing agent-based model of the Italian electricity market and we have investigated the rationality of the market participants by comparing three optimization methods.

We can conclude that in the reality the planning for managing GenCos follows a rational strategy which can be modeled thanks to an optimization method without reducing the action space. We can also conclude that the particle swarm optimization method is the method which allows to better simulate the behavior of the agents in the Italian electricity market — its results better fit with the historical PUN values. Therefore, our model can be used to predict the behavior of the Italian electricity market, for example by performing counterfactual analyses.

References

1. Betrò, B., Cugiani, M., Schoen, F.: Monte Carlo Methods in Numerical Integration and Optimization. Applied Mathematics Monographs CNR. Giardini, Pisa (1990)
2. Ela, E., et al.: Electricity markets and renewables. IEEE Power Energy Mag. **15**(27), 1540–7977 (2015)
3. Faia, R., Pinto, T., Vale, Z.A.: GA optimization technique for portfolio optimization of electricity market participation. In: 2016 IEEE Symposium Series on Computational Intelligence, SSCI 2016, Athens, Greece, 6–9 December 2016, pp. 1–7 (2016)

4. Giulioni, G., Hernández, C., Posada, M., López-Paredes, A. (eds.): Artificial Economics: The Generative Method in Economics. Lecture Notes in Economics and Mathematical Systems, vol. 631, 1st edn. Springer, Berlin (2009). https://doi.org/10.1007/978-3-642-02956-1
5. GME. http://www.mercatoelettrico.org/it/download/datistorici.aspx
6. Goldberg, D.E.: Genetic Algorithms in Search, Optimization & Machine Learning. Addison-Wesley, Reading (1989)
7. Guerci, E., Rastegar, M.A., Cincotti, S.: Agent-based modeling and simulation of competitive wholesale electricity markets. In: Rebennack, S., Pardalos, P.M., Pereira, M.V.F., Iliadis, N.A. (eds.) Handbook of Power Systems II. Energy Systems, pp. 241–286. Springer, Berlin (2010). https://doi.org/10.1007/978-3-642-12686-4_9
8. Holland, J.H.: Adaptation in Natural and Artificial Systems. The University of Michigan Press, Ann Arbor (1975)
9. Kennedy, J.: Particle swarm optimization. In: Sammut, C., Webb, G.I. (eds.) Encyclopedia of Machine Learning, pp. 967–972. Springer, Boston (2017). https://doi.org/10.1007/978-0-387-30164-8
10. Kennedy, J., Eberhart, R.: Particle swarm optimization. In: Proceedings of IEEE International Conference on Neural Networks, Part IV, pp. 1942–1948 (1995)
11. Pinto, T., Morais, H., Oliveira, P., Vale, Z., Praça, I., Ramos, C.: A new approach for multi-agent coalition formation and management in the scope of electricity markets. Energy **36**(8), 5004–5015 (2011)
12. Santos, G., Pinto, T., Praça, I., Vale, Z.: Mascem: optimizing the performance of a multi-agent system. Energy **111**((Supplement C)), 513–524 (2016)
13. Silva, F., Teixeira, B., Pinto, T., Santos, G., Praça, I., Vale, Z.: Demonstration of realistic multi-agent scenario generator for electricity markets simulation. In: Demazeau, Y., Decker, K.S., Bajo Pérez, J., de la Prieta, F. (eds.) PAAMS 2015. LNCS (LNAI), vol. 9086, pp. 316–319. Springer, Cham (2015). https://doi.org/10.1007/978-3-319-18944-4_36
14. Sioshansi, F.P.: Evolution of Global Electricity Markets: New Paradigms, New Challenges, New Approaches. Academic Press (2013)
15. Tribbia, C.: Solving the italian electricity power exchange (2015)
16. Urieli, D.: Autonomous trading in modern electricity markets. AI Matters **2**(4), 18–19 (2016)
17. Urieli, D., Stone, P.: Autonomous electricity trading using time-of-use tariffs in a competitive market. In: Proceedings of the Thirtieth AAAI Conference on Artificial Intelligence, 12–17 February 2016, Phoenix, Arizona, USA, pp. 345–352 (2016)
18. Vytelingum, P., Ramchurn, S.D., Voice, T., Rogers, A., Jennings, N.R.: Trading agents for the smart electricity grid. In: AAMAS, pp. 897–904. IFAAMAS (2010)

Analysis of Abstention in the Elections to the Catalan Parliament by Means of Decision Trees

Eva Armengol[(✉)] and Zaida Vicente

Artificial Intelligence Research Institute (IIIA-CSIC), Campus UAB,
Camí de Can Planes, s/n, 08193 Barcelona, Bellaterra, Spain
eva@iiia.csic.es

Abstract. Democracies are based on political parties and election systems allowing voters to put the confidence in representers of these political parties to defend their interests. There are many studies analyzing the results of elections with the goal of (1) explaining the results, and (2) trying to predict what will happens in future elections. However no many attention has received the abstention, why voters do not use their right to elect representers? Commonly, abstention has not been too significant, however in last years it has been increased in many countries and it could be of great interest to analyze the causes. Studies about elections, both voting and abstention, are commonly based on statistical methods. The current paper is focused on analyzing the abstention based on symbolic learning methods (decision trees). Particularly, we are interested on identifying the groups of potential voters that have decided to abstain. We worked on data of the elections to Catalan Parliament held in 2015.

Keywords: Electoral results · Ecological inference · Abstention ·
Explanations · Inductive learning methods · Decision trees

1 Introduction

One of the main rights in democracy is the free election of political representers. Representers belong to political parties that have a political program and voters choose the one that best fits their social, economic, and ethic values. Until recent times, political parties are aware to both predict the vote to their own party and, after the elections, explain why voters exhibited the resulting behaviour. Traditionally, predictions are made based on questionnaires before and during elections. In addition to directly ask for the vote, sometimes the voter is also questioned about aspects of his life. As McConway [13] explains, such questionnaires are difficult to elaborate and there is also a risk that the person asked do not answer what he actually believes (specially when asked directly for vote intention).

© Springer Nature Switzerland AG 2019
V. Torra et al. (Eds.): MDAI 2019, LNAI 11676, pp. 248–258, 2019.
https://doi.org/10.1007/978-3-030-26773-5_22

However, in addition to the interest in determining the vote to each political party, another important fact arises: the abstention. Lastly, the percentage of voters that abstain in elections increases in all the countries. Ferreira et al. [4] analyzed the causes of this increment and mainly they associate it to political disaffection and no representativeness of political parties. In fact it is increasing the percentage of population that does not identify theirselves with a particular party [8]. There are several reasons for this lack of representativeness, for instance, lack of confidence in democracy in general, corruption, etc. [1,2,14].

Cazorla et al. [2] analyze the causes of abstention of Spanish voters in European elections held in 2014. In this analysis the authors conclude that in the particular case of Spain, there are two main reasons for the increment of abstention: (1) the fragmentation of the space of political parties mainly due to the growing of two new parties, and (2) the demobilisation and dissatisfaction of the voters, mainly due to economic crisis, corruption cases and political scepticism. This analysis has been made using the results of a post-electoral questionnaire designed to provide relevant information about both abstentionists and supporters of the different political parties. By grouping and recoding variables they used *Structural Equation Modelling* to obtain a set of variables useful to explain abstention.

Ecological inference [10] is the process to use aggregated data to infer relationships at individual level when there are not information about individuals. The ecological inference handles two observed variables X_i and Y_i and two unobserved variables α_i and β_i for $1 \leq i \leq n$ being n the number of observations (aggregated units). Until 1997 there were two main approaches for solving ecological inference problems: the followers of the *method of bounds* proposed by Duncan y Davis [3], and the followers of the Goodman's approach, based on regression [7]. In 1997, King [9] introduced a new paradigm based on both approaches consisting on, given a set of constraints (see [10]) the goal is to obtain the straight lines that he calls *tomographies* that represent the space of possible solutions. Such space represents an estimation of the parameters α_i and β_i of the ecological inference.

Castela and Galindo-Villardón [1] use the ecological inference method proposed by King as a basis to use the HJ-Biplot method [19] to determine groups of population and their electoral behaviour from data of Portuguese elections held in 2002 and 2005. This work is interesting because it is possible to determine the evolution of votes. In our case we have only data belonging to one electoral date (the one of 2015 to Catalan Parliament).

A different approach is the one of Nwankwo et al. [14] that, from a questionnaire including socio-demographic questions, use *Principal Component Analysis* (PCA) to determine the main aspects that influence the abstention in South Eastern Nigeria. In particular, they obtained eight main components: socioeconomic status, lack of trust in the electoral process, social trust and unemployment, registration and demographic factor, corruption and inadequate security, deception and intimidation, indigene status and electoral manipulation and poverty.

A method that has provided interesting results is the one proposed by Flaxman et al. [6]. In particular, it has been used to explain the results of the E.E.U.U. elections held in 2012 [6] and 2016 [5]. The Flaxman's method exploits the connexion between Gaussian process regression [17] and kernel ridge regression [18] to use Bayesian approximation for learning and to include prior information.

In this paper we introduce a different approach: the use of symbolic learning methods to analyze the results of elections. Statistical methods give us valuable information such as correlations between variables. However, correlations take globally all the data and in our opinion to provide a more detailed exam of the data it is necessary to fragment them in significant groups that follow a pattern. This kind of pattern is the one that can be obtained using decision trees [15,16]. The advantage of decision trees is that their results are easily understood by experts and can be seen as explanations or descriptions of classes of objects. Notice that, in fact, the goal of methods such as the ones of Flaxman, King, Duncan and Davis, and Goodman is to solve the problem on ecological inference, i.e., to obtain particular data from general data. Instead, our goal is almost the converse one, we want to obtain significant patterns (i.e., general enough) to describe population groups having a similar behaviour from the point of view of the abstention. However, we could obtain patterns satisfied by reduced groups of electoral sections allowing to solve the ecological inference problem. In other words, if we allow overfitting when growing the decision tree, we could obtain patterns satisfied by only a few of electoral sections that is the goal of ecological inference.

In Catalonia, there are four different kind of elections: local, Catalan Parliament, Spanish Parliament and European Parliament. If we analyze separately the results of each one of them, the patterns can be easily compared and it will be possible to show the differences or similarities of voters according each kind of election. We focus our analysis in the abstention. Thus, our goal is to characterize electoral sections according to their abstention level. In the future we plan to extend this analysis to other elections.

The paper is organized as follows. In Sect. 2 there is a brief explanation of decision trees. In Sect. 3 there is the description of the data base used in the experiments. Section 4 contains a description and a discussion of the experiments carried on. Finally, Sect. 5 is devoted to conclusions and future work.

2   Decision Trees

A *Decision Tree* (DT) is a directed acyclic graph in the form of a tree. The root of the tree has not incoming edges and the remaining ones have exactly one incoming edge. Nodes without outgoing edges are called *leaf* nodes and the others are *internal* nodes. A DT is a classifier expressed as a recursive partition of the set of known examples of a domain [12]. The goal is to create a domain model predictive enough to classify future unseen domain objects.

Each node of a tree has associated a set of examples that are those satisfying the path from the root to that node. The leaves determine a partition of the original set of examples, since each domain object only can be classified following one of the paths of the tree. The construction of a decision tree is performed by splitting the source set of examples into subsets based on an attribute-value test. This process is repeated on each derived subset in a recursive manner. Figure 1 shows the ID3 algorithm [15,16] commonly used to grow decision trees. From a decision tree we can extract rules (i.e., patterns) giving descriptions of classes, since each path from the root to a leaf form a description of a class. When all the examples of a leaf belong to the same class such description is *discriminant*. Otherwise, the description is *no discriminant*.

ID3 (examples, attributes)
    create a *node*
    <u>if</u> all examples belong to the same *class* <u>return</u> *class* as the label for the node
    <u>otherwise</u>
        A ← best attribute
        <u>for</u> each possible value v_i of A
            add a new tree branch below *node*
            examples$_{vi}$ ← subset of examples such that A = v_i
            ID3(examples$_{vi}$, attributes - {A})
    <u>return</u> *node*

Fig. 1. ID3 algorithm for growing a decision tree.

A key issue of the construction of decision trees is the selection of the most relevant attribute to split a node. Each measure uses a different criteria, therefore the selected attribute could be different depending on it and thus the whole tree could also be different. In our experiments we used the López de Mántaras distance (LM) [11] that is an entropy-based normalized metric defined in the set of partitions of a finite set. It compares the partition induced by an attribute, say a_i, with the *correct partition*, i.e., the partition that classifies correctly all the known examples. The best attribute is the one inducing the partition which is closest to the correct partition. Given a finite set X and a partition $\mathcal{P} = \{P_1, \ldots, P_n\}$ of X in n sets, the entropy of $\mathcal{P}$ is defined as ($|\cdot|$ is the cardinality function):

$$H(\mathcal{P}) = -\sum_{i=1}^{n} p_i \cdot \log_2 p_i, \text{ where } p_i = \frac{|P_i|}{|X|}$$

and where the function $x \cdot \log_2 x$ is defined to be 0 when $x = 0$. The *López de Mántaras'* distance between two partitions $\mathcal{P} = \{P_1, \ldots, P_n\}$ and $\mathcal{Q} = \{Q_1, \ldots, Q_m\}$ is defined as:

$$LM(\mathcal{P}, \mathcal{Q}) = \frac{H(\mathcal{P}|\mathcal{Q}) + H(\mathcal{Q}|\mathcal{P})}{H(\mathcal{P} \cap \mathcal{Q})}, \tag{1}$$

where

$$H(\mathcal{P}|\mathcal{Q}) = -\sum_{i=1}^{n}\sum_{j=1}^{m} r_{ij} \cdot \log_2 \frac{r_{ij}}{q_j}, \quad H(\mathcal{Q}|\mathcal{P}) = -\sum_{j=1}^{m}\sum_{i=1}^{n} r_{ij} \cdot \log_2 \frac{r_{ij}}{p_i},$$

$$H(\mathcal{P} \cap \mathcal{Q}) = -\sum_{i=1}^{n}\sum_{j=1}^{m} r_{ij} \cdot \log_2 r_{ij},$$

$$\text{with } q_j = \frac{|Q_j|}{|X|}, \text{ and } r_{ij} = \frac{|P_i \cap Q_j|}{|X|}.$$

Decision trees can be useful for our purpose because their paths give us *patterns* describing classes of objects (electoral sections in our approach) in a user-friendly manner. One shortcoming of decision trees is *overfitting*, meaning that there are few objects in most of the leaves of the tree. In other words, paths are actually descriptions that poorly represent the domain. The main procedure to either avoid or reduce overfitting is by pruning the tree, i.e., under some conditions, a node is no longer expanded. However, this means that leaves can contain objects belonging to several classes and, therefore, paths do not represent discriminatory descriptions of classes, i.e., these descriptions are satisfied by objects of more than one class. In our approach, we managed overfitting by controlling the percentage of elements of each class. Let S_N be the set of objects associated with an internal node N, the stopping condition in expanding N (the *if* of the ID3 algorithm) holds when the percentage of objects in S_N that belong to the majority class decreases in one of the children nodes. In such a situation, the node N is considered as a leaf.

Notice, however, that allowing overfitting can produce results near to the goal of ecologic inference since the patterns obtained can be specific enough to almost identify particular objects.

3   The Data Base

Electoral landscape of Catalonia is formed by 5048 *electoral sections* each one of them composed of a minimum of 500 potential voters and a maximum of 2000. Most of times, each electoral section corresponds to an *electoral table* although this can variate if the number of voters of a table is either too high or the population is scattered throughout the territory represented by the electoral section. In such cases, an electoral section is divided in several electoral tables with the constraint that no electoral table can have less than 200 voters.

The data base we have is composed of 5048 records, each one of them corresponds to an electoral section in Catalonia. Each record has two kinds of information: socio-demographic data and electoral data. Socio-demographic data has been formed from an aggregation of the data coming from several public data bases. The result of such aggregation is the typology of families seen in Table 1 and it is the one contained in the data base Habits© from the AIS Group. The file containing the electoral results for all the Catalan electoral sections is public and contains (in absolute numbers and in percentage) the votes to each political party, the null votes, the blank votes, abstention and how many potential votes has an electoral section (*electoral census*). This file does not contain socio-demographic information associated to the electoral sections. Then, an ecological inference process has been carried out to assess how many families of each typology are in each electoral section. In https://www.ais-int.com/marketing-y-ventas/geomarketing-habits-big-data/ there is a description of the method used to assess how many families of each typology are in each electoral section. The resulting file containing the socio-demographic information related to each electoral section is the one we used in our experiments.

Table 1. Families typology as obtained as it is explained in https://www.ais-int.com/marketing-y-ventas/geomarketing-habits-big-data/

Type	Description
AF	Families with adolescents (until 18 years old) and children
BG	Families with young sons (from 18 to 35 years old)
CH	Families with children
DK	Families where the main salary is a pension
EI	Families with two working persons and without children (DINK)
J	Families with one or two members, no children, only one salary
LP	Families with one or two members, no children, one or two pensions
M	Singles older than 35 with a salary
N	The main person is a student, a housework or a permanent disability
O	The main person of the family is unemployed
Q	Family formed by only one person receiving a pension
R	The main person of the family is an immigrant
Expenses	Average of the expenses of a family
Income	Average of the income in a family

Table 2 shows the electoral attributes we used for each electoral section. Mainly there are percentages of both voters and votes to each political party, and we rejected all results corresponding to absolute numbers. We have take such decision in order to properly compare the results between political parties. The elections to the Catalan Parliament held in 2015 were somewhat special because of in addition to the traditional vote according right or left ideology, the independentist/no independentist aspect was very important. In fact, this aspect conducted to the creation of a new coalition and also was the focus of the electoral campaign. The main parties that concurred to the elections were the following:

- Cs (Ciudadanos). Party of right ideology and no independentist.
- JxSi (Junts pel Si). New independentist coalition formed by a right party and a left one.
- PSC (Socialist). Party of center-left ideology and no independentist.

- CSQEP (Catalunya Si Que Es Pot). Coalition of left ideology and mainly no independentist although some members of it could be favorable to independence.
- PP (Partit Popular). Party of right ideology and no independentist.
- CUP (Candidatures d'Unitat Popular). Party of extreme left and anti-system ideology, and independentist.

All the attributes have numerical values and we have discretized them. Such discretization has been done by computing the quartiles of each one and assessing a label to each one of the resulting intervals. Table 2 shows the quartiles used to define the discretization intervals of the electoral attributes and the label we associated with them. The information about the families of each electoral section has been discretized in a similar way.

Summarizing, the data base used in the experiments is composed of the attributes shown in Tables 1 and 2 each one of them discretized according the corresponding quartiles.

Table 2. Discretization intervals for some the atributes corresponding to electoral results, i.e., the percentage of votes for null votes, blank votes, abstention, number of voters, and votes to each relevant political party.

Attribute	Very low	Low	High	Very high
Census	[25, 796]	(796,1040]	(1040, 1310]	(1310, 3670]
Null	[0,0.15]	(0.15,0.32]	(0.32,0.54]	(0.54,5.37]
Blank	[0,0.29]	(0.29,0.49]	(0.49,0.72]	(0.72,4.03]
Abstention	[3.16,17.7]	(17.7,21.6]	(21.6,26.2]	(26.2,73.3]
Cs	[0,11]	(11,17.3]	(17.3,23.6]	(23.6,50]
JxSi	[2.45,24]	(24,40.6]	(40.6,54.2]	(54.2,91.3]
PSC	[0,7.19]	(7.19,11.2]	(11.2,17.5]	(17.5,58.3]
CSQEP	[0,4.67]	(4.67,8.37]	(8.37,12.4]	(12.4,27]
PP	[0,5.56]	(5.56,7.87]	(7.87,10.6]	(10.6,42.1]
CUP	[0,5.82]	(5.82,7.87]	(7.87,10.3]	(10.3,47.7]

4   Experiments

Because our focus is the analysis of the abstention in the elections to Catalan Parliament, we have growth a decision tree taking as class label the percentage of abstention of the electoral sections. In order to avoid overfitting we cut the node expansion when the percentage of the majority class decreases if a node was expanded (see Sect. 2). We only have manually analyzed the paths (patterns) corresponding to leaves that contain around the 25% of the population in the root of the tree. Most of significant patterns are not discriminant, i.e., they are satisfied by electoral sections having different abstention levels.

After growing a decision tree, two patterns involving around the 25% of the electoral sections have been found:

- P1: [[PSC = *very high*]]
- P2: [[PSC = *very low*]]

We expected that the abstention was related with socio-demographic attributes as it has been found for instance in [2,4]. However, our experiments show that in the particular case of the Catalan elections held in 2015 the abstention level is directly related to the percentage of votes to the PSC party. In fact, this is not a contradictory result with the ones in [2,4] since they only take into account the sociologic data coming from a questionnaire and no electoral data are included in their analysis. In our experiments, we included both socio-demographic data and electoral data. Moreover, our socio-demographic data come from real population data (census, register, and so on) instead of with a questionnaire.

Table 3. Level of abstention satisfied by the patterns P1, P2 and P3 and both number (#ES) and percentage (%ES) of electoral sections satisfying the patterns.

Pattern	#ES	%ES	Very low	Low	High	Very high
P1	1262	25.00	3.49	13.31	29.79	53.41
P2	1263	25.02	60.25	25.18	10.85	3.72
P3	976	19.33	15.06	32.58	31.97	20.39

Table 3 shows the support of each one of the patterns P1 and P2 to each abstention level. Notice that around the 83% of electoral sections that have had many votes for PSC have been high or very high abstention levels (pattern P1), the dual pattern (pattern P2) is also true: around the 85% of electoral sections that have had very few votes for PSC have been low or very low abstention levels. In other words, it seems that the level of abstention is directly correlated with the percentage of votes for PSC. In fact, this correlation has been proved statistically to be 0.580. Also we have seen that the percentage of abstention has an inverse correlation of -0.546 with the percentage of votes for JxSi. In our approach only the direct relations are explicit in the patterns, however a manual study of the patterns has show that the percentage of votes for PSC is inverse to the percentage of votes for JxSi. That is to say, electoral sections with low percentage of votes for PSC have high percentage of votes to JxSi (and low percentage of abstention). In our opinion the relatively low correlation between PSC and abstention is due to the fact that the percentage of votes for PSC explains well the abstention for the extreme values (very low and very high), whereas in the intermediate part we cannot found any relevant pattern because there is not a significant trend and is this part the one responsible of the relatively low correlation. With our approach we are capable to take several subsets with different behaviour.

Patterns including intermediate levels of votes (high or low) to PSC are not significant enough to represent a clear abstention level. For instance the pattern P3 = [PSC = *high*] satisfied by around the 19% of electoral sections has not a clear support to any of abstention levels (see Table 3). Notice that around the 52% of electoral tables satisfying P3 have high or very high abstention levels whereas the remaining 47.64% of electoral sections have low or very low abstention levels.

It has been surprising that information of families seems not to be relevant to explain the abstention level. In fact, we obtained some patterns relating the abstention with some family typology, for instance the following ones:

- $P4$: [[PSC = low], [R = $high$]]
- $P5$: [[PSC = low], [R = low]]

Notice that the patterns above relate low percentage of votes for PSC and the typology R (immigration) with a low level of abstention (around 65% and 77% respectively). Table 4 shows the support of the patterns $P4$ and $P5$ to each abstention level. These patterns are not significant enough due to low number of electoral sections that satisfy it, but we think that could be interesting as a basis to perform ecological inference. In the regression model also happens that the familiar information is not relevant to explain the abstention and we also found that families type R are the most correlated with the abstention. As future work, we have to check if this irrelevance of the family status is due to the way in which the family typologies has been obtained or conversely there is an actual independence between abstention and family status.

Table 4. Level of abstention satisfied by the patterns $P4$ and $P5$ and both number (#ES) and percentage (%ES) of electoral sections satisfying the patterns.

Pattern	#ES	%ES	Very low	Low	High	Very high
$P4$	237	4.70	20.68	44.30	26.58	8.44
$P5$	201	3.98	44.78	32.34	18.41	4.48

With the goal of trying to relate abstention to family typologies, we have growth a decision tree taking into account only the attributes describing family typologies (those in Table 1). Most of the patterns are not significant in the sense that they are satisfied by only a few electoral sections. However, there are two patterns satisfied by a significant number of electoral sections (more than 25%):

- [[R = $very\ low$]]: satisfied by 1262 electoral sections (25%) where the 44.14% of them have $very\ low$ abstention, the 25.59% low, the 18.86% $high$, and the 11.41% $very\ high$.
- [[R = $high$]]: satisfied by 1240 electoral sections (24.56%) where the 8.06% of them have $very\ low$ abstention, the 17.50% low, the 27.82% $high$, and the 46.61% $very\ high$.

Thus, patterns above shows some trend to relate electoral sections with very low or low percentage of families type R with low or very low abstention (69.73%) and electoral sections with high or very high percentage of families type R with high or very high abstention (74.43%). Both patterns seems to support the idea that immigrant collective tends to abstain.

Summarizing, the results provided by decision trees are consistent with those obtained by statistical models. However, our approach is more informative since in addition to explicit the correlation between variables also explain the intervals where such correlation is higher. Thus, from the patterns we have seen that the percentage of votes to PSC is directly correlated with the percentage of abstention, but only when the percentage of votes is in one of the extremes, i.e., it is $very\ low$ or $very\ high$, whereas in the intermediate values such correlation is not so clear. The statistical model gives a global perspective of the correlation between variables but it cannot focus on intervals as the patterns can do.

5   Conclusions and Future Work

In this paper we introduced a new approach to analyze electoral data: the decision trees. This kind of methods are commonly used to construct domain models useful for prediction. In the current application, the branches of trees can be interpreted as patterns, i.e., similarities between the electoral sections that have a particular behaviour. Our focus has been the abstention and the goal was to show if electoral sections with similar abstention levels have also some common pattern. Experimental results have shown that patterns allow to separate subgroups of electoral sections (differently than statistical methods that have a global vision of them) and support the experts in focusing on a particular group and investigate in depth the characteristics of it.

Our main conclusions are that abstention level in Catalan elections held in 2015 was directly related with the percentage of votes to the Catalan Socialist Party (PSC). When we focus on sociologic features, we found that the percentage of families belonging to the typology R (immigration) is directly correlated with abstention.

In the future we plan to use the same methods on data coming from different Catalan Parliament elections. The patterns of both elections will be easy to compare and the experts could determine similarities and differences of behaviour of the same subgroup of electoral sections in different elections. A different research line could be to use the patterns in combination with some ecological inference method, such as the one of G. King.

Acknowledgments. The authors acknowledge the AIS Group enterprise (https://www.ais-int.com/marketing-y-ventas/geomarketing-habits-big-data/) for having given us the Habits© Data Base in a selfless way. This research is funded by the project RPREF (CSIC Intramural 201650E044); and the grant 2014-SGR-118 from the Generalitat de Catalunya. Authors also thank to Àngel García-Cerdaña his helpful comments.

References

1. Castela, E., Galindo Villardón, M.P.: Inferencia ecológica para la caracterización de abstencionistas: El caso de portugal (in Spanish). Spat. Organ. Dyn. **3**, 6–25 (2011)
2. Cazorla-Martín, A., Rivera-Otero, J.M., Jaráiz-Gulías, E.: Structural analysis of electoral abstention in the 2014 european parliamentary elections. Revista Española de Investigaciones Sociológicas, pp. 31–50 (2017)
3. Duncan, O.D., Davis, B.: An alternative to ecological correlation. Am. Sociol. Rev. **18**, 665–66 (1953)
4. Ferreira, P., Dionisio, A.: Voters' dissatisfaction, abstention and entropy: analysis in european countries. https://www.researchgate.net/publication/23524254_Voters%27_dissatisfaction_abstention_and_entropy_analysis_in_European_countries, 022008
5. Flaxman, S., Sutherland, D., Wang, Y.X., Whye Teh, Y.: Understanding the 2016 US Presidential Election using ecological inference and distribution regression with census microdata. arXiv e-prints, page arXiv:1611.03787 (2016)
6. Flaxman, S., Wang, Y.X., Smola, A.J.: Who supported obama in 2012? ecological inference through distribution regression. In: Proceedings of the 21th ACM SIGKDD International Conference on Knowledge Discovery and Data Mining, KDD 2015, pp. 289–298. New York, ACM (2015)

7. Goodman, L.: Some alternatives to ecological correlation. Am. J. Sociol. - **64**, 05 (1959)
8. Holland, I., Miskin, S.: Interpreting election results in western democracies. Current Issues Brief no.2 2002–03. Politics and Public Administration Group. Parliament of Australia. https://www.aph.gov.au/About_Parliament/ Parliamentary_Departments/Parliamentary_Library/Publications_Archive/CIB/ cib0203/03CIB02,2002
9. King, G.: A Solution to the Ecological Inference Problem. Princeton University Press, New Jersey (1997)
10. King, G., Rosen, O., Tanner, M.A. (eds.): Ecological Inference: New Methodological Strategies. Cambridge University Press, New York (2004). http://gking. harvard.edu/files/abs/ecinf04-abs.shtml
11. López de Mántaras, R.: A distance-based attribute selection measure for decision tree induction. Mach. Learn. **6**, 81–92 (1991)
12. Maimon, O., Rokach, L. (eds.): Data Mining and Knowledge Discovery Handbook, 2nd edn. Springer, Berlin (2010). https://doi.org/10.1007/978-0-387-09823-4
13. McConway, K.: Explainer: how do you read an election poll? The Conversation. https://theconversation.com/explainer-how-do-you-read-an-election-poll-41204 (2015)
14. Nwankwo, C., Okafor, U., Asuoha, G.: Principal component analysis of factors determining voter abstention in south eastern nigeria. J. Pan Afr. Stud. **10**, 249–273 (2017)
15. Quinlan, J.R.: Discovering rules by induction from large collection of examples. In: Michie, D. (ed.) Expert Systems in the Microelectronic Age, pp. 168–201. Edimburg Eniversity Press, Edinburgh (1979)
16. Quinlan, J.R.: Induction of decision trees. Mach. Learn. **1**(1), 81–106 (1986)
17. Rasmussen, C.E., Williams, C.K.I.: Gaussian Processes for Machine Learning. The MIT Press, Cambridge (2006)
18. Saunders, C., Gammerman, A., Vovk, V.: Ridge regression learning algorithm in dual variables. In: Proceedings ot the 15th International Conference on Machine Learning, ICML (1998)
19. Galindo Villardón, M.P.: Una alternativa de representación simultánea: Hj-biplot (in Spanish). Questiio **10**, 13–23 (1986)

Making Decisions with Knowledge Base Repairs

Rafael Peñaloza[✉]

University of Milano-Bicocca, Milan, Italy
rafael.penaloza@unimib.it

Abstract. Building large knowledge bases (KBs) is a fundamental task for automated reasoning and intelligent applications. Needing the interaction between domain and modeling knowledge, it is also error-prone. In fact, even well-maintained KBs are often found to lead to unwanted conclusions. We deal with two kinds of decisions associated with faulty KBs. First, which portions of the KB (and their conclusions) can still be trusted? Second, which is the correct way to repair the KB? Our solution to both problems is based on storing all the information about repairs in a compact data structure.

1 Introduction

In logic-based knowledge representation, the goal is to encode the relevant knowledge of an application domain through a collection of axioms, which intuitively restrict the way in which the symbols used may be interpreted, so as to provide them with an unambiguous and clear meaning. Such a collection of axioms is known as a knowledge base (KB). Historically, many knowledge representation languages have been proposed; most notably, perhaps, are propositional logic [6], constraint systems [1], and description logics (DLs) [2]. Their success has led to the creation of more and larger KBs.

It should come as no surprise that constructing a KB, which requires a combination of domain and modeling knowledge, is an arduous and error-prone task. As a consequence, it is not uncommon to detect errors even in well-maintained KBs. Unfortunately, after an error has been detected, it is also quite difficult to identify the main sources, and select the adequate correction of the fault. Moreover, KB updates are typically subject to a production cycle that prevents them from publishing corrected versions on-demand. For example, SNOMED CT [15]—a very large KB about medical terms—publishes two updated versions every year. Thus, even when a KB is known to be faulty, it may be necessary to wait for several months before a corrected version is available.

Since KBs are not static entities to be observed, but rather tools necessary for automated reasoning in practical scenarios, users cannot simply drop a faulty KB and wait for the next version to appear. At the same time, they cannot ignore the fact that some of the knowledge that they contain is wrong. Hence, they need to be able to use the KB, but in a way that preserves some guarantees of correctness.

V. Torra et al. (Eds.): MDAI 2019, LNAI 11676, pp. 259–271, 2019.
https://doi.org/10.1007/978-3-030-26773-5_23

In this paper, we propose a method for reasoning about consequences that takes into account the *repairs* of the KB. In a nutshell, the repairs correspond to the ways in which all errors may be removed, by deleting a minimal amount of axioms from the KB. Our approach consists in encoding all repairs through a Boolean function over the axioms in the KB. This function can be easily updated if new errors are encountered, and can then be used to decide whether a consequence follows from one or all repairs. Thus, our method can be used throughout deployment time, collecting all known errors as they are detected, and providing guarantees over all reasoning results.

At some point, the knowledge engineer will take control of the KB, and will need to decide which of the many potential repairs is the right one; that is, which axioms are indeed wrong. We emphasise that this task cannot be automated because it needs the expert human knowledge to discern correctness. Rather than making them verify each axiom, we devise a method that suggest to them which axioms to check first, in order to reduce the search space efficiently. With the help of this method, finding the right repair requires minimal human effort, which is the most expensive resource in the repairing scenario.

2   Preliminaries

We consider an abstract logic-based knowledge representation language, where explicit knowledge is expressed via a set of constraints (or axioms), and logical entailments derive other implicitly encoded knowledge. For simplicity, we consider that axioms and implicit consequences have the same shape, although the results can be easily generalised to avoid this restriction (see e.g. [4,11]).

Formally, a *knowledge representation language* is defined by an infinite set $\mathfrak{A}$ of (well-formed) *axioms* and an *entailment relation* $\models\ \subseteq \mathscr{P}_{\mathsf{fin}}(\mathfrak{A}) \times \mathfrak{A}$, where $\mathscr{P}_{\mathsf{fin}}(\mathfrak{A})$ denotes the class of all finite subsets of $\mathfrak{A}$. In general, we call any finite subset of $\mathfrak{A}$ a *knowledge base* (KB). Thus, a knowledge representation language provides the syntax of the axioms that form a knowledge base, and the semantics are given through the entailment relation expressing what consequences can be derived from which KBs. We use infix notation for the entailment relation; i.e., $\mathcal{K} \models \alpha$ expresses that the KB $\mathcal{K}$ entails the axiom α. In this work, we are only interested in *monotonic* knowledge representation languages; these are those where the entailment relation is monotonic in the sense that for every two KBs $\mathcal{K}, \mathcal{K}'$, and axiom α, if $\mathcal{K} \models \alpha$ and $\mathcal{K} \subseteq \mathcal{K}'$, then $\mathcal{K}' \models \alpha$.

For the sake of building an example and improving understanding of the results presented here, we consider a very simple knowledge representation language consisting of directed graphs and reachability between nodes. In this case, given an infinite set $\mathcal{V}$ of *nodes*, an axiom is of the form $v \to w$, where $v, w \in \mathcal{V}$. Intuitively, this axiom expresses that w is reachable from v. A KB, that is, a finite set of axioms, can be represented as a finite graph. The graph $\mathcal{K}$ entails the axiom $v \to w$ iff the node w is reachable from v in $\mathcal{K}$ (see Fig. 1).

An important operation for KB update and belief revision is known as *contraction*. The goal of this operation is to remove a consequence from a KB; in other words, given the KB $\mathcal{K}$ and the axiom α, the contraction $\mathcal{K} - \alpha$ should yield a KB $\mathcal{K}'$ such that $\mathcal{K}' \not\models \alpha$. While it is possible to define many different kinds of contractions, we focus on one that provides minimal syntactic changes to the original KB, which is based on the notion of a repair.

Definition 1 (repair). *A repair of $\mathcal{K}$ w.r.t. α is a sub-KB $\mathcal{K}' \subseteq \mathcal{K}$ such that (i) $\mathcal{K}' \not\models \alpha$ and (ii) for all $\mathcal{K}' \subset \mathcal{L} \subseteq \mathcal{K}$, $\mathcal{L} \models \alpha$.*

In words, a repair is a maximal sub-KB of $\mathcal{K}$ that does not entail α. It is important to notice that repairs are not unique. In fact, removing a single consequence from a KB can produce exponentially many such repairs [10]. To try to reduce the number of options, some heuristics can be proposed; for example, to consider only the repairs with the largest cardinality. However, in general, this is still insufficient to identify one single solution. In order to find only one (adequate) solution, human intervention is needed to provide the expert domain knowledge that distinguishes the axioms that are in fact incorrect w.r.t. the current knowledge and should hence be removed.

As it is well known in the area of belief revision, the problem of choosing the right repair becomes even more crucial in the presence of an iterated contraction [8]; that is, when more than one consequence is to be removed in successive steps. Obviously, making a wrong choice of repair in any given step will negatively affect the following contraction steps. For example, suppose that we are only interested in finding a repair of maximum cardinality that removes a set of consequences from $\mathcal{K}$. If we simply choose one maximum cardinality repair at each step, we are not guaranteed to end up with a solution of maximum cardinality. If human intervention is necessary, then asking an expert to choose the right repair at every single contraction step becomes excessively expensive in terms of human expert resources.

Henceforth, it is useful to consider the dual notion of a repair, called a justification, that corresponds to a minimal sub-KB entailing a consequence.

Definition 2 (justification). *A justification of a consequence α w.r.t. the KB $\mathcal{K}$ is a set $\mathcal{K}' \subseteq \mathcal{K}$ such that $\mathcal{K}' \models \alpha$ and for all $\mathcal{L} \subset \mathcal{K}'$ it holds that $\mathcal{L} \not\models \alpha$.*

It is well known (see [16]) that repairs and justification are dual in the sense that are repair can be obtained by removing at least one axiom from every justification, and justifications are obtained similarly from the axioms that are removed to form repairs. This duality will allow us to exploit efficient methods developed for finding justifications to deal with repairs as well.

In the following sections, we first show how to deal with iterative contractions automatically without losing any valuable information, and then provide a method for helping a human expert to choose the right solution among the potentially exponentially many available, minimising the need of human effort.

3 Iterative Contractions

We are interested in an iterative process for repairing a KB. Starting from a given KB $\mathcal{K}$, we assume that a user is detecting a sequence of erroneous consequences

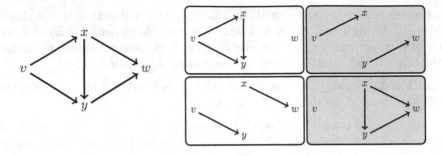

Fig. 1. A KB, and its repairs w.r.t. $v \to w$ and $v \to y$.

that they try to bypass, while waiting for an official correction by the knowledge experts. In the meantime, the KB is still operational and new errors may be derived. At any point in this process, the knowledge expert may attempt to find the *correct* repair, for which automated support should be given.

In order to preserve all the information needed for computing the correct repair, *all* possible solutions should be stored; otherwise, we risk removing the best option from the search space. As mentioned already, even at the first contraction, the number of repairs may become exponential on the size of the KB. Moreover, suppose that we have already a set of repairs obtained after a sequence of contractions. When the following erroneous consequence is detected, it is necessary to contract each of these repairs. In practice, this means potentially computing an exponential number of new KBs for each previously known repair, with the additional cost of verifying that no set appears twice, and that they are all actual repairs. For example, Fig. 1 depicts a simple KB and its repairs w.r.t. the consequence $v \to w$ (all the squared KBs), and w.r.t. the two errors $v \to w, v \to y$ (with a grey background). Notice that the set of axioms obtained by removing $v \to y$ from the lower-left repair is not a repair, because it is strictly contained in the lower-right repair.

To solve this issue, it was proposed in [14] to succinctly encode the class of all repairs by associating with each axiom in the KB a propositional formula expressing to which repairs this axiom belongs, and to which it does not. Although effective, this approach has several issues; most importantly, there is a trade-off between the succinctness of the representation, and the readability of the repairs. That is, reducing the size of the representation comes at the cost of making it harder to know the axioms that belong to each repair.

We propose an alternative approach, encoding the class of all repairs by a Boolean function over the original KB, called a repair function.

Definition 3 (repair function). *Let $\mathcal{K}$ be a KB and C a set of consequences. A function $\mathsf{rep}_C : \mathscr{P}(\mathcal{K}) \to \{0, 1\}$ is a repair function of $\mathcal{K}$ w.r.t. C iff for every $\mathcal{K}' \subseteq \mathcal{K}$ it holds that $\mathsf{rep}_C(\mathcal{K}') = 0$ iff there is an $\alpha \in C$ such that $\mathcal{K}' \models \alpha$.*

Intuitively, a repair function takes as input a sub-KB $\mathcal{K}'$ of $\mathcal{K}$, and returns 1, if $\mathcal{K}'$ does not entail any of the (erroneous) consequences in C (that is, if $\mathcal{K}'$ gets rid of all errors), and 0 otherwise. Notice, however, that $\mathsf{rep}_C(\mathcal{K}') = 1$ does not mean that $\mathcal{K}'$ is a repair, since the maximality criterion is not being considered in the definition of this function, but it is easy to detect the repairs from it. When clear by the context, we remove the subscript C from the name rep.

If we see every axiom as a propositional variable, then rep is simply a propositional formula over the variables in $\mathcal{K}$. Hence, we can encode this function using any of the existing datatypes for compact representation of formulas, such as circuits [17], binary decision diagrams (BDDs) [9], or the more recent sentential decision diagrams (SDDs) [7]. For the sake of an example, and to be consistent with previous work (as should become clear later in this paper), we consider BDDs. In a nutshell, a BDD is a directed acyclic graph with one root node and two terminal nodes (called 0 and 1) such that every non-terminal node node is labelled with a propositional variable and has exactly two successors called the *low* and the *high* branch, and in every path from the root to a terminal node the same variable can appear at most once. A valuation is checked by the BDD by traversing the graph starting from the root and following the low branch if the variable is set of false, and the high branch otherwise. The valuation is a model of the BDD iff this traversal leads to the terminal 1.

To achieve our goal, we build a sequence of repair functions, updating the last one whenever a new fault is found. At the beginning, we have only a KB $\mathcal{K}$, and no errors; that is, $C = \emptyset$. Hence, we have a trivial repair function that maps every subset of $\mathcal{K}$ to 1. In terms of propositional formulas, we represent it as a tautology, which corresponds to the BDD having only one node 1.

Suppose now that we have already detected a set of faults C, and that we have computed a repair function rep for it. For simplicity, we observe this function as a propositional formula. Our goal now is, given a new unwanted consequence α, to compute a repair function rep' w.r.t. $C \cup \{\alpha\}$. One could, of course, compute this new function from scratch, ignoring the properties of rep. We instead opt to update rep to exclude those sub-KBs of $\mathcal{K}$ that entail α from being accepted.

Recall that a justification is a minimal sub-KB that entails α. Intuitively, if $\mathcal{M}$ is a justification, then any set containing $\mathcal{M}$ will entail α. If $\mathcal{M}$ was the *only* justification, then minimality would imply that excluding at least one axiom from $\mathcal{M}$ would also get rid of the consequence α; i.e., if $\mathcal{M} \setminus \mathcal{K}' \neq \emptyset$, then $\mathcal{K}' \not\models \alpha$ holds for all $\mathcal{K}' \subseteq \mathcal{K}$. Since there may exist more than one justification, we need to ensure that none of them is contained in a set to guarantee that α is not entailed. We do this with the help of a *pinpointing formula* [3].

Definition 4 (pinpointing formula). *Let $\mathcal{K}$ be a KB, and α a consequence. A pinpointing formula for α w.r.t. $\mathcal{K}$ is a Boolean function $\mathsf{pin}_\alpha : \mathscr{P}(\mathcal{K}) \to \{0,1\}$ such that, for every $\mathcal{K}' \subseteq \mathcal{K}$ it holds that $\mathsf{pin}_\alpha(\mathcal{K}') = 1$ iff $\mathcal{K}' \models \alpha$.*

Notice that the pinpointing formula is the analogous notion to the repair function, when speaking about justifications instead of repairs. The name is chosen to preserve consistency with existing terminology. *Pinpointing* refers to the task

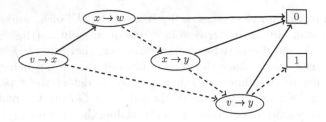

Fig. 2. A BDD for the repairs of Fig. 1.

of computing all justifications, and the notion of *formula* arises from considering all axioms as propositional variables inside the Boolean function pin, as we did before for the repair function. It is perhaps worth noting that Definition 4 is simpler than Definition 3 mainly because the former refers to only one consequence, while the later must consider a whole set of them. It is easy to see that for all α and all C, $\text{rep}_{\{\alpha\}} = \neg\text{pin}_\alpha$ and $\text{rep}_C = \bigwedge_{\alpha \in C} \neg\text{pin}_\alpha$.

Effective methods for constructing and dealing with pinpointing formulas have been studied and implemented for different representation languages; most notably for DLs [18]. The bottom line is that it is possible to efficiently construct a pinpointing formula for many different cases. In particular, one can even build a BDD encoding this formula for expressive representation languages.

As mentioned already, the pinpointing formula can be used to update the repair function to exclude one more consequence. Remember that, from each repair, we need to exclude every justification. That is, the new repair function should map a sub-KB $\mathcal{K}'$ to 1 if the original rep did so (i.e., $\text{rep}(\mathcal{K}') = 1$), but the pinpointing formula did not ($\text{pin}(\mathcal{K}') = 0$). If we see these functions as formulas, this means that rep' should be $\text{rep} \land \neg\text{pin}$.

Theorem 5. *Let $\mathcal{K}$ be a KB, rep a repair function w.r.t. a set of consequence C, and pin a pinpointing formula w.r.t. the consequence α. Then $\text{rep}' := \text{rep} \land \neg\text{pin}$ is a repair function w.r.t. $C \cup \{\alpha\}$.*

Proof. Let $\mathcal{K}' \subseteq \mathcal{K}$. We need to show that $\text{rep}'(\mathcal{K}') = 0$ iff there is a $\beta \in C \cup \{\alpha\}$ such that $\mathcal{K}' \models \beta$. We see that $\text{rep}'(\mathcal{K}') = 0$ iff $\text{rep}(\mathcal{K}') = 0$ or $\text{pin}(\mathcal{K}') = 1$. The former holds iff there is a $\beta \in C$ such that $\mathcal{K}' \models \beta$, while the latter holds iff $\mathcal{K}' \models \alpha$. Hence, overall, we get the desired result.   □

This means that the repair functions can be iteratively constructed by conjoining the negations of the pinpointing formulas of the unwanted consequences that are encountered during the use of the KB. At any moment, we can try to extract some meaningful information out of this repair function. Before we describe the kind of information that can be extracted, and the methods for doing so, we note that the correctness of Theorem 5 depends fundamentally on the fact that the repair function captures all sub-KBs that do not entail the consequences in

C, and not only the repairs (i.e., not only the subset-maximal ones). Figure 2 depicts a BDD for the repairs w.r.t. $\{v \to w, v \to y\}$ constructed this way.

Recall now that we are using, during production, a KB that is known to contain some errors. Still, until it is completely repaired, we want to be able to extract some meaningful information out of this KB. In particular, we would like to be able to derive consequences with a guarantee, or at least a hint, of correctness. This motivates the use of cautious and brave consequences.

Definition 6 (cautious, brave consequences). *Let $\mathcal{K}$ be a KB, C a set of consequences, and α a consequence. $\mathcal{K}$ cautiously entails α w.r.t. C ($\mathcal{K} \models_c \alpha$) iff for every repair $\mathcal{K}'$ of $\mathcal{K}$ w.r.t. C it holds that $\mathcal{K}' \models \alpha$. $\mathcal{K}$ bravely entails α w.r.t. C ($\mathcal{K} \models_b \alpha$) iff there exists a repair $\mathcal{K}'$ of $\mathcal{K}$ w.r.t. C such that $\mathcal{K}' \models \alpha$.*

Intuitively, a cautious consequence is guaranteed to hold regardless of which repair we choose, and hence we are certain that, after the errors have been fixed, this consequence will still hold. Thus, a user can safely use this consequence, which circumvents all the known errors. On the other hand, brave consequences only need to hold in at least one repair. Note that this could, in fact, be the correct repair, in which case this consequence would still hold after the KB is fixed. However, there is no guarantee that this will be the case. Although brave consequences are relatively weak, and do not preserve some basic logical closure properties (e.g., $\mathcal{K} \models_b a \to b$ and $\mathcal{K} \models_b b \to c$ does not imply that $\mathcal{K} \models_b a \to c$), they can be useful to know that the consequence is still possible after repairing. If this is a wanted consequence, we can use this information to guide the search for the right repair. Or, if it is an unwanted consequence, knowing that it is not bravely entailed allows us to avoid making an additional (but irrelevant) contraction step. We first see that brave consequences can be decided by operating over the repair function and the pinpointing formula.

Proposition 7. *Let $\mathcal{K}$ be a KB, rep a repair function w.r.t. C, and pin a pinpointing formula for α. $\mathcal{K}$ bravely entails α w.r.t. C iff rep does not entail ¬pin.*

Proof. If $\mathcal{K} \models_b \alpha$, then there exists a repair $\mathcal{K}'$ w.r.t. C such that $\mathcal{K}' \models \alpha$. By definition, this means that $\mathsf{rep}(\mathcal{K}') = 1 = \mathsf{pin}(\mathcal{K}')$. So, it cannot hold that rep $\Rightarrow$ ¬pin. Conversely, if $\mathcal{K} \not\models_b \alpha$, then for every repair $\mathcal{K}'$ we know that $\mathcal{K}' \not\models \alpha$. By monotonicity of the consequences, the same is true for all subsets of a repair. That is, for every $\mathcal{L} \subseteq \mathcal{K}$, if $\mathsf{rep}(\mathcal{L}) = 1$ then $\mathcal{L} \not\models \alpha$. In terms of the pinpointing formula, this means that if $\mathsf{rep}(\mathcal{L}) = 1$, then $\mathsf{pin}(\mathcal{L}) = 0$, or alternatively, $\neg\mathsf{pin}(\mathcal{L}) = 1$. Hence rep $\Rightarrow$ ¬pin. □

Dealing with cautious consequences requires a slightly more complex analysis. To verify that a consequence is not cautiously entailed, it does not suffice to simply check whether there is a set accepted by rep that is rejected by pin (i.e., a set that avoids C and also rejects α). This is because rep also accepts all subsets of the repairs. In particular, the empty KB is such that $\mathsf{rep}(\emptyset) = 1$ and $\mathsf{pin}(\emptyset) = 0$, assuming that α and the consequences in C are not tautologies. Thus, we really need to be careful that the maximality of the repairs is taken into consideration.

Proposition 8. *Let $\mathcal{K}$ be a KB,* rep_C *a repair function w.r.t. C, and* pin_α *a pinpointing formula for α. $\mathcal{K}$ cautiously entails α w.r.t. C iff for every set $\mathcal{K}'$ satisfying* $\mathsf{rep}_C \wedge \neg\mathsf{pin}_\alpha$ *there exists some $\mathcal{L} \supseteq \mathcal{K}'$ that satisfies* $\mathsf{rep}_C \wedge \mathsf{pin}_\alpha$.

Proof. Suppose that $\mathcal{K} \models_c \alpha$, and let $\mathcal{K}'$ be such that satisfies $\mathsf{rep}_C \wedge \neg\mathsf{pin}_\alpha$. By definition of pin, $\mathcal{K}' \not\models \alpha$. Since α is cautiously entailed, $\mathcal{K}'$ cannot be a repair, but since $\mathsf{rep}_C(\mathcal{K}') = 1$, it should be a subset of a repair. Let $\mathcal{L}$ be any repair containing $\mathcal{K}'$. By construction, $\mathcal{L}$ satisfies $\mathsf{rep}_C \wedge \mathsf{pin}_\alpha$. Conversely, if $\mathcal{K} \not\models_c \alpha$, then there exists some repair $\mathcal{K}'$ such that $\mathcal{K}' \not\models \alpha$. Again, by construction $\mathcal{K}'$ satisfies $\mathsf{rep}_C \wedge \neg\mathsf{pin}_\alpha$. But since $\mathcal{K}'$ is a repair, for every $\mathcal{L} \supseteq \mathcal{K}'$ we know that $\mathsf{rep}_C(\mathcal{L}) = 0$. Hence, there exists no $\mathcal{L} \supseteq \mathcal{K}'$ that satisfies $\mathsf{rep} \wedge \mathsf{pin}$. □

For the sake of completeness, we consider a third kind of entailment that has been studied in the literature. A KB *IAR entails* the consequence α w.r.t. a set C iff α follows from the intersection of all repairs of $\mathcal{K}$ w.r.t. C [5]. In this case, the name IAR is an acronym for *intersection of all repairs*. Note that when dealing with IAR entailments, the structure of the repairs becomes irrelevant, and the only important information is which axioms appear in all of them. In fact, if we know the intersection of all the repairs, then IAR entailment becomes just a standard entailment over this sub-KB.

To solve IAR entailments, it suffices to identify the axioms that appear in the intersection. This can be done in two ways: either, on demand, whenever an IAR entailment test is required, or during the construction of the repair function, while preserving an additional data structure. We can see that the intersection of all repairs is the complement of the union of all justifications. Hence, as we construct rep, every time that we compute the set of all justifications, we can simply remove their union from the set of still active axioms. This approach becomes more efficient if many IAR entailment tests are expected, since the intersection is computed only once, and used for all tests.

We have so far focused on the problem of computing the repair function, which encodes all the repairs for a sequence of unwanted consequences from a KB, and how to use this function, together with the pinpointing formula, to perform meaningful reasoning tasks that avoid the known errors. In the next section, we consider a different problem; namely, helping a knowledge engineer to find the right repair (that is, identify which are the truly faulty axioms) from a potentially exponential class of options without having to analyse all of them independently, and trying to minimise their effort.

4   Choosing the Right Repair

So far we have worked under the assumption that a user of the KB, while capable of detecting errors in the consequences observed, is not knowledgeable enough— or is not authorised—to definitely fix the KB. As we have seen, there may exist many different repairs, and each of them is a potential solution to get rid of all the errors. However, not all of them are equally valid. Suppose for example that our KB refers to a concept hierarchy, where $a \rightarrow b$ means that every a is a b. In this setting, we can consider the KB that has two axioms human $\rightarrow$ mammal and mammal $\rightarrow$ plant. A consequence of this KB is that every human is a plant, which is obviously unwanted. Each of the two axioms human $\rightarrow$ mammal and mammal $\rightarrow$ plant is a repair, but clearly, only the former one is a valid repair w.r.t. the domain of the KB. The task of the knowledge engineer is to identify this right repair from the sea of all potential repairs.

It should be clear that it is completely infeasible for the knowledge engineer to explore all the repairs one by one, and decide which is the correct one. First, there are many such repairs, making this an overwhelming task for a human resource. Moreover, each repair is expected to be large and, as in the original KB, it may be very hard to spot an error in an axiom hidden within thousands of other axioms. Our proposal is to make the knowledge engineer verify only one axiom at a time, rather than a full repair. Once again, one could think of checking each axiom in the KB, which is clearly infeasible for very large KBs. Instead, we select first those axioms that are more likely to lead to the correct repair in the least amount of steps, regardless of the answer provided by the knowledge engineer. More precisely, we try to find an axiom that belongs to half of all repairs, or as close as possible to that.

Definition 9 (cut axiom). *Let $\mathcal{K}$ be a KB, and $\mathcal{R}$ the set of all repairs of $\mathcal{K}$ w.r.t. a set of consequences C. Given an axiom $\alpha \in \mathcal{K}$, we define the sets*

$$\mathcal{R}_\alpha^+ := \{\mathcal{L} \in \mathcal{R} \mid \alpha \in \mathcal{L}\},$$
$$\mathcal{R}_\alpha^- := \{\mathcal{L} \in \mathcal{R} \mid \alpha \notin \mathcal{L}\}.$$

The axiom α is called a cut axiom iff for every $\beta \in \mathcal{K}$ it holds that

$$||\mathcal{R}_\alpha^+| - |\mathcal{R}_\alpha^-|| \le ||\mathcal{R}_\beta^+| - |\mathcal{R}_\beta^-||.$$

The idea behind the cut axiom is that, by verifying its correctness, we can immediately cut the search space (almost) in half. Specifically, if α is correct, then we know that the right repair is among $\mathcal{R}_\alpha^+$, and if it is wrong, we should focus only on $\mathcal{R}_\alpha^-$. Hence, the first question is how to compute such an axiom. Unfortunately, it turns out that deciding whether an axiom is a cut axiom is coNP-hard already for the very simple representation language that we are using as an example, and only one unwanted consequence is known.

Theorem 10. *Given a graph $\mathcal{K}$, an edge $v \rightarrow w \in \mathcal{K}$, and an unwanted reachability entailment $x \rightarrow y$, deciding whether $v \rightarrow w$ is a cut axiom is coNP-hard.*

Proof. We prove this by a reduction from the following coNP-hard problem: is there a repair $\mathcal{L}$ for $\mathcal{K}$ w.r.t. $x \rightarrow y$ such that $v \rightarrow w \notin \mathcal{L}$? [12]. Let $n := |\mathcal{K}|$. We notice that there can exist at most 2^{n-1} repairs that contain $v \rightarrow w$ and at most 2^{n-1} that do not contain this axiom. Assuming w.l.o.g. that there are at least two repairs, we construct the new KB $\mathcal{M}$, which extends $\mathcal{K}$ by adding $2n$ new vertices $z_1, \ldots, z_{2n}$, and the edges

$$E' := \{v \rightarrow z_i, z_i \rightarrow w \mid 1 \le i \le 2n\}.$$

Clearly, the size of $\mathcal{M}$ is linear on the size of $\mathcal{K}$.

This new KB has the following property. For every repair $\mathcal{K}'$ of $\mathcal{K}$ w.r.t. $x \to y$, (i) if $v \to w \in \mathcal{K}'$ then $\mathcal{K} \cup E'$ is the only repair of $\mathcal{M}$ w.r.t. $x \to y$ that contains $\mathcal{K}'$ (that is, there is a one-to-one correspondence between the repairs of $\mathcal{K}$ and of $\mathcal{M}$ that contain $x \to y$); and (ii) if $v \to w \notin \mathcal{K}'$, then there exist 2^{2n} different repairs of $\mathcal{M}$ w.r.t. $x \to y$ that contain $\mathcal{K}'$; in particular, exactly 2^{2n-1} of those repairs contain the axiom $v \to z_1$.

In particular, if $\mathcal{K}$ has m repairs containing $v \to w$ and ℓ not containing this axiom, then $\mathcal{M}$ will have $m + \ell \cdot 2^{2n-1}$ repairs containing $v \to z_1$ and $\ell \cdot 2^{2n-1}$ repairs not containing the axiom. Moreover, every other axiom of $\mathcal{K}$ will appear in at most 2^{n-1} repairs of $\mathcal{M}$, and all the axioms in E' will be in exactly the same number of repairs as $v \to z_1$. Thus, $x \to z_1$ is a cut axiom w.r.t. $\mathcal{M}$ iff $\ell \geq 1$; that is, iff there is at least one repair of $\mathcal{K}$ that does not contain $v \to w$. □

What this theorem shows is that it is in general hard to detect which axiom is the best option to verify first in order to guarantee a reduction of the search space. In fact, it is very unlikely that the exact complexity coincides with this lower bound. It is known that counting the number of justifications, and the number of justifications to which an axiom belongs are #P-complete problems [13]. Although such hardness results have not been shown, to the best of our knowledge, also for the problem of counting repairs, the duality between both problems strongly suggests that this is the case as well.

On the other hand, notice that not all knowledge representation languages are as inexpressive as our example graph language. Indeed, even if we restrict to decidable cases, there are some mainstream languages where reasoning is at least NExpTime-hard [2]. In such cases (and indeed in any language where reasoning is PSpace-hard), finding a cut axiom is as expensive, in terms of computational complexity, as merely deciding whether a consequence follows.

Suppose that we have found a cut axiom α. The next step is to use it to prune the search space in a way that the process can be iterated until the right repair is found. After we propose this axiom to the knowledge engineer, they will respond either that the axiom is (i) correct, or that it is (ii) incorrect. In the second case, we know that α cannot appear in the right repair. Thus, we simply eliminate it from the KB $\mathcal{K}$. In practical terms, this means updating the repair function to ignore α, considering it as being always false. If the repair function is expressed as a BDD, this operation corresponds simply to removing the nodes representing α from the diagram, and updating every edge pointing to those nodes to now point to their lower children—that is, to the node reached when α is evaluated to false at that node.

In the case (i)—i.e., when α is marked as correct—then we should ignore all repairs that do not contain α, and focus only on those that include it. In principle, we could do as before and remove all α nodes, but now assuming that it is always true within the repair function. However, we want to preserve the knowledge that α must appear in all repairs. Hence, rather than removing the variable from the formula, we enforce the repair function to exclude any set that does not contain α. In a BDD, this is achieved by substituting the lower branch of every α node with the 0-terminal.

Importantly, after these operations the resulting structure is still a repair function, but which now accepts only those repairs that comply with the information provided by the knowledge engineer. Hence, we can repeat the process, reducing at each step the total number of repairs remaining. Note that once that an axiom has been analysed by the knowledge engineer, it will never be proposed by the system again. Hence, in the worst case, it will need as many tests as there are axioms in the KB; i.e., it is not worse than the naïve approach of testing all axioms in order. In reality, far less axioms will be tested. First, notice that all axioms that appear either in all repairs, or in none of them will never be proposed by the method: the only way they become cut axioms is if there is only one repair left, in which case the process has already finished. Second, every time that we test an axiom, we reduce the class of remaining repairs, which in turn increases both, the class of axioms that belong to all repairs, and the class of axioms that belong to none. Thus, more axioms will be excluded from testing.

5 Conclusions

We have proposed a methodology for helping knowledge engineers to make decisions in relation to faulty knowledge bases. The premise of this work is based on the idea of preserving all the information regarding all the possible ways in which errors, which may be detected while a KB is in use, may be avoided. For this purpose, we propose to store a Boolean function, which accepts only those sub-KBs that are free of all known errors. If new faults are encountered, this Boolean function—called the repair function—can be updated using known techniques from the area of consequence explanation (or axiom pinpointing).

The importance of the repair function is that it can be used to decide different properties of the consequences of the original KB. Specifically, one can verify whether a consequence can still be derived from any possible repair (cautious reasoning), or from at least one of them (brave reasoning). As we have no precise notion *a priori* of the actual repair that will be obtained after the knowledge engineer has verified the correctness of the axioms, brave consequences have also a place as ones that can potentially remain.

In addition, we introduced the notion of a cut axiom, which is the one that appears in half the repairs, or as close to that as possible. Although finding a cut axiom is computational hard, it is an effective tool for guiding the search for the right repair that fixes all the faults encountered. The computational effort needed to find the cut axiom is a good investment in reducing the human cost of verifying the correctness of a repair. We notice that a full repair plan (e.g. a decision tree) can be computed offline before any intervention by the human expert, so that they are given a sequence of questions without delay.

In summary, our proposal is to preserve all the information of the potential repairs during deployment to allow users to make informed decisions that sidestep known errors, and to use this information to minimise the number of queries made to a human expert to identify the actual correct knowledge base that should be used in the following publishing cycle.

One line of future research is to try to extend the notion of a cut axiom into a *cut consequence*; that is, using potentially complex consequences to better separate the space of repairs. The challenge in this direction is to adequately restrict the class of consequences that may be used, and to develop an effective method that does not need to enumerate them all. Another goal is to improve the notion of a cut axiom so that it also takes into account the improvements on the following steps; for instance, to consider those that minimise the average number of questions that need to be asked overall. Finally, we will implement a prototypical system for a well-known knowledge representation language, and test the effectiveness of our methods on realistic KBs.

References

1. Apt, K.: Principles of Constraint Programming. Cambridge University Press, New York (2003)
2. Baader, F., Calvanese, D., McGuinness, D., Nardi, D., Patel-Schneider, P. (eds.): The Description Logic Handbook: Theory, Implementation, and Applications, 2nd edn. Cambridge University Press, New York (2007)
3. Baader, F., Peñaloza, R.: Automata-based axiom pinpointing. In: Armando, A., Baumgartner, P., Dowek, G. (eds.) IJCAR 2008. LNCS (LNAI), vol. 5195, pp. 226–241. Springer, Heidelberg (2008). https://doi.org/10.1007/978-3-540-71070-7_19
4. Baader, F., Peñaloza, R.: Axiom pinpointing in general tableaux. J. Logic Comput. **20**(1), 5–34 (2010). https://doi.org/10.1093/logcom/exn058
5. Bienvenu, M., Rosati, R.: Tractable approximations of consistent query answering for robust ontology-based data access. In: Proceedings IJCAI 2013, pp. 775–781. AAAI Press (2013)
6. Biere, A., Heule, M., van Maaren, H., Walsh, T. (eds.): Handbook of Satisfiability, Frontiers in Artificial Intelligence and Applications, vol. 185. IOS Press, Amsterdam (2009)
7. Darwiche, A.: SDD: a new canonical representation of propositional knowledge bases. In: Proceedings IJCAI 2011, pp. 819–826. IJCAI/AAAI (2011)
8. Darwiche, A., Pearl, J.: On the logic of iterated belief revision. In: Proceedings TARK 1994, pp. 5–23 (1994)
9. Drechsler, R., Becker, B.: Binary Decision Diagrams - Theory and Implementation. Springer, Berlin (1998)
10. Ludwig, M., Peñaloza, R.: Error-tolerant reasoning in the description logic $\mathcal{EL}$. In: Fermé, E., Leite, J. (eds.) JELIA 2014. LNCS (LNAI), vol. 8761, pp. 107–121. Springer, Cham (2014). https://doi.org/10.1007/978-3-319-11558-0_8
11. Peñaloza Nyssen, R.: Axiom pinpointing in description logics and beyond. Ph.D. thesis, Technische Universität Dresden, Germany (2009)
12. Peñaloza, R.: Inconsistency-tolerant instance checking in tractable description logics. In: Costantini, S., Franconi, E., Van Woensel, W., Kontchakov, R., Sadri, F., Roman, D. (eds.) RuleML+RR 2017. LNCS, vol. 10364, pp. 215–229. Springer, Cham (2017). https://doi.org/10.1007/978-3-319-61252-2_15
13. Peñaloza, R., Sertkaya, B.: Understanding the complexity of axiom pinpointing in lightweight description logics. Artif. Intell. **250**, 80–104 (2017)

14. Peñaloza, R., Thuluva, A.S.: Iterative ontology updates using context labels. In: Proceedings JOWO 2015. CEUR Workshop Proceedings, vol. 1517. CEUR-WS.org (2015)

15. Price, C., Spackman, K.: Snomed clinical terms. Br. J. Healthc. Comput. Inf. Manag. **17**(3), 27–31 (2000)

16. Reiter, R.: A theory of diagnosis from first principles. AIJ **32**(1), 57–95 (1987)

17. Shannon, C.E.: The synthesis of two-terminal switching circuits. Bell Syst. Tech. J. **28**(1), 59–98 (1949)

18. Zese, R., Bellodi, E., Riguzzi, F., Cota, G., Lamma, E.: Tableau reasoning for description logics and its extension to probabilities. AMAI **82**(1–3), 101–130 (2018)

k-Medoids Clustering
Based on Kernel Density Estimation
and Jensen-Shannon Divergence

Yukihiro Hamasuna$^{1(\boxtimes)}$, Yuto Kingetsu^2, and Shusuke Nakano^2

^{1} Department of Informatics, School of Science and Engineering, Kindai University,
3-4-1 Kowakae, Higashiosaka, Osaka 577-8502, Japan
yhama@info.kindai.ac.jp
^{2} Graduate School of Science and Engineering, Kindai University,
3-4-1 Kowakae, Higashiosaka, Osaka 577-8502, Japan
{1933340439u,shusuke.nakano}@kindai.ac.jp

Abstract. Several conventional clustering methods consider the squared L_2-norm which is calculated from objects coordinates. To extract meaningful clusters from a set of massive objects, it is required to calculate the dissimilarity from both objects coordinates and other features such as objects distribution. In this paper, JS-divergence based k-medoids (JSKMdd) is proposed as a novel method for clustering network data. In the proposed method, the dissimilarity that is based on objects coordinates and an object distribution is considered. The effectiveness of the proposed method is verified through numerical experiments with artificial datasets which consist non-linear clusters. The influence of the parameter in the proposed method is also described.

Keywords: Clustering · k-medoids · Kernel density estimation ·
Jensen-Shannon divergence

1 Introduction

Clustering is one of the data analysis methods that divide a set of objects into groups called clusters [1,2]. Objects classified in the same cluster are considered similar, whereas those in different clusters are considered dissimilar. A similarity or dissimilarity which are defined between objects is considered in the clustering procedure. The squared L_2-norm is a typical dissimilarity. When the squared L_2-norm is considered as the dissimilarity, a cluster partition forms a Voronoi diagram. Because of the property of the squared L_2-norm, it is difficult to cluster a set of objects into non-linear clusters. Kernel method is actively used in pattern analysis to handle complex datasets and extract important properties [1,3,4]. The kernel method is well-known as the significant method to handle complex datasets which consist non-linear cluster boundary. The dissimilarity between object and cluster center is considered in high dimensional feature space in the

© Springer Nature Switzerland AG 2019
V. Torra et al. (Eds.): MDAI 2019, LNAI 11676, pp. 272–282, 2019.
https://doi.org/10.1007/978-3-030-26773-5_24

kernelized clustering method. Several conventional clustering methods consider the squared L_2-norm which is calculated from objects coordinates.

To extract meaningful clusters from a set of massive objects, it is required to calculate the dissimilarity from both objects coordinates and other features such as objects distribution. For example, the Mahalanobis distance that is based on the variance-covariance matrix is a typical example of above [1,5]. In addition, noise clustering [6], tolerance [7], and even-sized [8] approaches are typical examples of clustering methods that considers typical cluster structures into optimization framework. Moreover, local outlier factor (LOF) which is known to as a useful anomaly detection method uses k-distance to calculate the density of objects [9]. These methods and their utility suggest that objects coordinates, cluster structure, and data distribution are effective to cluster a set of massive objects.

In this paper, JS-divergence based k-medoids (JSKMdd) is proposed as a novel method for clustering network data. In the proposed method, the dissimilarity that is based on objects coordinates and an object distribution is considered. First, the probability density function is estimated from an object and its neighbors by kernel density estimation (KDE) [11]. Second, the dissimilarity between the distributions obtained by KDE is calculated by Jensen-Shanon divergence (JS-divergence) [12]. Third, k-medoids (KMdd) [10] is executed to obtain cluster partition by using the dissimilarity based on JS-divergence.

Further, the effectiveness of the proposed method is verified through numerical experiments with artificial datasets which consist non-linear clusters. The influence of the parameter in the proposed method is also described.

The remainder of this paper is organized as follows: In Sect. 2, we introduce the notation, KMdd, KDE, and JS divergence. In Sect. 3, we propose JS-divergence based KMdd (JSKMdd). In Sect. 4, we describe the conducted experiments to demonstrate the effectiveness of the proposed method. In Sect. 5, we provide some concluding remarks regarding this research.

2   Preliminaries

A set of objects to be clustered is given, and it is denoted by $X = \{x_1, \ldots, x_n\}$, where x_k $(k = 1, \ldots, n)$ is an object. In most cases, each object x_k is a vector in the p-dimensional Euclidean space $\Re^p$, that is, an object $x_k \in \Re^p$. A cluster is denoted by G_i, and a collection of clusters is given by $\mathcal{G} = \{G_1, \ldots, G_c\}$. The membership degree of x_k belonging to G_i and a partition matrix is denoted as u_{ki}, and $U = (u_{ki})_{1 \leq k \leq n, \ 1 \leq i \leq c}$.

2.1   k-medoids

k-medoids is a variant of k-means clustering. The cluster center is used as a cluster representative in k-means [2]. In contrast, an object in each cluster is chosen as a cluster representative in k-medoids [10]. An objective function of k-medoids is as follows:

$$J_{\mathrm{KMdd}}(U, W) = \sum_{i=1}^{c} \sum_{k=1}^{n} \sum_{l=1}^{n} u_{ki} w_{li} r_{kl}. \tag{1}$$

Here, r_{kl} represents a measure of relationship between objects and $W = (w_{li})_{1 \leq l \leq n, \, 1 \leq i \leq c}$ is a variable called prototype weight. In many cases, r_{kl} is considered as a dissimilarity between objects. An algorithm of k-medoids is based on the alternating optimization with u_{ki} and w_{li} under the constraints on u_{ki} and w_{li} as follows:

$$\mathcal{U}_h = \left\{ (u_{ki}) : u_{ki} \in \{0, 1\}, \ \sum_{i=1}^{c} u_{ki} = 1, \ ^{\forall}k \right\}, \tag{2}$$

$$\mathcal{W}_h = \left\{ (w_{li}) : w_{li} \in \{0, 1\}, \ \sum_{l=1}^{n} w_{li} = 1, \ ^{\forall}i \right\}. \tag{3}$$

The lth-object that takes $w_{li} = 1$ is the representative in a cluster. The important feature of k-medoids is that it handles relational data denoted as a table of distances between objects such as network data.

The optimal solutions for u_{ki} and w_{li} are as follows:

$$u_{ki} = \begin{cases} 1 & \left(i = \arg\min_{s} \sum_{l=1}^{n} w_{ls} r_{kl} \right), \\ 0 & (\text{ otherwise }) \end{cases} \tag{4}$$

$$w_{li} = \begin{cases} 1 & \left(l = \arg\min_{t} \sum_{k=1}^{n} u_{ki} r_{kt} \right). \\ 0 & (\text{ otherwise }) \end{cases} \tag{5}$$

The medoid of G_i is denoted in another form as follows:

$$\mathrm{Mdd}(G_i) = \arg\min_{x_k \in G_i} \sum_{x_l \in G_i} d(x_k, x_l). \tag{6}$$

(5) and (6) mean the same optimal solution. By considering the optimization problem of J_{kd}, the optimal solution of w_{li} is described in (5). Further, (6) is considered by considering k-medoids in the algorithmic procedure. The algorithm of KMdd is summarized as follows:

Algorithm 1. KMdd

KMdd1 Set initial medoids $w_{li} \in W$.
KMdd2 Calculate $u_{ki} \in U$ by (4).
KMdd3 Calculate $w_{li} \in W$ by (5).
KMdd4 If convergence criterion is satisfied, stop. Otherwise go back to **KMdd2**.

The number of repetitions, the convergence of each variable, or the convergence of an objective function is used as the convergence criterion in **KMdd4**.

2.2   Kernel Density Estimation

KDE is one of the methods to estimate probability density function [11]. KDE is a nonparametric method and does not require a specific density function. KDE estimates the probability density function from given a set of objects automatically.

Let x_k be an independent, identically distributed random variable. KDE estimates a probability density function according to the following function:

$$p(x) = \frac{1}{nh} \sum_{k=1}^{n} K\left(\frac{x - x_k}{h}\right). \tag{7}$$

where $K(\cdot)$ is a kernel function and $h > 0$ is the bandwidth parameter which controls the smoothness.

The kernel function $K(\cdot)$ satisfies the following properties:

$$K(x) \geq 0, \quad \int K(x)dx = 1, \quad \int xK(x)dx = 0, \quad \int x^2 K(x)dx > 0.$$

A common example of $K(x)$ is the Gaussian kernel:

$$K(x) = \frac{1}{\sqrt{2\pi}} \exp\left(-\frac{x^2}{2}\right).$$

A multivariate density function estimated from KDE is as follows:

$$p(x) = \frac{1}{n} \sum_{k=1}^{n} \prod_{j=1}^{p} \frac{1}{h_j} K\left(\frac{x^j - x_k^j}{h_j}\right). \tag{8}$$

where, h_j is also the bandwidth parameter which controls the smoothness for each dimension.

2.3   Jensen-Shannon Divergence

JS-divergence is a method that measures the dissimilarity between two probability distribution functions [12]. JS-divergence is based on the Kullback-Leibler divergence (KL-divergence) [13]. KL-divergence between two probability distribution functions $p(x)$ and $q(x)$ is as follows:

$$KL(p \parallel q) = \int_{-\infty}^{+\infty} p(x) \log \frac{p(x)}{q(x)} dx.$$

It is known that KL-divergence is not symmetric. Then, JS-divergence has been proposed to realize symmetry. JS-divergence is as follows:

$$JS(p \parallel q) = \frac{1}{2} KL(p \parallel m) + \frac{1}{2} KL(q \parallel m), \tag{9}$$

where $m(x) = \frac{1}{2}(p(x) + q(x))$.

3　Proposed Method

JSKMdd is proposed as a novel method that considers objects coordinates and distribution to calculate the dissimilarity. In JSKMdd, KDE estimates an object distribution of x_k by using its neighbors. JS-divergence gives the dissimilarity between two distributions.

First, the procedure of estimating the object distribution is described. The neighbor objects x_k is defined as follows:

$$N(x_k) = \{x \in X \mid d(x_k, x) \leq D\} \tag{10}$$

where $D > 0$ is a parameter which is a radius of hyper-sphere and $d(x_k, x)$ is a distance of two objects like L_2-norm. It means that objects x within the range from the object x to D are included to $N(x_k)$. Another definition is as follows:

$$N(x_k) = \{x \in X \mid d(x_k, x) \leq d(x_k, x_{q(t)})\} \tag{11}$$

Here, $q(t)$ means the object numbers sorted ascending order of the distance. First, the distance between x_k to $x \in X$ is calculated.

$$d(x_k, x_1), \ldots, d(x_k, x_n).$$

The above distance is sorted in ascending order as follows:

$$d(x_k, x_{q(1)}) \leq \ldots \leq d(x_k, x_{q(n)}),$$

where $q(t) \in \{1, \ldots, n\}$ means the object number with t-th smallest distance. $q(t)$ is the number replaced in order of the distance. Compared with (10), $N(x_k)$ defined by (11) consists t neighbors.

The object distribution $p(x_k)$ is calculated by KDE using the $N(x_k)$ defined by (10) or (11) as follows:

$$p(x_k) = \frac{1}{|N(x_k)|} \sum_{x \in N(x_k)} \prod_{j=1}^{p} \frac{1}{h_j} K\left(\frac{x^j - x_k^j}{h_j}\right). \tag{12}$$

where $|N(x_k)|$ means the number of objects in $N(x_k)$. $p(x_k)$ is the data distribution calculated by its neighbors $N(x_k)$. The dissimilarity is calculated by JS-divergence with the data distribution $p(x_k)$ according to (9).

The optimization problem of JSKMdd is as follows:

$$J_{\text{JSKMdd}}(U, W) = \sum_{i=1}^{c} \sum_{k=1}^{n} \sum_{l=1}^{n} u_{ki} w_{li} r'_{kl},$$

$$\mathcal{U}_h = \left\{ (u_{ki}) : u_{ki} \in \{0, 1\}, \ \sum_{i=1}^{c} u_{ki} = 1, \ ^{\forall}k \right\},$$

$$\mathcal{W}_h = \left\{ (w_{li}) : w_{li} \in \{0, 1\}, \ \sum_{l=1}^{n} w_{li} = 1, \ ^{\forall}i \right\}.$$

where the objective function and constraints are the same as (1), (2), and (3). The difference of KMdd and JSKMdd is that r'_{kl} is the JS-divergence calculated by (9) under (11).

The algorithm of JSKMdd is summarized as follows:

Algorithm 2. JSKMdd

JSKMdd1 Set parameter t or D. Calculate $p(x_k)$ by (12).
JSKMdd2 Calculate r_{kl} by (9).
JSKMdd3 Set initial medoids.
JSKMdd4 Calculate $u_{ki} \in U$ by (4).
JSKMdd5 Calculate $w_{li} \in W$ by (5).
JSKMdd6 If convergence criterion is satisfied, stop.
    Otherwise go back to **JSKMdd4**.

In this paper, k-medoids type algorithm is considered because a large calculation time is required to calculate JS-divergence. If JS-divergence is considered in k-means type algorithm, the dissimilarity between object and cluster centers is required to update in each iteration. JS-divergence based k-means takes much computation costs. To avoid much computation costs, k-medoids type algorithm is considered.

4 Experiments

We conducted numerical experiments with two datasets to verify the effectiveness of JSKMdd. First, we describe the calculation conditions of the numerical experiments. Second, we describe the results by JSKMdd, k-means, KMdd and spectral clustering [14]. Third, we summarize the results and the features of the proposed method.

4.1 Experimental Setup

The abovementioned methods are compared using the Polaris and Double circle dataset and evaluated the value of adjusted rand index (ARI) [15]. ARI is a measure of similarity between two cluster partitions, and it takes the best value of 1 when two cluster partitions are exactly the same.

The Polaris dataset comprises 3 clusters of 51 objects in total, and it has two attributes in original form. There are 13, 15, and 23 objects in each cluster. This dataset is linearly separable. The Double circle dataset comprises 2 clusters of 150 objects in total, and it has two attributes in original form. There are 100 objects in outer circle and 50 ones in inner ball. This dataset is not linearly separable and known to as one of the benchmark dataset which has nonlinear cluster boundary. Figures 1 and 2 are illustrative examples of adequate cluster partition.

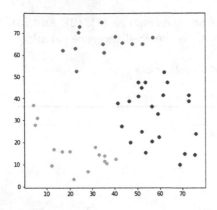

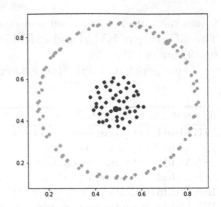

Fig. 1. Porlais dataset.   **Fig. 2.** Double circle dataset.

4.2   Experimental Results

First, the results of ARI of each method are summarized in Table 1. Table 1 shows
that several methods including the proposed method obtain adequate cluster
partition. The result of JSKMdd for the Polaris dataset is when $|N(x_k)| = 6$
and for each $h_j = 2.000$. The result of JSKMdd for Double circle dataset is
when $|N(x_k)| = 4$ and for each $h_j = 0.001$. The best ARI are shown by k-means
and KMdd from the 100 trials.

Figures 3 and 4 are illustrative examples of cluster partition by spectral clustering and k-medoids, respectively. Spectral clustering fail to obtain adequate
cluster partition that one object which is the center bottom is clustered into right
cluster from Fig. 3. KMdd can not obtain adequate cluster partition because of
the Double circle dataset consists non-linear cluster from Fig. 4.

Table 1. Results of ARI.

	Polaris	Double cicle
JSKMdd	1.000	1.000
k-means	1.000	0.008
KMdd	1.000	0.005
Spectral clustering	0.935	1.000

Next, the graphs showing how the value of ARI changes with Double circle
dataset are presented. In Figs. 6, 7, 8, 9, 10, 11 and 12, the vertical and horizontal
axes indicate the value of ARI and the value of h_j, respectively. In these graphs,
all h_j are the same value and are set between 0.001 and 0.030. The value of
ARI increases ranges in 0.001 and around 0.008 and tales 1.0. The value of ARI
decreases as the value of h_j increases. From these Figs., JSKMdd are robust with
a wider interval of h_j in case of $|N(x_k)| = 15$.

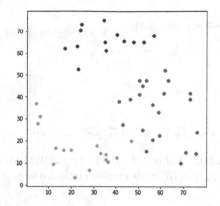

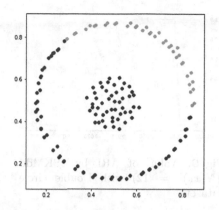

Fig. 3. Result of Porlais dataset by spectral clustering.

Fig. 4. Result of Double circle dataset by KMdd.

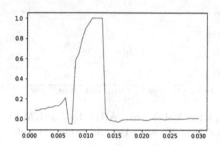

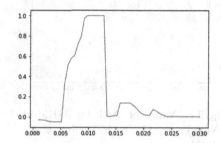

Fig. 5. Value of ARI by JSKMdd ($|N(x_k)| = 3$) with Double circle dataset.

Fig. 6. Value of ARI by JSKMdd ($|N(x_k)| = 5$) with Double circle dataset.

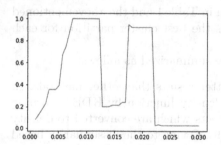

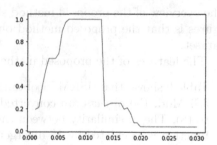

Fig. 7. Value of ARI by JSKMdd ($|N(x_k)| = 7$) with Double circle dataset.

Fig. 8. Value of ARI by JSKMdd ($|N(x_k)| = 10$) with Double circle dataset.

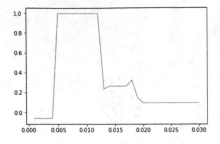

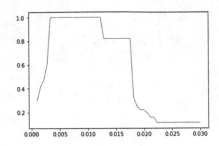

Fig. 9. Value of ARI by JSKMdd ($|N(x_k)|$ = 12) with Double circle dataset.

Fig. 10. Value of ARI by JSKMdd ($|N(x_k)| = 15$) with Double circle dataset.

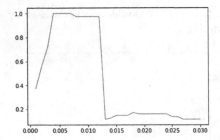

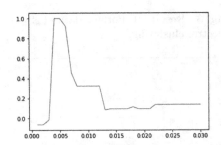

Fig. 11. Value of ARI by JSKMdd ($|N(x_k)|$ = 17) with Double circle dataset.

Fig. 12. Value of ARI by JSKMdd ($|N(x_k)| = 20$) with Double circle dataset.

4.3 Discussions

The overview of the proposed method shown in Table 1 and the abovementioned figures is that the proposed method obtains the best cluster partition for each dataset.

The features of the proposed method are summarized as follows:

- Table 1 shows that JSKMdd obtains better results than other methods. In JSKMdd, the objects are converted to density function by KDE according to (8). The dissimilarity between the objects which are converted to density function is calculated by JS-divergence (9). Owing to the dissimilarity based on both objects coordinates and distribution, JSKMdd obtains better results for Polaris and Double circle datasets..
- JSKMdd uses two parameters $|N(x)|$ and h_j. Figures 5, 6, 7, 8, 9, 10, 11 and 12 show relationship between ARI and parameters. Except in the case of $|N(x)|$ = 20 described in Fig. 12, ARI is 1 near h_j = 0.010 described in Figs. 5, 6, 7, 8, 9, 10 and 11. In the case of $|N(x)|$ = 7 described in Fig. 7, ARI increases or decreases between 0.010 and 0.020. These results show that determining two parameters $|N(x)|$ and h_j is important for JSKMdd.
- JSKMdd obtains both linear and non-linear cluster partition. Applying KDE and JS-divergence requires much calculation cost. Simplification and

approximation for calculating the dissimilarity are important to apply JSKMdd to real world datasets. To clarify the impact of those procedures to cluster boundary is also important.

The strength of JSKMdd is obtaining linear and non-linear cluster boundary. To obtain better cluster partition and clarify the cluster boundary, automatic determination mechanism for parameters is required.

5 Conclusions

In this paper, JSKMdd was proposed. The proposed method is different from KMdd in that, it uses KDE and JS-divergence to calculate dissimilarity. Moreover, the effectiveness of the proposed method is verified through numerical experiments. These experiments indicate that JSKMdd can be used in a wider range than *k*-means, KMdd and Spectral clustering.

In future works, we will compare the proposed method with other clustering methods with several benchmark datasets. Comparison with other estimation based on other density function is also possible future work. We will also construct new clustering methods for network data [16] according to the procedure of JSKMdd.

Acknowledgments. This work was partly supported by JSPS KAKENHI Grant Numbers JP19K12146. This work was also partly supported by Telecommunications Advancement Foundation.

References

1. Miyamoto, S., Ichihashi, H., Honda, K.: Algorithms for Fuzzy Clustering. Springer, Heidelberg (2008). https://doi.org/10.1007/978-3-540-78737-2
2. Jain, A.K.: Data clustering: 50 years beyond *K*-means. Pattern Recogn. Lett. **31**(8), 651–666 (2010)
3. Girolami, M.: Mercer kernel-based clustering in feature space. IEEE Trans. Neural networks **13**(3), 780–784 (2002)
4. Endo Y., Haruyama H., Okubo T., On some hierarchical clustering algorithms using kernel functions. In: Proceedings of IEEE International Conference on Fuzzy Systems (FUZZ-IEEE2004), pp. 1513–1518 (2004)
5. Gustafson, D.E., Kessel, W.C.: Fuzzy clustering with a fuzzy covariance matrix. In: Proceedings of IEEE Conference on Decision and Control, pp. 761–766 (1979)
6. Davé, R.N., Krishnapuram, R.: Robust clustering methods: a unified view. IEEE Trans. Fuzzy Syst. **5**(2), 270–293 (1997)
7. Hamasuna, Y., Endo, Y.: On semi-supervised fuzzy *c*-means clustering for data with clusterwise tolerance by opposite criteria. Soft Comput. **17**(1), 71–81 (2013)
8. Endo, Y., Hamasuna, Y., Hirano, T., Kinoshita, N.: Even-sized clustering based on optimization and its variants. J. Adv. Comput. Intell. Intell. Inform. (JACIII) **22**(1), 62–69 (2018)
9. Breunig, M.M., Kriegel, H.-S., Ng, R.T., Sander, J.: LOF: identifying density-based local outliers. In: Proceedings of the 2000 ACM SIGMOD International Conference on Management of data (SIGMOD 2000), pp. 93–104 (2000)

10. Kaufman, L., Rousseeuw, P.J.: Finding Groups in Data: An Introduction to Cluster Analysis. Wiley, New York (1990)
11. Epanechnikov, V.A.: Non-parametric estimation of a multivariate probability density. Theory Probab. Appl. **14**, 153–158 (1969)
12. Fuglede, B., Topsoe, F.: Jensen-Shannon divergence and Hilbert space embedding. In: Proceedings of the International Symposium on Information Theory (ISIT 2004) (2004)
13. Kullback, S., Leibler, R.A.: On information and sufficiency. Ann. Math. Stat. **22**, 79–86 (1951)
14. von Luxburg, U.: A tutorial on spectral clustering. Stat. Comput. **17**(4), 395–416 (2007)
15. Hubert, L., Arabie, P.: Comparing partitions. J. Classif. **2**(1), 193–218 (1985)
16. Blondel, V.D., Guillaume, J.-L., Lambiotte, R., Lefebvre, E.: Fast unfolding of communities in large networks. J. Stat. Mech. Theory Exp. P10008 (2008)

Biases Affecting Human Decision Making in AI-Supported Second Opinion Settings

Federico Cabitza[✉]

Dipartimento di Informatica, Sistemistica e Comunicazione,
University of Milano–Bicocca, Viale Sarca 336, 20126 Milano, Italy
federico.cabitza@unimib.it

Abstract. In this paper we focus on a still neglected consequence of the adoption of AI in diagnostic settings: the increase of cases in which a human decision maker is called to settle a divergence between a human doctor and the AI, i.e., second opinion requests. We designed a user study, involving more than 70 medical doctors, to understand if the second opinions are affected by the first ones and whether the decision makers tend to trust the human interpretation more than the machine's one. We observed significant effects on decision accuracy and a sort of "prejudice against the machine", which varies with respect to the respondent profile. Some implications for sounder second opinion settings are given in the light of the results of this study.

Keywords: Human decision making · Second opinion ·
ECG overreading · Decision support systems

1 Introduction

Many approaches to the design of computer-based decision support systems take a cognitivist perspective: according to this stance, a decision maker is an agent that must be supported in reaching the optimal decision in light of the available evidence or data. However, studies in the field of *naturalistic decision making* [9] have observed that human decision making seldom occurs in single-agent settings but rather in social and collaborative ones. In real-world settings, decision makers can rely on the support of colleagues in case they deem themselves unable to reach a definitive decision autonomously. In medicine, these situations are common and are called "second opinion". When a second opinion is pursued by a doctor (and not by a patient to get an alternative point of view) this involves a colleague to get support on the interpretation of a complex case.

In this paper we focus on particular (and still prospective) second-opinion scenario, where this opinion is requested by a human doctor to resolve a discordance with a diagnostic AI. We report the main results of a user study performed in the application domain of Artificial Intelligence-assisted Electrocardiogram (ECG) reading. AI-supported ECG reading will likely improve this clinical task both in terms of efficiency (i.e., higher throughput) and in terms of effectiveness

© Springer Nature Switzerland AG 2019
V. Torra et al. (Eds.): MDAI 2019, LNAI 11676, pp. 283–294, 2019.
https://doi.org/10.1007/978-3-030-26773-5_25

(i.e., high diagnostic accuracy). Preliminary researches seem to back this prediction: In [1] the authors describe a deep learning algorithm capable to identify asymptomatic left ventricular dysfunction from traditional 12-lead ECGs with a better performance with respect to the common BNP blood test. Similarly, [19] claimed to have developed a deep learning algorithm capable to detect and classify myocardial infarctions in traditional ECGs at the performance level of human cardiologists; [13] claimed to have developed an algorithm that *even exceeds* the performance of board-certified cardiologists in detecting a wide range of heart arrhythmias from electrocardiograms recorded with a single-lead wearable monitor. The deep learning system described in [17] exhibited an accuracy of 92% in detecting a wide range of abnormalities in emergency situations, while the system described in [5] reached an accuracy of 98% in detecting myocardial infarctions.

Furthermore, the need for this kind of support is clear: indeed, it is known that cardiologist proficiency at ECG interpretation is far from being perfect, not only in case of residents [16], when almost one ECG out of two can be misread, but also in case of expert cardiologists, when accuracy rates ranging from 53% to 96% [14] also depending on the difficulty of the cases considered. This is also reflected by the relatively low agreement that is observed among cardiologists who interpret the same ECG [3].

For all these reasons, it is a reasonable guess that in the next future most cardiological practices will be supported by new-generation, AI-driven computer supported ECG interpretation systems. It is likewise reasonable to believe that current recommendations will be valid even when these systems spread in cardiological practice, and that therefore AI-based decision support systems will remain adjuncts to the clinician, and that "all computer-based reports [will] require physician overreading" [10].

To our knowledge, no study has yet focused on the potential biases occurring in second opinion settings[1], that is the process in which a previous decision is either confirmed or rejected by a second decision maker. In particular, in this paper we will focus on two constructs: *conformity bias*[2], which is the tendency to conform to, and hence confirm, the opinion of others, thus considering them more expert in regard to some matter than they actually are; and the balance between authority bias and automation bias, that is any tendency of second-opinion decision makers to prefer either the opinion of the human expert (considered a sort of authority) or the advice by the machine [7]. In our study, this means to address whether overreading (or second opinion) is affected by, respectively, a tendency to confirm the first opinion (cf. conformity bias) or by a tendency to agree with either an expert colleague or any machine that is perceived as highly accurate (authority vs automation bias).

---

[1] These kind of settings in the cardiological domain is also called *ECG overreading*.

[2] Other contributions call this bias "truth bias" [18], which is defined as the tendency of someone to believe that "others are telling the truth more often than they actually are" and hence to confirm what they say, mainly to spare themselves feelings of discomfort.

2   Methods

Methodologically we have adopted the business simulation game approach, which is an established method to understand how people interact with technologies and are affected by the business models and policies underlying and enacted by those technologies [4]. For this study, we conceived a simulation game where we invited a panel of clinicians with varying proficiency in ECG reading to formulate informed opinions on three clinical cases, described succinctly and accompanied with a high-resolution 12-lead ECGs.

We asked the participants to get themselves into a part in the following business scenario: they were told that a colleague of theirs, a senior cardiologist with 26 years of experience in ECG reading, had disagreed with the interpretation that an advanced and certified Artificial Intelligence (AI) system supplied him, in regard to three non-trivial cases; therefore he requested each participant to act as an *arbitrator* and express a 'second opinion' by over-reading the ECGs at hand. The participants were also told that the above AI had been recently acquired by their hospital and had shown a diagnostic accuracy on border-line cases of approximately 96–97%.

To help design this scenario, two cardiologists extracted three cases from the *ECG Wave-Maven*[3] database, which is an online repository of more than 500 ECGs of various difficulty developed for the self-assessment of students and clinicians in ECG reading proficiency. The ECG Wave-Maven database categorizes cases according to a 5-level scale of difficulty (from 1, the simplest cases to 5, the most difficult ones): the three cases were extracted randomly from the 94 cases denoted as of difficulty 4 (namely, cases no. 6, 18, 73), in order to avoid extremely difficult cases and at the same time to make the need to rely on over-reading and second opinion credible.

An online, multi-page questionnaire was then prepared on the Limesurvey platform, to be administered to a sample of clinicians. These gave their consent to the anonymous and aggregate use of the responses to address our research questions. In the online questionnaire, each of the above three cases was presented in a different page, with a brief clinical description and single ECG. This was rendered as a raster color image of 675×450 pixels (see Fig. 1), but a floating magnifying lens feature was available to let respondents look at ECG details at a twofold resolution. The form asked the respondents to either choose among two alternative diagnoses, one given by the AI and the other one by the human colleague, or, in case of disagreement with these interpretations, to propose a third one, different from the former two. Respondents were also requested to assess the *plausibility* of each diagnosis proposed to them (on a semantic differential rating scale, from 1 'very low' to 4 'very high'). To assess preferences, since we were interested in assessing how conformity bias affects human decision in case of erroneous advice, *both* the proposed diagnoses were wrong

---

[3] ECG Wave-Maven, Copyright (c) 2001–2016 Beth Israel Deaconess Medical Center. All rights reserved. https://ecg.bidmc.harvard.edu/maven/mavenmain.asp Last Accessed: 17th May 2018.

(although variously plausible). In order to avoid order bias, half of the sample read the diagnosis proposed by the AI before the diagnosis by the human colleague, while the other half read the diagnoses in the opposite order. Also the two wrong diagnoses associated to each case were randomly assigned to either the AI or the human colleague so that each diagnosis was seen associated with both kinds of first opinions an even number of times at sample level. In its last page, the online questionnaire asked the respondents some profiling information (namely gender, main specialty, work experience in that specialty, and self-assessed proficiency in ECG reading on a 3-level rating scale, from 1, basic skills, to 3 advanced skills). This would allow to create three further strata (besides gender): cardiologists (vs. non-cardiologists), long-experienced clinicians (vs. clinicians with less than 10 years of practice), and advanced proficiency readers (vs. lower proficiency readers).

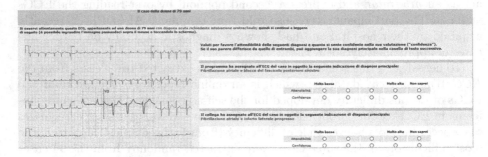

Fig. 1. Screenshot from the questionnaire page presenting the case no. 18 '79-year-old woman with shortness of breath requiring intubation' and its ECG (Copyright, 2005 Beth Israel Deaconess Med Ctr.). The magnifying lens feature is superimposed on the ECG raster image. The alternative options were displayed to each study participant in random order.

Since we would not be able to detect those cases in which neither of the two first opinions were really misleading the reader but, so to say, one of the advices just confirmed an autonomous (wrong) interpretation by the human reader, we also conceived a second survey, where the same panel of experts were supposed to consider the same three ECGs (in different order) after a due wash-out period but without receiving any prior advice, neither by AI nor a human colleague. Since the respondents were never given the right diagnoses, nor received any feedback on their interpretation, we consider four months as an adequate lapse for the wash-out period.

3   Results and Analyses

In July 2017, a panel of 157 potential participants were personally invited by email to complete participate in the simulation game. Invitations were personal

and the questionnaires had restricted (tokenized) access to avoid multiple com-
pilations and allow for one gentle reminder that was sent after two weeks after
the initial invitation. When we closed the survey two additional weeks later,
we collected 246 unique interpretations by 75 clinicians (see Fig. 2) who had
completed the overall task of reading 3 ECG and giving their opinion in approximately 10 min and 7 s on average (M = 607, SD = 538 s).

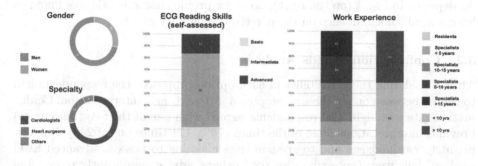

Fig. 2. Descriptive statistics of the profile of the participants of the second opinion
study (N = 73, as two participants did not fill in the profile items).

3.1 Plausibility Analysis

In regard to the perceived plausibility of the first opinions, not surprisingly, those
who discarded these options (mean rank = 148) considered these significantly less
plausible that those who trusted them (mean rank = 243, Mann whitney test,
U = 10684, Z = −7.54, P < .000). The former respondents did not found significantly different the two interpretation with which they disagreed (Mann Whitney test, U = 1999, Z = −.41, P = .68). Conversely, the latter respondents found
the responses given by the human colleague *significantly more plausible* (mean
rank = 164, see Fig. 4) than those given by the AI (mean rank = 134) although,
we remind the reader, these were the same (and equally wrong) at sample level
(Mann Whitney test, U = 8863, Z = −3.12, P = .002). As a whole, the respondent
sample deemed the interpretations proposed by the human colleague significantly more plausible than those proposed by the AI (Mann Whitney test: mean
ranks = 231 vs. 201, U = 20149, Z = −2,5, P = .011). Moreover, in absolute terms
respondents found the AI advice to be generally of low plausibility significantly
($Median_{AI} = 2$ CI = [2,1], frequency of plausibility values equal or lower than 2:
.42 CI = .7; and equal or greater than 3: .58 CI = .7, proportion test: P = .009),
while statistical significance was not found with respect to the average perceived
accuracy of the human colleague ($Median_{Human} = 3$ CI = [3,2]). Even among
the respondents who confirmed the interpretation by the AI, the majority (76
vs. 70) attached to this interpretation a plausibility of 1 or 2 (Median = 2, in a
4-point scale).

The average plausibility attached to the options given by the AI can be considered associated with trust in the AI support. Gender and proficiency in ECG reading were found to affect this perceived accuracy of AI: Performing a Mann Whitney test we found that female readers tended to find the AI advice significantly more accurate/reliable than male readers (U = 3299, P = .045, means = 2.27 vs 2.62, medians = 2 vs 4), and the same holds for the less skilled readers (U = 2213, P = .048, means = 2.25 vs 2.0, medians = 2 vs 2). As depicted in Fig. 4 (on the right), no other profile characteristic was found to have a significant influence on the perceived reliability of AI.

3.2   Confirmation Trends Analysis

After considering the plausibility of each option proposed, the respondents had to choose between one of these or propose a *different, new* interpretation. Confirmation rate was high: The respondents agreed with one of the two (inaccurate) previous interpretations most of the times (70%, 171 times out of 246). Not surprisingly, cardiologists and the respondents claiming to possess advanced ECG reading skills were those who discarded others' advice significantly more often (more precisely, cardiologists rejected the advice twice more often than non cardiologists, 35% vs 18%, P = .016; the same holds for the more skilled readers with respect to the less skilled ones, 50% vs 25%, P = .005). Thus, observations allow to claim that specialization and (self-perceived) reading proficiency influence reliance on others' opinion. Conversely, gender and work experience did not affect this behavior. Indeed, the more experienced readers discarded the advices as many times as the less experienced ones (29% vs 29% P = 1), and the male readers did it slightly more often than the female ones (31% vs 24%, P = .33) but not significantly so.

Interestingly (and in accordance with the results on plausibility mentioned above), the participants chose the human interpretation significantly more often than the AI one (57%4, 43%4) although, as recalled above, the given advices had been randomized so that both the AI and the human colleague were giving the very same interpretations. This result (all together with the finding about the perceived plausibility) amounts to what we can call a *prejudice against the machine*. This prejudice was stronger in the cardiologists than in other medical specialists and in those claiming an advanced ECG reading competency in comparison to those readers who claimed to have a less than advanced skill.

Respondents preferred the interpretation formulated by the human colleague more often than the diagnosis given by the AI (97 vs. 74). This difference is statistically significant: The one-tailed probability of this ratio (or a more extreme one) assuming that frequencies should be equal because the two sources (i.e., human and AI) offered the very same diagnoses is P = .044 (z = 1.70). As depicted in Fig. 3, the cardiologists, the clinicians with less than 10 years of ECG reading practice, and clinicians claiming to possess advanced ECG reading skills were the categories of respondents that preferred the human advice over the AI's one more often. In the former case the difference in the proportions of choice is statistically significant.

3.3   Impact on Accuracy Analysis

To assess the impact of first opinions on accuracy, we refer to the results of the subsequent questionnaire, sent in December 2017. As said in Sect. 2, we recontacted the respondents who had actually participated in the comparative study and asked to report their most plausible diagnosis for each case in a free text field (each case was presented in random order). As hinted above, while the case description and ECG were the same with respect to the second opinion study, in this second study no diagnostic alternative or previous interpretation was proposed.

Fewer people responded to this second invitation (N = 21) and completed the three reading tasks in approximately 10 min (M = 611.27 s, SD = 361.64). Although different in cardinality, the samples in the two studies *were not significantly different* in regard to: *work experience* (Fisher's exact test on the experience dichotomized in terms of proportions of ECG readers with less than 10 years of experience vs. readers with at least 10 years, P = .79, OR = 0.81, CI = 0.29 to 2.29); in regard to the proportion between cardiologists and non-cardiologists involved (Fisher's exact test P = .78 OR = 1.2941, CI: .42 to 4); and in regard to the *self-assessed proficiency* in ECG reading (Median = 2, Mann-Whitney test on mean ranks, 53.2 vs. 45.4, P = .21, Z = −1.26, U = 586): in particular, the proportions of clinicians declaring to possess at least intermediate skills were approximately the same in the two studies, i.e., three quarters of the sample (Fisher's exact test P = 1, OR = 1.0345, CI = 0.3 to 3.56). This is important to allow for the comparison of the results.

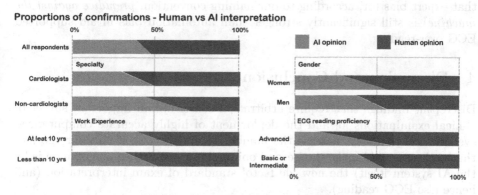

Fig. 3. Stacked bar charts indicating the proportions of respondents who confirmed the external advice provided by either the human (blue side) or the AI (red side). Since the proportion of respondents who confirmed one advice is the complement of the proportion of those who chose the other, if a confidence interval (denoted as the slanting border between the two sides of the bar) does not cross the 50% vertical line, the corresponding estimates of the proportions of choice are significantly different. (Color figure online)

3.4   Conformity Bias Analysis

If we combine the results from both stages of the study, we can address whether the respondents' opinions were affected by conformity bias and, if so, what kind of opinion source was prominent. As said above, all the times the respondents confirmed a first opinion in the first part of the study (either by the AI or the expert colleague) they actually confirmed an inaccurate diagnosis. Even when they differed with the external opinions, the participants proposed an accurate[4] diagnosis only in one fourth of cases (19 out of 75, 25%). Therefore (see Fig. 5), the overall error rate was surprisingly high, 92% (19 right diagnoses out of 245). It is worth noting that the three cases were labelled as a level 4 difficulty in the ECG Wave-Maven database (on a 1 to 5 scale), that is challenging but not extremely difficult. Fisher's exact test revealed that the error rate in readers claiming to have at least 10 years of experience and the error rate of those with less than 10 years of experience were not statistically different (P = .62, odd ratio = 1.53, CI = .5662 to 4.1429); likewise, no significant difference was found with respect to error rate between cardiologists and non cardiologists (P = .42, OR = .78, CI = .2976 to 2.0857). Conformity bias affects clinicians irrespective of their work experience (38–89 vs. 27–64, P = 1[5]). Cardiologists were the professional stratum less prone to conformity bias and who refused the others' advice more often (53–98 vs. 9–52, P = .003). The same holds for those who claimed to have advanced EG reading skills (18-18 vs. 48–135, P = .009). Lastly, we have observed that cardiologists, when provided with two likewise plausible diagnoses by an expert colleague and a state-of-the-art artificial intelligence, still prefer significantly the human advice, even when this is wrong: we can therefore claim that expert bias (or, according to our naming convention, *prejudice against the machine*) is still significantly stronger than automation bias in aid supported ECG over-reading.

4   Discussion and Conclusion

Discrepant findings that require arbitration constitute an important subset of clinical examinations [6] and the deployment of highly accurate computerized systems will probably increase the number of divergences, if only for the fact that their use could make the 'second opinion' protocol (when one opinion is by the AI system itself) the new 'de facto' standard of exam interpretation (and hence also ECG reading).

In this paper we have presented a user study aimed at investigating the impact of inaccurate decision support in ECG reading and whether respondents

---

[4] Accuracy was judged by two cardiologists according to whether the diagnosis was either the same given by the gold standard (i.e., the official diagnosis associated with the ECG), or it was somehow close to it and would have informed an appropriate treatment or management of the case at hand.

[5] We present the data in a 2×2 contingency table and the P value associated with a Fisher's exact test. The first figure in the pair represents the number of clinicians who discarded the given advice.

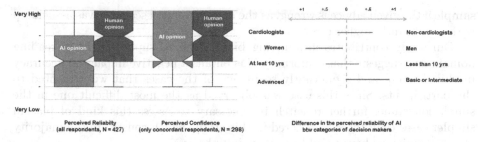

Fig. 4. On the left: Box plots of the perceived reliability of the first opinions, for all respondents (on the left), and for the respondents who agreed with one of the proposed interpretations (on the right). Responses were given on an ordinal 4-value scale from 'very low' to 'very high'. Notches indicate the 95% confidence intervals of the medians (no overlap indicates statistically significant difference). N indicates the number of plausibility values recorded, not the number of respondents who gave those values. On the right: Confidence intervals of the difference of the average perceptions of reliability of the AI advice. If an interval does not contain the vertical line denoted as 0, the corresponding means are significantly different. The interval is represented closer to the category of respondents who attached the higher reliability to the AI advice. Thicker lines are associated with statistical significance.

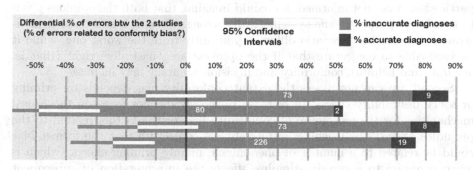

Fig. 5. Bar charts indicating the interpretation performance in the second opinion study: the 3 cases above and the total error rate at the bottom bar. Accuracy has been assessed with respect to the Gold Standard. The percentage of errors that can be related to the conformity bias is conjectured by comparing the error rate of the second study undertaken by the same sample of respondents on the same cases after due washout period. The confidence intervals are calculated for the difference of proportions between the two experimental conditions (opinions given vs. no opinion given). Whenever the confidence intervals does not cross the 0% line the observed difference is statistically significant and not due to chance.

could be "primed" by knowing what interpretation would be given by what advice source (either a human colleague or a specialized AI). We define *conformity bias* as 'the extent the second opinion of a decision maker is misled by a first opinion', irrespective of the fact that this latter opinion is given by a human or an AI system. We observed a significant conformity bias in every strata of our sample if the given advice is wrong, as the second opinion scenario was associated with significantly higher error rates.

Our study contributes in studying bias in the AI-supported ECG reading domain and suggests that conformity bias entails a negative impact on accuracy in this domain, and significantly so in one of the cases that we presented to the participants. Since this case was observed as the most difficult one in the simulation game, further research is necessary to assess this kind of bias in simpler cases than those considered in this study, which constitute the majority of cases clinicians have to address on a daily basis.

In light of our findings, a clear indication from this study is the following one: first opinions should not be displayed to doctors involved in second opinion services before they produce their own interpretation. Only afterwards, the previous diverging opinions can be exposed with indication of who/what produced them, to avoid conformity and prejudices.

Our study presents some limitations: first, the experiment design presented in this paper was motivated by the intention to use the full power of the study to assess effects on accuracy due to bias. Thus, one could argue that interpreters were purposely misled by the opinions given and this inflated the role of conformity bias. However, the effect observed should not be considered in hindsight: participants were not informed, nor could imagine, that both the opinions given to them were purposely and systematically wrong. Thus, it is still to be demonstrated that the best scenario differs significantly from the worst one, while it is reasonable to conjecture that if the opinions are "basically" right (that is, accurate and helpful), conformity and decision accuracy may increase.

Second, we can not assess the extent conformity bias depends on priming or social desirability effects. In fact, we purposely did not focus on the causal mechanisms for the emergence of conformity bias or the factors related to the prejudice against the machine: intuitively, we conjecture that the former bias could be related to a number of phenomena, mainly priming effects (which is when exposure to a certain stimulus affects the interpretation of subsequent stimuli), but also other factors in various degrees: availability heuristic (which occurs when one relies on immediate examples coming to one's mind when evaluating a specific decision); informational social influence, social loafing [12] and group conformity bias [8] (which all occur when one relies on the options proposed within a group by peers and colleagues whom are felt to belong to the same social group).

Interestingly, we found that the participants to our study generally found the AI opinion less reliable than the human one (which, nevertheless, gave the very same wrong advice), as depicted in Fig. 4 (on the left). In particular, cardiologists chose to follow the human advice more often than the AI one (see Fig. 3): this probably reflects a general distrust of these professionals in the reliability of most of currently available tools of computerized interpretation of ECGs [15]. Among those who chose the others' advice, we did not find a significant propensity towards either the human advice or the AI's one. However, if we consider the clinicians who most likely could consider a competent opinion helpful (that is, in our sample, non-cardiologists, people with less than 10 years of work experience or claiming to possess not higher than intermediate ECG reading skills), conformity is common, and reliance on AI is greater than on the human colleague. Thus, if a study involving a greater sample size confirmed the higher reliance by this kind of clinician toward the AI aid instead of more specialized or expert colleagues, the adoption of this kind of tools should undergo further scrutiny, not only for its misleading potential, but also for its potential to undermine human collaboration among different specialists and the young.

To conclude: the main original contribution of this study is the characterization and assessment of the effect we called *prejudice against the machine*. This study has also a confirmatory value; in general, this regards the general phenomenon of the *automation bias* and its impact on decision accuracy. More in particular, this work confirms previous research about *conformity bias* and the tendency of people toward accepting, rather than refuting, diagnostic hypotheses [11]. Also in our study physicians found it difficult to reject incorrect diagnoses and had their accuracy affected by confirmatory tendencies. We also confirm the findings by [20], who found that clinicians (non cardiologists) are influenced significantly by the incorrect advice of computerized aid in ECG reading. Furthermore, we can generalize this effect to different specialists (including cardiologists) and to any external aid; to this latter regard, we also found that the error-inducing effect is even stronger in case of human support (at least, to date). We can also confirm the findings of a recent study [2] that considered only automated aids and involved fewer experts (30 vs. 75), but was carried out on more ECG interpretations (9000 vs. 311[6]): incorrect advices from automated aids reduce diagnostic accuracy and hence bring to automation bias; non-cardiologists tend to agree more with the automated aids in comparison to cardiologists and are hence affected more by automation bias. These studies and our contribution thus suggest caution in the deployment of powerful AI-based decision support systems and to consider the potential negative effects of biases, like conformity bias, authority bias and automation bias, in second-opinion settings.

Acknowledgments. The author wishes to thank Dr. Raffaele Rasoini and Dr. Camilla Alderighi, cardiologists, for their help in the design and dissemination of the survey, and for the valuable suggestions after reviewing a preliminary draft of the manuscript.

References

1. Attia, Z.I., et al.: Application of artificial intelligence to the standard 12 lead ECG to identify people with left ventricular dysfunction. J. Am. Coll. Cardiol. **71**(11), A306 (2018)
2. Bond, R.R., et al.: Automation bias in medicine: the influence of automated diagnoses on interpreter accuracy and uncertainty when reading electrocardiograms. J. Electrocardiol. **51**(6), S6–S11 (2018)

---

[6] To be precisely, we collected 249 interpretations in the first part of the study and 62 in the following one.

3. Brailer, D.J., Kroch, E., Pauly, M.V.: The impact of computer-assisted test interpretation on physician decision making: the case of electrocardiograms. Med. Decis. Making **17**(1), 80–86 (1997)
4. Brauner, P., et al.: A game-based approach to raise quality awareness in ramp-up processes. Qual. Manag. J. **23**(1), 55–69 (2016)
5. Dohare, A.K., Kumar, V., Kumar, R.: Detection of myocardial infarction in 12 lead ECG using support vector machine. Appl. Soft Comput. **64**, 138–147 (2018)
6. Duijm, L.E., Groenewoud, J.H., Hendriks, J.H., de Koning, H.J.: Independent double reading of screening mammograms in the Netherlands: effect of arbitration following reader disagreements. Radiology **231**(2), 564–570 (2004)
7. Goddard, K., Roudsari, A., Wyatt, J.C.: Automation bias: empirical results assessing influencing factors. Int. J. Med. Inform. **83**(5), 368–375 (2014)
8. Kaba, A., Wishart, I., Fraser, K., Coderre, S., McLaughlin, K.: Are we at risk of groupthink in our approach to teamwork interventions in health care? Med. Educ. **50**(4), 400–408 (2016)
9. Klein, G.: Naturalistic decision making. Hum. Factors **50**(3), 456–460 (2008)
10. Kligfield, P., Gettes, L.S., Bailey, J.J., Childers, R., Deal, B.J., Hancock, E.W., Van Herpen, G., Kors, J.A., Macfarlane, P., Mirvis, D.M., et al.: Recommendations for the standardization and interpretation of the electrocardiogram: part I. J. Am. Coll. Cardiol. **49**(10), 1109–1127 (2007)
11. Mannion, R., Thompson, C.: Systematic biases in group decision-making: implications for patient safety. Int. J. Qual. Health Care **26**(6), 606–612 (2014)
12. Parasuraman, R., Manzey, D.H.: Complacency and bias in human use of automation: an attentional integration. Hum. Factors **52**(3), 381–410 (2010)
13. Rajpurkar, P., Hannun, A.Y., Haghpanahi, M., Bourn, C., Ng, A.Y.: Cardiologistlevel arrhythmia detection with convolutional neural networks. Nat. Med. **25**(1), 65–69 (2019). https://doi.org/10.1038/s41591-018-0268-3
14. Salerno, S.M., Alguire, P.C., Waxman, H.S.: Competency in interpretation of 12-lead electrocardiograms: a summary and appraisal of published evidence. Ann. Intern. Med. **138**(9), 751–760 (2003)
15. Schläpfer, J., Wellens, H.J.: Computer-interpreted electrocardiograms: benefits and limitations. J. Am. Coll. Cardiol. **70**(9), 1183–1192 (2017)
16. Sibbald, M., Davies, E.G., Dorian, P., Eric, H.: Electrocardiographic interpretation skills of cardiology residents: are they competent? Can. J. Cardiol. **30**(12), 1721–1724 (2014)
17. Smith, S.W., et al.: A deep neural network learning algorithm outperforms a conventional algorithm for emergency department electrocardiogram interpretation. J. Electrocardiol. **52**, 88–95 (2019)
18. Street, C.N., Masip, J.: The source of the truth bias: Heuristic processing? Scand. J. Psychol. **56**(3), 254–263 (2015)
19. Strodthoff, N., Strodthoff, C.: Detecting and interpreting myocardial infarction using fully convolutional neural networks. Physiol. Measur. **40**(1), 015001 (2019)
20. Tsai, T.L., Fridsma, D.B., Gatti, G.: Computer decision support as a source of interpretation error: the case of electrocardiograms. J. Am. Med. Inform. Assoc. **10**(5), 478–483 (2003)

Applications of Different CNN Architectures for Palm Vein Identification

Szidónia Lefkovits[1], László Lefkovits[2], and László Szilágyi[2,3(✉)]

[1] Department of Computer Science, University of Medicine, Pharmacy,
Sciences and Technology, Tg. Mureş, Romania
szidonia.lefkovits@umfst.ro
[2] Department of Electrical Engineering, Sapientia Hungarian University
of Transylvania, Tg. Mureş, Romania
{lefkolaci,lalo}@ms.sapientia.ro
[3] John von Neumann Faculty of Informatics, Óbuda University, Budapest, Hungary
szilagyi.laszlo@nik.uni-obuda.hu

Abstract. In this paper a palm vein identification system is presented, which exploits the strength of convolutional neural network (CNN) architectures. We built and compared six different CNN approaches for biometric identification based on palm images. Four of them were developed by applying transfer learning and fine-tuning techniques to relevant deep learning architectures in the literature (AlexNet, VGG-16, ResNet-50 and SqueezeNet). We proposed and analysed two novel CNN architectures as well. We experimentally compared the identification accuracy and training convergence of these models. Each model was trained and evaluated using the PUT palm vein near infrared image database. To increase the accuracy obtained, we investigated the influence of some image quality enhancement methods, such as contrast adjustment and normalization, Gaussian smoothing, contrast limited adaptive histogram equalization, and Hessian matrix based coarse vein segmentation. Results show high recognition accuracy for almost every such CNN-based approach.

Keywords: Convolutional neural networks · Transfer learning · Biometric identification · Palm vein recognition

1 Introduction

The rapid development of smart devices, cloud computing and home automation brings up various security and privacy problems. People gradually recognize the importance of guarding their personal data stored in networks. Access to resources is usually granted based on certain control credentials that must

The research was partially supported by Sapientia Foundation – Institute for Scientific Research, and Domus Hungarica Research Grant. L. Szilágyi is János Bolyai Fellow of the Hungarian Academy of Sciences.

© Springer Nature Switzerland AG 2019
V. Torra et al. (Eds.): MDAI 2019, LNAI 11676, pp. 295–306, 2019.
https://doi.org/10.1007/978-3-030-26773-5_26

represent unique identifiers possessed or memorized by the user. Traditional authentication and identification methods (e.g. tokens, ID cards, etc.) are losing ground, as they are relatively easy to intercept or reproduce.

The most secure automated human authentication methods rely on biometric data, which represent physical or behavioral characteristics of the individual that are difficult to forge or steal: e.g. face, fiducial points, fingerprints, veins, palm and dorsal hand veins, finger veins, iris, retina, ears, voice and the DNA. Biometrics based on vascular system patterns have recently attracted increasing attention [1].

It is known that the blood vessel network, which is already formed in embryo state, substantially differs from one individual to another. On the contrary, it is not known what is the reason of this uniqueness. Technically, palm vein detection is usually performed by Near Infrared (NIR) or Far Infrared (FIR) cameras because of the different absorption of IR radiation in the vessels and the surrounding tissues. The acquisition protocol is not standardized and differs from one database to another. Images are usually characterized by very low resolution, low contrast, and grayscale intensities of narrow spectrum. Vein-based approaches can be classified into three different types:

1. Shape based methods rely on the vascular structure and extract lines or curves as basic representation. Meraoumia et al used Discrete Wavelet Transform (DWT) along the Laplacian of Gaussian (LoG) operator to extract the veins and eliminate noise [2]. The Hamming distance was applied for the matching step. The line-based methods also called geometric methods are based on the extraction of short line-segments that contribute to the approximation of the veins. The algorithms from this category make use of minutiae, bifurcation, endpoint and crossing point detection. Soliman et al detected minutiae with SIFT feature points and used Linear Vector Quantization for classification [3]. Kang et al combined not only SIFT points, but also SURF and Affine-SIFT points [4]. The most important drawback of geometric methods is the difficult comparison of several skeleton-like structures in order to obtain their similarity coefficient. They require registered images and a standardized alignment of the vessel structure not only the input image. In our previous article we proposed an alignment and registration method using 3D rigid transformation for dorsal hand veins [5].
2. The subspace methods reduce the data dimensionality. The most used data reduction methods are the principal component analysis (PCA) that extracts the highest variation components like in [6], or the linear discriminant features [2] determining the components that maximize the he ratio of the variance between the classes to the variance within the classes.
3. Local or texture based methods usually involve descriptors like local binary patterns (LBP) [7]. This type of local feature was used along with DWT and isometric projection obtaining reduced feature vectors and applying a sum based matching [8]. On the other hand, the most used descriptor types are based on different combinations of Gabor jets with optimized parameters [9].

The most recent research area of convolutional neural networks is very rarely used in biometric identification, especially in case of palm vein recognition. The first CNN based approach is presented in [10]. Here the authors use an extremely simple CNN with only two convolution layers and two fully connected layers of 7 activations each. The second method applying CNN is the so called Deep Hashing Palm Network (DHPN) [11]. Here the architecture relies on the pre-trained weights of VGG-16 and 5 convolutional and 3 FC layers.

This paper proposes two CNN architectures that will be deployed detection in palm vein network identification, and provides a comparison with four existing pre-trained CNN architectures fine-tuned via transfer learning.

The remainder of the paper is structured as follows: Sect. 2 describes the database used, followed by Sect. 3 in which the transfer learned and fine-tuned CNNs are presented. Section 4 describes our proposed CNN architectures used in the experimental setup. Finally, in Sect. 5 the experiments carried out are presented, putting an accent on the comparison of architectures and the influence of different image processing steps on the identification accuracy. The paper ends with conclusions and future perspectives (Sect. 6).

2   The Database

The presented approach was trained and evaluated on the PUT Vein database of the Poznan University of Technology [1]. The vein pattern database consists of 2400 images of human veins of the hand. Half of them 1200 are wrist vein patterns and the other half palmar vein patterns. Obviously, we have used only the palmar images. The near infrared images of 50 different subjects were acquired both for the left and right hand. For each subject there are 12 images taken in three series of four snapshots each with one-week interval between them. The images were aligned by positioning the palm according to a line that had to be right at the base of their fingers. The images have a resolution of 1024×768 pixels with a depth of 24 bit. Owing to the low-cost acquisition equipment the images captured for this database have small illumination changes, a narrow intensity histogram with a high noise level. There are only small changes in translation, rotation and illumination because of the physical image adjustment step.

3   Transfer Learning

Given a certain CNN architecture, its weights can be fine-tuned via transfer learning, to make it suitable for new input data. Transfer learning proved to enhance the accuracy and efficiency, compared to retraining the CNN from the beginning. Two of the CNN network architectures participating in this study were built and trained from scratch, while the other four existing networks enumerated below were fine-tuned starting from their pre-trained state:

1. AlexNet was the winner architecture at the ImageNet Challenge in 2012 [12], the first one to implement parallelism in network training. Originally it

worked with 3-channel RGB images of size 224×224. It consists of five convolutional layers, three overlapping max-pool layers, and two fully connected (FC) layers at the end. In our approach, we fine-tuned the trained weights for ImageNet and changed the final FC layer to be adequate for our 100-class classification problem.

2. VGG is a much deeper model than AlexNet, containing 16 (VGG-16) or 19 (VGG-19) layers [13]. The idea was to stack several convolutional layers. The VGG-16 model that is involved in our study, has 13 conv-layers, 5 max-pooling layers and 3 FC layers. The other design novelty in this architecture was the doubling of filter depth after reducing the input layers size to half in both dimensions, thus reducing the number of parameters. This network has about 138 million parameters and was extremely hard to train. In our approach, we retrained the initial weights of this network obtained from ImageNet and replaced the last FC layer according to our 100 class scores.

3. ResNet was introduced with the intent to optimize the training and improve the decision accuracy of very deep convolutional networks [14]. The authors formulated a residual learning framework in which layers are learning residual functions with respect to layer inputs, and provided empirical evidence to prove the superiority of the residual training technique. In our experiments, we used the ResNet-50 architecture consisting of fifty layers and fine-tuned the pre-trained weights of this architecture using the PUT palm vein database, added data augmentation and refined the weights and the dense layers according to the palm vein network identification problem.

4. The SqueezeNet architecture is a CNN with a highly decreased number fo weights (of 1.2 million only), but with performances in ImageNet similar to AlexNet [15]. The main ideas at the foundation of this network are the so-called fire modules in conjunction with the same activation map sizes and by reducing their spatial sizes later and fewer times in the architecture. The SqueezeNet is built of a conv-layer, max-pool for downsampling, three fire modules, max-pool again, four fire modules, max-pool, a final conv-layer and average pooling at the end, i.e., a total of 11 layers based on convolutions. In our application, we refined the vanilla SqueezeNet without bypasses, retrained the above described network and adapted the FC layer at the end to the 100-class classification problem.

4   The Proposed Approach

The aim of this paper is the design, application and evaluation of two end-to-end trained CNN networks with the purpose of palm vein based identification. These two networks are built of simple CNN layers like convolution and fully connected ones, thus they are relatively easy to train. Their performance is compared to well-known CNNs such as AlexNet, VGG-16, ResNet-50 and SqueezeNet, described above. The first network proposed and trained end-to-end is a 6-layered CNN formed of 4 convolutional and 2 dense layers. The input image was resized with bilinear interpolation to the standard image size used for network training at the ImageNet competition, i.e., to $224 \times 224 \times 3$. The number of training datasets is very small for an adequate training from scratch without using transfer learning. The most widespread method used to extend the training image set is the data augmentation technique. Instead of twelve, several thousand images are generated applying different affine transformations in every epoch. In our case the most relevant augmentation steps were random rotation by $\pm 10°$. We have used centre-crop based on background extraction to consider only the palm region. Random rotation of the images by $\pm 10°$ and random translation in both X and Y directions of maximum 10% horizontally and 5% vertically were also applied. The last step was the normalization of images to a mean of 0 and a standard deviation of 1.

In this research, our intension was to build a simple CNN architecture with a reduced number of parameters in order to be trained easily, which leads to identification accuracy comparable to other pre-trained CNNs described in the literature.

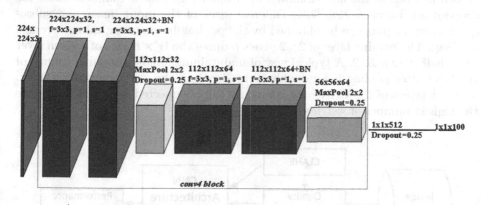

Fig. 1. *conv4-fc2* architecture trained from end-to-end

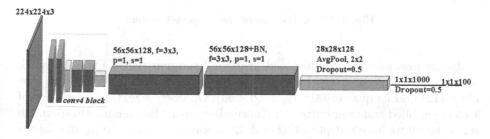

Fig. 2. *conv6-fc2* architecture trained from end-to-end

First we have proposed to build a simpler network with only 4 convolutional layers for considerably fewer number of parameters. The convolutional kernel in case of every layer was 3×3. As shown in [13], the activation map of larger filters of size 5, 7, 9, 11 pixels can be obtained by multiple layers of kernel 3×3. The bigger the number of filters, the higher is the number of weights between consecutive layers. The number of parameters (weights) between two layers with a convolutional filter of size $a \times a$ and a 3D input layer of $W \times H \times D_I$ (W-width, H-height, D_I-depth) is $W \times H \times D_I \times D_O + D_O$, where D_O is the number of activation maps, and the additive term D_O stands for the bias. The output of the conv-layer is of size $W \times H \times D_O$. The output size of a layer is preserved to the next layer only if the adequate padding and stride parameters are chosen. In case of 3×3 convolution the stride has to be $s = 1$ and the zero-padding also 1. The activation function of every layer is a rectifying linear unit (ReLU), which is more suitable for CNNs than tanh or sigmoid activations.

In order to make the training process more efficient, batch normalization (BN) is applied. This has a regularization effect and trainable parameters for normalization. Dropout is used in training to deactivate a number of randomly chosen neurons in the layer followed by dropout. The most common values for dropout are between 25%–50%. Different sizes of the input image and subsequent layer outputs can be obtained by the pooling layers (max, min or average pooling). The pooling layer of 2×2 pixels reduces the $W \times H$ size of a given layer to its half $W/2 \times H/2$. A typical way of maintaining almost the same volume of the layer after pooling is by doubling the depth of the next convolution layer. The last layers of the CNNs are commonly fully connected ones which produce the highest number of weights.

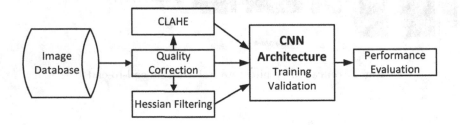

Fig. 3. Block diagram of the proposed system

In the first architecture proposed *conv4-fc2* summarized in Fig. 1. we work with three $W \times H$ sizes: the input size ($size_1 = 224 \times 224$), its half ($size_{1/2} = 112 \times 112$) and its quarter ($size_{1/4} = 56 \times 56$). On every size, two conv-layers of 3×3 are applied and their outputs are normalized using batch normalization. On $size_1$ the filters have a depth of 32 and 32, respectively. On $size_{1/2}$ the depths of filters are doubled for both layers to 64. Finally, on $size_{1/4}$ the layer-depth is 64. We use a dropout right after the first reduction in size with a probability of 0.25 and with 0.5 after the second reduction from $size_{1/2}$ to $size_{1/4}$. The last pooling layer is global average pooling, because this type of layer reduces the overfitting before the first dense layer. The 4 conv-layers are followed by 2 dense layers (FC). The first FC has an input of $56 \times 56 \times 64$ and an output of 512. The last FC has on output equal to the number of persons in the dataset. The probability responses after this last layer are obtained by a softmax classifier.

The second variant of our CNN (*conv6-fc2*) is built from 6 convolutional layers and two fully connected layers. The 4 convolutional layers from the previous architecture are completed by another two on the 1/4 size of the original dimension. So the input of conv-layer5 is $56 \times 56 \times 64$ producing 128 activations and its output will be $56 \times 56 \times 128$. The sixth layer has the 5th layer as input and produces $56 \times 56 \times 128$ output. In this case the max pooling is used between layers 4–5 and the average pooling is applied only before the two dense layers. The first FC has 1000 neurons. It makes the mapping form from 100352 to 1000 and the second from 1000 to the number of classes (persons). The input-output dimensions and kernel sizes for this network are summarized in Fig. 2.

5   Results and Experiments

This section presents our experiments about palm vein recognition with different types of convolutional networks. Beside comparing some well-known CNNs with our proposed two models, we have also carried out some experiments related to the influence of several image processing steps on the identification accuracy and training convergence of the presented networks.

The block diagram of the proposed system is shown in Fig. 3. Our system uses the PUT database images for training all the desired networks. The image processing steps are followed by the fundamental building block in our approach, namely, the CNN architecture used for palm vein based biometric identification.

In our study we have carried out different experiments with various setups. At first, we have compared our two networks trained from scratch and the 4 networks with pre-trained weights fine-tuned via transfer learning approach. We have considered three datasets: the training set, validation set and test set. Out of 12 images per person we have randomly selected 6 into the training set, 3 to the validation and the remaining 3 to the test set. With data augmentation we have multiplied ten times the training set and five times the other two sets. For all the proposed and fine-tuned networks we have compared the recognition accuracy using the same number of epochs (400) in training. Obviously, all the other hyperparameters were also set to the same value. The error between the last layer and the target was expressed as cross entropy loss. The optimization of the weights used common mini-batch stochastic gradient descent. The learning rate was set to 0.01 with a decay of 10 every 40 epochs. Training and validation convergence is better in case of *conv6-fc2* network compared to *conv4-fc2*, however, the total number of parameters in both cases are approximately the same. Figure 4 shows a better decrease for the 6 layered network for the training and the validation data as well. The accuracy of these networks is 92.5% and 95% respectively (Table 1).

Our next experiment was related to the learning rate and optimization methods of the networks. We trialed different types of learning rates between 10^{-3}

and 10^{-1}, with and without learning rate decay. Finally, the best learning rate value found $lr = 0.1$, with learning rate decay of 10 in every 35–45 epochs. The next experiment analysed the influence of different optimization methods. We compared Adam (with $\beta_1 = 0.9$ and $\beta_2 = 0.999$), SGD ($lr = 0.1$), mini-batch SGD ($batch\_size = 32$) and RMSProp ($lr = 0.1$, $\alpha = 0.9$). The best optimization method in our case was the mini-batch SGD with learning rate decay. We also found that the same value of validation loss cannot be achieved by all the 6 trialed networks. Thus, their performance can be compared based on the limitation of the number of epochs.

We also performed experiments with the fine-tuning of the most well-known CNNs from the state-of-the-art, detailed in Sect. 4. We managed to modify them and adjust the pre-trained weights to be adequate for palm vein based biometric identification. We studied not only the final accuracy of the networks, but also the behavior of the networks in training and validation phases. Results show (Figs. 4 and 5) best performances for the ResNet-50 network, over 99.8% in recognition rate. It is somehow evident, because it contains the highest number of conv-layers (50) and FC layers at the end. Our *conv6-fc2* architecture obtains similar results to the outcome of VGG-16, about 96%–97%. The advantage of our network compared to VGG is its smaller number of layers (only 8 instead of 16) and its reduced number of parameters i.e. weights. Our *conv4-fc2* architecture with only 4 convolutional layers has about 3–4% worse accuracy than *conv6-fc2*. Figure 5 shows a good training loss decrease for all these four networks. On the other hand, the same figure draws the attention of the SqueezeNet and AlexNet, the losses of which drop much slower. Moreover, AlexNet has an oscillating behaviour. This worse training process is in connection with the lower accuracy results (between only 84–92%) see Table 1.

Our second group of experiments analysed the effect of different kind of image processing steps, while the experiments described above were related to the CNN architectures and the variation of hyperparameters. We applied the presented 6 CNN not only on the original images, but also to processed images with the following approaches: inhomogeneity correction, contrast limited adaptive histogram equalization (CLAHE), and Hessian affine detector.

The prepocessing steps were necessary because of the low-quality images in the database described in Sect. 2. According to the authors of the database the images are acquired with a low-cost USB web camera and the system uses IR LED lamp for a better vein imaging. The images are having 24 bit depth, still they cannot be considered colored. Some of them are redish others greenish and the majority is simply grayscale (Fig. 6(a)). This variety of colors, but with single channel intensity information is disturbing the training process. The other artefact is the misalignment of the palms. However, the images are somehow aligned with a quasi-rigid positioner, which does not impose a fixed hand position. The misalignment was corrected with translation and rotation in the data augmentation phase of our experiments.

The first image processing step is the conversion to grayscale and normalization of image intensities on a single color channel. Histogram equalization was performed to enhance the contrast of the image. In this way the intensity range was extended from the original interval to [0, 255]. This is followed by the noise filtering of the images with a 3 × 3 Gaussian kernel to reduce the noise and blurriness (Fig. 6(b)). This way we managed to obtain images with quite uniform luminance.

The next step, in order to increase image quality and local contrast, was the CLAHE algorithm. We evaluated the histograms in blocks of 127 × 127 pixels. The clipping limit was set to a maximum slope-size of 7.

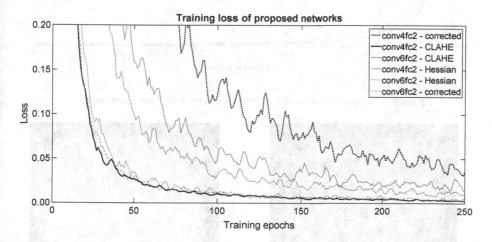

Fig. 4. Comparison of the two proposed networks trained from scratch

The Hessian matrix can be used to locally detect line-like, plate-like and blob-like image structures. Each element of the Hessian matrix is a regularized derivative of the input image. It is obtained by convolving the image with the derivative filters of a Gaussian kernel at scale σ. The eigenvalues and eigenvectors are computed for the Hessian matrix. The relations between the eigenvalues of the Hessian lead to the detection of different structures. In particular, a pixel belonging to a vessel region or a tubular structure will be signaled by large eigenvalues and the corresponding eigenvectors indicate the direction along the vessel. In this way, different vessel widths can be detected at their corresponding scales. The filter applied obtains the largest eigenvalues of the Hessian matrix with a smoothing scale of $\sigma = 8$ for every image point. Vessels and tubular structures of the skin are displayed in high intensities [16]. The effect of Hessian filtering on a palm image is given in Fig. 6(d). The obtained tube-like structures may support a visual comparison and a checkup of the test results. Moreover, we have tried to use these types of images as input for CNNs.

The above presented CNN networks were trained, validated and tested on the corrected images, on CLAHE images and on the Hessian coded images, as well. The identification accuracies for each type of image are presented in Table 1.

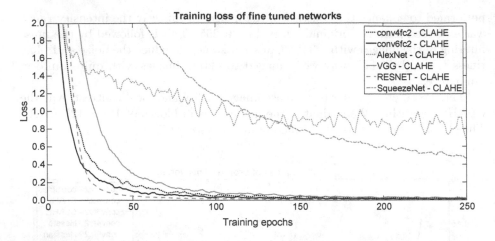

Fig. 5. Comparison of training losses of the fine-tuned networks

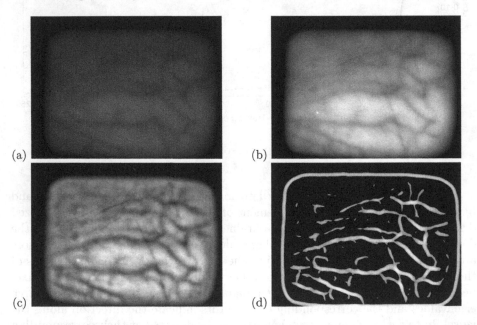

Fig. 6. Image quality enhancement: (a) original; (b) corrected; (c) CLAHE; (d) Hessian.

Table 1. Identification accuracy obtained using the test set

Image Enhancement	CNN network					
	conv4-fc2	*conv6-fc2*	AlexNet	VGG16	ResNet-50	SqueezeNet
Corrected	0.9083	0.9366	0.8916	0.9650	0.9983	0.8750
CLAHE	0.9250	0.9516	0.9216	0.9733	0.9983	0.9166
Hessian	0.9100	0.9416	0.8466	0.9400	0.9966	0.8633

According to the results measured, best identification accuracies are obtained for the CLAHE images, which overpass by 3–4% the outcome using the corrected images. The Hessian images representing tubular structures are not beneficial in this case. The objective function in the networks applied in our case was the cross entropy loss, adequate in case of classification problems. The tubular structure extraction with the Hessian method will come in handy for our future experiments of vein segmentation.

Overall, the identification accuracies of the transfer-learning approach and the end-to-end training are comparable and similar. The best results, regardless the preprocessing step, obtains the ResNet-50 CNN with over 99% for all the three preprocessed image types. It is followed by VGG-16 with 97% and the our network trained from scratch reaches 95%. The AlexNet and SqueezeNet have 10% lower identification rate for the same test set (Table 1).

An even higher, about 100% identification, rate can be achieved if we combine the softmax output of two or three of the proposed networks. This approach is based on the linear opinion pool that combines of probabilistic opinions calculating a weighted linear average of the individual responses.

6   Conclusion and Future Work

In this research we compared different convolutional neural network architectures with the purpose of palm vein based biometric identification. We managed to create and train two simple architectures from scratch and compared their performances with four well-known state-of-the-art CNNs, initially trained for ImageNet Challenge. The best results were obtained for the largest networks with pre-trained weights and transfer learning techniques. Although ResNet-50 obtains almost 100% in accuracy, it is considerable difficult to train. In this work we demonstrate that a simple architecture, even if it is trained from scratch, is also suitable for palm vein based biometric identification. The most important advantage of networks with smaller number of parameters is not only their training in reasonable time, but also the possibility of embedding them in low-resource systems.

For the future we propose to validate our method not only on the PUT palm database, but on other larger and higher quality image sets. Another interesting experiment would be a comparison and combination of palm and dorsal hand [17] or even finger vein based identification of the same subjects.

References

1. Kabaciński, R., Kowalski, M.: Human vein pattern segmentation from low quality images - a comparison of methods. In: Choraś, R.S. (ed.) Image Processing and Communications Challenges 2. Advances in Intelligent and Soft Computing, vol. 84, pp. 105–112. Springer, Berlin, Heidelberg (2010). https://doi.org/10.1007/978-3-642-16295-4_12

2. Meraoumia, A., Chitroub, S., Bouridane, A.: Are infrared images reliable for palm-print based personal identification systems? In: 2013 Saudi International Electronics, Communications and Photonics Conference, pp. 1–6 (2013)

3. Soliman, H., Mohamed, A.S., Atwan, A.: Feature level fusion of palm veins and signature biometrics. Int. J. Video Image Process. Network Secur. **12**(1), 28–39 (2012)

4. Kang, W., Liu, Y., Wu, Q., Yue, X.: Contact-free palm-vein recognition based on local invariant features. PLoS ONE **9**(5), 1–12 (2014)

5. Lefkovits, S., Emerich, S., Szilágyi, L.: Biometric system based on registration of dorsal hand vein configurations. In: Satoh, S. (ed.) PSIVT 2017. LNCS, vol. 10799, pp. 17–29. Springer, Cham (2018). https://doi.org/10.1007/978-3-319-92753-4_2

6. Lee, Y.P.: Palm vein recognition based on a modified $2D^2$ LDA. SIViP **9**(1), 229–242 (2015)

7. Fronitasari, D., Gunawan, D.: Palm vein recognition by using modified of local binary pattern (LBP) for extraction feature. In: 15th International Conference on Quality in Research (QiR), International Symposium on Electrical and Computer Engineering, Nusa Dua, Indonesia, pp. 18–22 (2017)

8. Al-Juboori, A.M., Bu, W., Wu, X., Zhao, Q.: Palm vein verification using multiple features and isometric projection. Int. J. Signal Process. Image Process. Pattern Recogn. **7**(1), 33–44 (2014)

9. Ma, X., Jing, X., Huang, H., Cui, Y., Mu, J.: Palm vein recognition scheme based on an adaptive gabor filter. IET Biometrics **6**(5), 325–333 (2017)

10. Du, D., Lu, L., Fu, R., Yuan, L., Chen, W., Liu, Y.: Palm vein recognition based on end-to-end convolutional neural network. J. South. Med. Univ. **39**(2), 207–214 (2019)

11. Zhong, D., Liu, S., Wang, W., Du, X.: Palm vein recognition with deep hashing network. In: Lai, J.-H., et al. (eds.) PRCV 2018. LNCS, vol. 11256, pp. 38–49. Springer, Cham (2018). https://doi.org/10.1007/978-3-030-03398-9_4

12. Krizhevsky, A., Sutskever, I., Hinton, G.E.: Imagenet classification with deep convolutional neural networks. In: Pereira, F., Burges, C.J.C., Bottou, L., Weniberger, K.Q. (eds.) Advances in Neural Information Processing Systems vol. 25, pp. 1097–1105 (2012)

13. Simonyan, K., Zisserman, A.: Very deep convolutional networks for large-scale image recognition. arXiv preprint arXiv:1409.1556 (2014)

14. He, K., Zhang, X., Ren, S., Sun, J.: Deep residual learning for image recognition. In: IEEE Conference on Computer Vision and Pattern Recognition (CVPR), pp. 770–778 (2016)

15. Iandola, F.N., Han, S., Moskewicz, M.W., Ashraf, K., Dally, W.J., Keutzer, K.: Squeezenet: alexnet-level accuracy with 50x fewer parameters and <0.5 mb model size. arXiv preprint arXiv:1602.07360 (2016)

16. Frangi, A.F., Niessen, W.J., Hoogeveen, R.M., van Walsum, T., Viergever, M.A.: Model-based quantitation of 3-d magnetic resonance angiographic images. IEEE Trans. Med. Imaging **18**(10), 946–956 (1999)

17. Lefkovits, S., Lefkovits, L., Szilágyi, L.: CNN approaches for dorsal hand vein based identification. In: 27th International Conference in Central Europe on Computer Graphics, Visualization and Computer Vision. Computer Science Research Notes (2019, in press)

An Infinite Replicated Softmax Model
for Topic Modeling

Nikolas Alexander Huhnstock[1]([✉]), Alexander Karlsson[1], Maria Riveiro[1,2],
and H. Joe Steinhauer[1]

[1] School of Informatics, University of Skövde, Skövde, Sweden
{nikolas.huhnstock,alexander.karlsson,joe.steinhauer}@his.se
[2] School of Engineering, Jönköping University, Jönköping, Sweden
maria.riveiro@ju.se

Abstract. In this paper, we describe the infinite replicated Softmax model (iRSM) as an adaptive topic model, utilizing the combination of the infinite restricted Boltzmann machine (iRBM) and the replicated Softmax model (RSM). In our approach, the iRBM extends the RBM by enabling its hidden layer to adapt to the data at hand, while the RSM allows for modeling low-dimensional latent semantic representation from a corpus. The combination of the two results is a method that is able to self-adapt to the number of topics within the document corpus and hence, renders manual identification of the correct number of topics superfluous. We propose a hybrid training approach to effectively improve the performance of the iRSM. An empirical evaluation is performed on a standard data set and the results are compared to the results of a baseline topic model. The results show that the iRSM adapts its hidden layer size to the data and when trained in the proposed hybrid manner outperforms the base RSM model.

Keywords: Restricted Boltzmann machine · Unsupervised learning · Topic modeling · Adaptive Neural Network

1 Introduction

One important task of data analysis is clustering which usually addresses the problem of grouping data points that share similar conceptual characteristics into groups. Analyzing textual data can be seen as a special case of clustering. Here, clusters, often referred to as topics, consist of words that are frequently occurring together.

Topic modeling algorithms are statistical methods for analyzing co-occurrences of words within text documents to discover topics that can be further used for categorization purposes within different application scenarios [1]. One of the most widely employed topic model is the latent Dirichlet allocation (LDA) [10], introduced by Blei and Jordan (2003) which is able to discover the thematic structure within large archives of text [1]. Each document within such

V. Torra et al. (Eds.): MDAI 2019, LNAI 11676, pp. 307–318, 2019.
https://doi.org/10.1007/978-3-030-26773-5_27

a document corpus, as explained in [10, pp. 5–6], can be regarded as a bag-of-words that has been produced by the mixture of topics that the document's author intended to discuss. Each topic is hence represented by a distribution over all words that can be found in the document corpus. Abstractly speaking, when a document is generated, the author would repeatedly pick a topic, then a word belonging to that topic and places it in the bag until a document is complete. The objective of topic modeling is then to find the statistical parameters of such a process that is likely to have generated the corpus [10, pp. 5–6].

Topic modeling algorithms work unsupervised and do not usually require any prior annotations or labeling of the documents since topics emerge from the original texts under analysis [1]. However, most topic modeling methods rely on manually setting important initial input parameters, such as the number of topics that is to be expected to be found in the document corpus [3]. The estimation of this rather crucial parameter is challenging and usually requires a certain level of knowledge about the content of document corpus that sometimes could be provided by human experts.

In this work, we attempt to overcome the challenge of determining the number of topics manually and propose a neural network-based approach to topic modeling that is able to self adapt the number of topics within a corpus of text. This method utilizes the combination of two recently developed extensions to the restricted Boltzmann machine (RBM) [13]: the replicated Softmax model (RSM) [6] which adapts the RBM to be usable for topic modeling, and the infinite restricted Boltzmann Machine (iRBM) [2] which is an adaptation of the RBM able to self identify the number of clusters needed for a traditional clustering problem. We combine these two different extension of the RBM into the infinite replicated Softmax model (iRSM) that is capable to self identify the number of topics within a corpus of text documents.

The remainder of the paper is structured as follows: In order to describe the aforementioned approach to topic modeling, we first provide some formal information and preliminaries regarding RBM, RSM and iRBM in Sect. 2. This is followed by the presentation of relevant related work in Sect. 3 after which the proposed model is introduced in Sect. 4. Section 6 describes the empirical and qualitative evaluations of our approach and our results are presented in Sect. 7. We conclude the paper with a brief summary of our findings and conclusions drawn, in Sect. 9.

2   Preliminaries

Two main problems arise when trying to model the contents (represented by topics) of a corpus of textual documents with a RBM. Firstly, the number of words within documents may vary from one document to another and secondly, to infer topics from documents the number of topics that the model is able to represent has to be set in advance which requires knowledge about the corpus' contents which a user cannot be guaranteed to have.

In this section, the two methods the iRSM is based on are introduced: the RBM and the two different adaptions to it, namely the RSM and the iRBM.

2.1   RBM

The RBM [13] can be described as an undirected bipartite graphical model composed of one visible layer $\mathbf{v}$ and one hidden layer $\mathbf{h}$. A weight W_{ij} is associated with each connection between units v_i and h_j of the two layers. Given a binary RBM with n visible and m hidden units we can describe the energy of the model for a given state $(\mathbf{v}, \mathbf{h})$ as:

$$E(\mathbf{v}, \mathbf{h}) = -\mathbf{c}^T\mathbf{h} - \mathbf{h}^T\mathbf{W}\mathbf{v} - \mathbf{b}^T\mathbf{v} \tag{1}$$

Due to its bipartite structure, states of visible and hidden units are only dependent on the other layers' units. The conditional distributions of the layers are therefore described by:

$$p(h_k = 1 \mid \mathbf{v}) = \sigma\left(c_k + \sum_{i=1}^{n} W_{ki}v_i\right), \tag{2}$$

$$p(v_k = 1 \mid \mathbf{h}) = \sigma\left(\sum_{j=1}^{m} h_j W_{jk} + b_k\right), \tag{3}$$

where $\sigma(x) = (1 + \exp(-x))^{-1}$.

2.2   RSM

The replicated Softmax model (RSM), proposed by Salakhutdinov and Hinton [6], has been used to enable the RBM to model documents of words. The RSM addresses the problem of a varying number of words within documents by allocating one visible unit per word in the document while sharing parameters (weights) over all visible units. Hence, it allows the RSM to model arbitrarily sized documents while decoupling the number of free parameters from the document length. This comes, however, at the cost of disregarding the order in which the words occur within the document.

When deploying an RSM, a document is modeled as binary matrix $\mathbf{U} \in \{0,1\}^{V,D}$, where V is the number of words in the dictionary and D is the number of words in the document. The matrix $\mathbf{U}$ defines the observed state of the visible units $\mathbf{v}$ such that $U_k^\eta = 1$ is equal to the kth unit taking value η ($v_k = \eta$). The energy of the RSM given a state $(\mathbf{v}, \mathbf{h})$ is described by:

$$E(\mathbf{v}, \mathbf{h}) = -D\mathbf{c}^T\mathbf{h} - \sum_{i=1}^{n=D} \mathbf{h}^T\mathbf{W}_{\cdot, v_i} - \sum_{i=1}^{n} b_{v_i} \tag{4}$$

To balance the offset that has been introduced through the varying number of visible units that are contributing to the model's energy, the hidden bias term $\mathbf{c}^T\mathbf{h}$ is scaled according to the document's length D.

Since the bipartite structure of the RBM is preserved, the conditional distributions of hidden and visible units are given by:

$$p(h_k = 1 \mid \mathbf{v}) = \sigma \left(Dc_k + \sum_{i=1}^{n=D} W_{k,v_i} \right), \tag{5}$$

$$p(v_k = v^* \mid \mathbf{h}) = \frac{\exp \left(\sum_{j=1}^{m} h_j W_{j,v^*} + b_{v^*} \right)}{\sum_{t=1}^{V} \exp \left(\sum_{j=1}^{m} h_j W_{j,i}^t + \sum_{i=1}^{n} b_i^t \right)}. \tag{6}$$

2.3 iRBM

The infinite restricted Boltzmann machine [2] extends the RBM by enabling it to adapt the size of its hidden layer. This behavior is achieved by introducing a, in theory infinitely large, hidden layer $\mathbf{h}$ of that only a subset $\{h_j \mid j \leq z\}$ is considered. The number of hidden units describing this subset is given by the value of the introduced random variable z. The weights and biases associated with the hidden units $\{h_j \mid j > z\}$ are assumed to have a value of 0 and the energy of a given binary iRBM is given by:

$$E(\mathbf{v}, \mathbf{h}, z) = - \sum_{j=1}^{z} (c_j h_j - \beta_j) - \sum_{j=1}^{z} h_j \mathbf{W}_{j,\cdot} \mathbf{v} - \mathbf{b}^T \mathbf{v}. \tag{7}$$

To counteract the growth of z, Salakhutdinov and Hinton [2] introduced a penalty term β_j, which penalizes the accumulation of untrained units. The penalty term is parametrized on each hidden unit's bias with a global penalty β as $\beta_j = \beta \, \text{soft}_+(c_j)$. With $\text{soft}_+(x) = ln(1 + \exp(x))$

With the introduced random variable z the conditional distributions of the model are given by:

$$p(h_k = 1 \mid \mathbf{v}) = \begin{cases} \sigma(c_k + \mathbf{W}_{j,\cdot} \mathbf{v}), & k \leq z \\ 0, & \text{otherwise} \end{cases} \tag{8}$$

$$p(v_k = 1 \mid \mathbf{h}) = \begin{cases} \sigma \left(\sum_{j=1}^{m} h_j W_{jk} + b_k \right), & k \leq z \\ 0, & \text{otherwise} \end{cases} \tag{9}$$

$$p(z \mid \mathbf{v}) = \frac{\exp -F(\mathbf{v}, z)}{\sum_{z^*}^{\infty} \exp -F(\mathbf{v}, z^*)} \tag{10}$$

It can be shown that the infinite sum occurring in the denominator of (10) can be reformulated into a sum over a term of trained hidden units and a finite geometric series that can be a computed analytically, given that β is greater than 1 [2].

3   Related Work

Inspired by the RSM's weight sharing technique, Larochelle and Laury [8] extended the neural autoregressive distribution estimator (NADE) [9] and enabled the model to represent documents. The so-called DocNade inherits the advantageous characteristic of computing the gradient of the negative log-likelihood over the data without requiring approximation. The DocNode uses a hierarchy of binary logistic regressions to represent the distribution of words, which results in a sublinear scaling with V when sampling the probability of an observed word. Although the DocNade architecture corresponds to several parallel hidden layers, i.e. one for each input word, with tied weights the number of units in each layer needs to be defined manually and is static.

Based on the RSM, Srivasta et al. [14] developed the Over-Replicated Softmax model, which belongs to the family of Deep Boltzmann Machines, i.e. Boltzmann Machines that contain at least two hidden layers. The Over-Replicated Softmax has softmax visible units and binary hidden units in the first layer and on top of that another softmax hidden layer. This is supposed to provide a more flexible prior over the hidden representations.

Srivasta et al. introduced this second hidden layer without the usual increase in model parameters, by reusing the weights that connect the visible layer to the first hidden layer, for the connections between the first and second hidden layer. This allows the Over-Replicated Softmax model to be trained as efficiently as the RSM despite the presence of an additional layer.

Even though the Over-Replicated Softmax model and the DocNade model achieved better results than the RSM model, both models require manual setting of the hidden layer(s), which is the fundamental issue that will be resolved within the proposed iRSM.

4   Proposed Model

The proposed model is a combination of the RSM [6] and the iRBM [2] which we refer to as the infinite replicated Softmax model (iRSM). It combines the capability of the RSM as an undirected topic model while, at the same time, it adapts to the number of represented topics automatically.

The iRSM can be trained on documents of varying length due to the use of the RSM's weight sharing technique, allowing it to replicate input units depending on the input document's length. Furthermore, the iRBM's hidden layer's growing behavior has been adopted by introducing a theoretical infinite hidden layer together with a growing penalty. Figure 1 shows a graphical illustration of the proposed model.

The energy function of the iRSM takes the following from:

$$E(\mathbf{v}, \mathbf{h}, z) = -D \sum_{j=1}^{z} (c_j h_j - \beta_j) - \sum_{j=1}^{z} \sum_{i=1}^{n=D} h_j W_{j,v_i} - \sum_{i=1}^{n} b_{v_i}, \qquad (11)$$

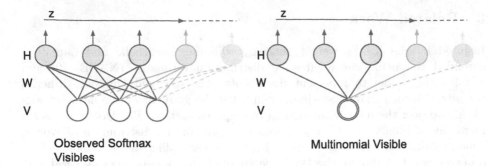

Fig. 1. Graphical representation of the iRSM. (left) An iRSM with three visible softmax units. (right) Visible softmax units replaced with a single multinomial unit which is sampled D times. The shaded hidden units indicate that these are added based on the state of z.

with $\beta_j = \beta \, \text{soft}_+(c_j)$. The growing penalty β_j, defined in the same way as for the iRBM, enables the model to adapt its hidden layer size according to the inputs. In addition to the scaling of the hidden term of the RSM, the growing penalty β is scaled by the size of the document in order to maintain balance among terms.

Given an iRSM with binary hidden units, the conditional distributions are given by:

$$p(h_k = 1 \mid \mathbf{v}) = \begin{cases} D(c_k - \beta_k) + \sum_{i=1}^n n\mathbf{W}_{j,v_i}, & k \le z \\ 0, & \text{otherwise} \end{cases} \tag{12}$$

$$p(v_k = 1 \mid \mathbf{h}) = \begin{cases} \dfrac{\exp(\sum_{j=1}^m h_j W_{j,v*} + b_{v*})}{\sum_{t=1}^V \exp(\sum_{j=1}^m h_j W_{j,i}^t + \sum_{i=1}^n b_i^t)}, & k \le z \\ 0, & \text{otherwise} \end{cases} \tag{13}$$

$$p(z \mid \mathbf{v}) = \frac{\exp -F(\mathbf{v}, z)}{\sum_{z*}^\infty \exp -F(\mathbf{v}, z*)} \tag{14}$$

The iRSM's learning parameters are obtained through the application of gradient descent on the model's negative log-likelihood (NLL) over documents. For a single document v this takes the form:

$$\frac{\partial - log(p((v)))}{\partial \theta} = \mathbb{E}_{\mathbf{h}, z \mid v}\left[\frac{\partial}{\partial \theta} E(v, \mathbf{h}, z)\right] - \mathbb{E}_{\mathbf{v}, \mathbf{h}, z}\left[\frac{\partial}{\partial \theta} E(\mathbf{v}, \mathbf{h}, z)\right] \tag{15}$$

The computation of the second expectation term is considered to be too expensive since it involves the sum over all possible states of the network. Instead, Contrastive Divergence (CD) [5] can be used to approximate the gradient of the NLL by running a short Markov Chain wherein sampling is alternated between $z \sim p(z \mid \mathbf{v})$, $\mathbf{h} \sim p(\mathbf{h} \mid \mathbf{v}, z)$ and $\mathbf{v} \sim p(\mathbf{v} \mid \mathbf{h}, z)$.

5   Hybrid Training

Additionally to the iRSM model introduced previously we propose a hybrid training approach. The idea of this training method is to combine the iRSM and RSM into a two phase training procedure, where the former determines the networks hidden layer size and the latter is used to improve performance. The motivation behind this procedure is that the iRSM is, even in later stages of the training process, still slightly adjusting its hidden layer size. This behavior was as well observed in previous work for the iRBM in the context of clustering [7]. This leads to the situation, in which some of the already limited amount of information is continuously devoted to the task of adjusting the hidden layer's size. In order to leverage as much as possible from the sparse information, we decided to discard the adaptive behavior of the iRSM at a point in the training process where the iRSM has had sufficient time to develop its hidden layer. From this point forward the training is solely focused on optimising weight and bias parameter to learn the representation of the data as good as possible. By making the size of the hidden layer static at a given point in time, the parameters will be fine-tuned to that size of the hidden layer. The training process begins by training an iRSM; after some predefined time, e.g. half the total training time, this iRSM is transformed into a RSM and training continues until termination.

6   Experiment Design

In this section, we describe the empirical and qualitative experiments we conducted. Since Salakhutdinov and Hinton [6] showed that the base RSM is able to outperform Latent Dirichlet Allocation (LDA) we do not further go into the comparison with LDA and focus on comparing the RSM with iRSM and the iRSM trained in the proposed hybrid manner.

The first experiment quantitatively analyses the influence of the regularization parameter beta on the behavior of the iRSM and makes the comparison with the base RSM, we report mean and standard deviation of 10 trials. The second experiment provides an analysis of the top words per topic identified by an hybridly trained iRMS for different parameter settings. The "Reuters-21578, Distribution 1.0" corpus contains 10,788 news documents totaling 1.3 million words and was compiled by David Lewis[1]. The data set is split into 7,769 training documents and 3019 test documents. Common stopwords are removed from the data and the words are stemmed. To effectively reduce the dimensionality of the problem space, we only consider the 2000 most frequent words similar to what Salakhutdinov and Larochelle [6,8] suggested by.

We use per word perplexity as a metric to assess the models generative performance through:

$$exp\left(-\frac{1}{N}\sum_{i=1}^{N}\frac{1}{D_i}\log p(\mathbf{v}_i)\right), \tag{16}$$

---

[1] Available at: http://www.daviddlewis.com/resources/testcollections/reuters21578/ [Accessed 22 May 2019].

where D_i represents the word count of the i-th document. The perplexity is evaluated in a similar fashion as Salakhutdinov [6] over 50 randomly held out test documents. Computing the probability of held-out documents exactly is intractable for undirected models, such as the RBM, since it requires to enumerate of over an exponential number of terms. Therefore, annealed importance sampling (AIS) [6] is deployed to obtain $p(\mathbf{v})$ of the RSM and iRSM by averaging over 100 runs using 1,000 in $[0, 1]$ uniformly spaced temperatures β.

To allow comparison between the models: all models processed the data in batches of size 100 and were trained for an equal amount of epochs. During training, the adaptive gradient algorithm ADAGRAD [4] with an initial learning rate of 0.05 is deployed.

7 Results

The results of the first experiment are depicted in Fig. 2. The plot shows average perplexity scores and final hidden layer sizes for the iRSM trained and hybridly trained iRSM (iRSM_hybrid) for beta values from 1.1 to 2. Additionally, average RSM score are indicated as lines in the plot for models with hidden layer sizes of 25, 50, 100 and 250 hidden units.

The plots show that a higher beta value results in smaller sized hidden layers. This is the expected behavior since a higher penalty term increases the model's growing threshold. For lower beta values, i.e., lower than 1.2, the size of the hidden layer falls between 75 and 100 hidden units, whereas for higher values of beta, i.e., greater than 1.5, hidden layer size average between 25 an 50 hidden units. Overall, the range of hidden layer sizes of the iRSM seems to be in a reasonable range considering that the best performing RSM has as well 50 hidden units. Hidden layer sizes of iRSM and iRSM_hybrid is very similar which is not surprising considering that the iRSM does not tend to change its hidden layer size much in later stages of the training and since iRSM_hybrid is an iRSM for the first half of the training process its hidden layer size is almost equal to hidden layer sizes of iRSM models.

The top plot of Fig. 2, depicting average perplexity scores, shows that the iRSM does not reach the performance of any of the RSM models. The iRSM scores improve with higher values of beta which seems to be correlated with the resulting smaller sized hidden layers. It becomes as well apparent that the iRSM suffers from slightly higher variances (indicated by the shaded areas) than the RSM, which is certainly caused by the non-static hidden layer sizes. The plot shows as well that the hybridly trained iRSM (IRSM_hybrid) performs better than the iRSM and as well better than all RSMs, for the whole range of evaluated beta values.

Table 1 illustrates the influence of training time on the performance of the different models. It can be seen that at epoch 200 RSM models perform better than the iRSM based models. Among the base RSM models, the RSMs with 50 hidden units show the best performance of all throughout the course of the training.

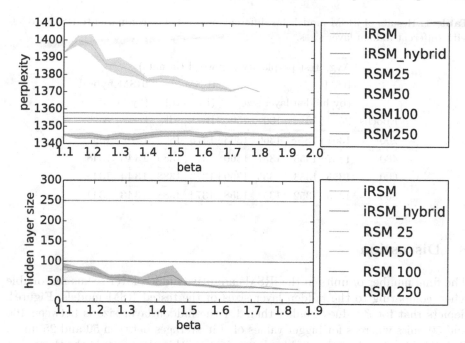

Fig. 2. (top) The plot depicts the average perplexity results of several RSM models with hidden layer sizes of 25, 50, 100 and 250; mean and standard deviation of iRBM models trained with different β settings as well as mean and standard deviation of hybridly trained iRSMs. Mean (line) and standard deviation (shade) based on 10 evaluations per configuration are plotted. (bottom) Mean (line) and standard deviation (shade) of the hidden layer sizes. Results of 10 evaluations per setting. All models have been trained for 800 epochs.

In the interval from epoch 200 to epoch 400 the hybridly trained iRSM has had the largest improvement of all models.

From epoch 200 on all RSM models gradually increase their performance scores as training progresses. Although, the best iRSM models do as gradually increase their performance as the RSM models its performance wrt. the average overall considered beta value decreases. This is causes by the poor performance of the very low valued beta settings which accumulate too many hidden units during the course of the training.

Figure 3 shows the perplexity for 50 randomly selected test documents of RSM, iRSM and iRSM trained in the proposed hybrid manner to give a closer look at how models compare with to each other. The given iRSM and hybrid iRSM models were trained with beta set to 1.5, the RSM model has a hidden layer size of 50 and all models were trained for 800 epochs. The left plot shows that the iRSM perplexity scores are consistently higher than the RSM for all test documents. Similarly, the right plot shows that the hybridly trained iRSM reaches lower perplexity scores than the RSM model.

Table 1. Results of iRSM models for different beta settings compared to RSM models with different hidden layer sizes.

	Avg. Test perplexity per word (in nats)							
	RSM (by hidden layer size)				iRSM (by beta)		iRSM_hybrid (by beta)	
Epochs	25	50	100	250	Best	All	Best	All
200	1361	**1358**	1360	1363	1375	1390	1375	1390
400	1357	1355	1357	1360	1373	1385	**1346**	1348
600	1355	1353	1355	1359	1373	1388	**1344**	1345
800	1353	1352	1354	1358	1374	1388	**1343**	1344

8 Discussion

The final number of units in the iRSMs adaptive hidden layer seems reasonable when comparing to the hidden layer sizes of the tested RSM models. Figure 2 depicts that for β values smaller than 1.4 the hidden layer size is between 100 and 50 units whereas for bigger values of β it averages between 50 and 25 units. Remarkably, this change in hidden layer size is within the range of the three best performing RSM models under test. These results show that the iRSM is able to adequately adapt its hidden layer size to the documents and reach a reasonable hidden layer size for a broad range of β values.

The results of the conducted experiments summarized in Table 1 and Fig. 2 show that the RSM models do reach better perplexity scores early into the training process than the adaptive iRSM models. This is most likely caused by the fact that the iRSMs first have to adapt the size of their hidden layers, from initially 1 unit, on by gradually growing their hidden layers in the first epochs of the training process. Therefore, they suffer from a slow start with respect to representation learning compared to the RSM based models which have all hidden units available to train from the very start. Especially, the hybridly trained iRSM does quickly surpass the performance of the RSM models.

The increase in performance is well depicted in Fig. 2. The hybrid iRSM achieves superior scores than both the iRSM and the RSM. A possible explanation for this might be that the RSM models do make steady but small improvements throughout the learning process. The hybridly trained iRSM, on the other hand, does not have this steady monotonous perplexity improvements: it starts rather slow, as already discussed above, but as soon as the transformation to an ordinary RSM takes place the model is able make a big leap wrt. its perplexity scores, see Table 1. The ordinary RSM seems to be able to improve upon the essentially, by the iRSM, pretrained model much better than by starting from a normal initialized RSM model, given that the iRSM learned a reasonable sized hidden layer.

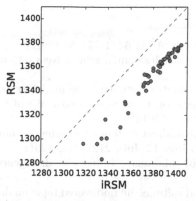

(a) Document-wise perplexity comparison of an RSM with 50 hidden units and an iRSM.

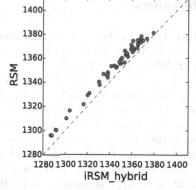

(b) Document wise perplexity comparison of an RSM with 50 hidden units and an hybridly trained iRSM.

Fig. 3. Perplexity score comparison of RSM, iRSM and hybridly trained iRSM on each of the 50 randomly selected test documents. All models were trained for 800 epochs.

Despite the difference in performance values, Fig. 3 as well depicts that all three models seem to, represent each document almost equally well relative to their individual performance realm, which is indicated by the fact that the dots are arranged along an imaginary straight line. One would maybe expect different kind of models to showcase differing representational behavior here, e.g. being able to represent some pattern better than others, and therefore expect a more diffused score pattern. But considering that all the iRSM models inherited especially their representational characteristics from the base RSM, this behavior seems reasonable.

9   Conclusion

In order to adapt automatically the number of topics found in a corpus of text, this paper presents a novel combination of two RBM based methods: the RSM and the iRBM. The resulting iRSM inherits the topic modeling properties of the RSM as well as the iRBM's adaptive hidden layer, which obviates the need to set the size of the hidden layer manually. In addition, we also introduce a hybrid training procedure to effectively increase the performance of the iRSM over the standard training procedure. We conducted empirical experiments to showcase the functioning of the proposed method.

In upcoming work, we are interested in comparing the proposed model with already existing topic models that are as well able to adapt the number of topics, such as the Hierarchical Dirichlet Process Model [15]. For future extensions of this method we are interested in moving from a flat representational structure to structures consisting of several layers of units, that could enable a beneficial interaction between topic features, as already discussed by Salkhutdinov and Hinton [12]. Similarly, Peng et al. developed the infinite deep Boltzmann machine (IDBM) by stacking a fixed number of iRBMs [11].

References

1. Blei, D.M.: Probabilistic topic models. Commun. ACM **55**(4), 77–84 (2012)
2. Côté, M.A., Larochelle, H.: An infinite restricted Boltzmann machine. Neural Comput. **28**(7), 1265–1288 (2016)
3. DiMaggio, P., Nag, M., Blei, D.: Exploiting affinities between topic modeling and the sociological perspective on culture: application to newspaper coverage of US government arts funding. Poetics **41**(6), 570–606 (2013)
4. Duchi, J., Hazan, E., Singer, Y.: Adaptive subgradient methods for online learning and stochastic optimization. J. Mach. Learn. Res. **12**(Jul), 2121–2159 (2011)
5. Hinton, G.E.: Training products of experts by minimizing contrastive divergence. Neural Comput. **14**(8), 1771–1800 (2002)
6. Hinton, G.E., Salakhutdinov, R.R.: Replicated softmax: an undirected topic model. In: Advances in Neural Information Processing Systems, pp. 1607–1614 (2009)
7. Huhnstock, N.A., Karlsson, A., Riveiro, M., Steinhauer, H.J.: On the behavior of the infinite restricted Boltzmann machine for clustering. In: Proceedings of the 33rd Annual ACM Symposium on Applied Computing, pp. 461–470. ACM (2018)
8. Larochelle, H., Lauly, S.: A neural autoregressive topic model. In: Advances in Neural Information Processing Systems, pp. 2708–2716 (2012)
9. Larochelle, H., Murray, I.: The neural autoregressive distribution estimator. In: Proceedings of the Fourteenth International Conference on Artificial Intelligence and Statistics, pp. 29–37 (2011)
10. Mohr, J.W., Bogdanov, P.: Introduction-topic models: what they are and why they matter. Poetics **41**(6), 545–569 (2013). topic Models and the Cultural Sciences
11. Peng, X., Gao, X., Li, X.: An infinite deep Boltzmann machine. In: Proceedings of the 2nd International Conference on Compute and Data Analysis. ICCDA 2018, pp. 36–41. ACM, New York (2018)
12. Salakhutdinov, R., Hinton, G.: Semantic hashing. Int. J. Approximate Reasoning **50**(7), 969–978 (2009)
13. Smolensky, P.: Information processing in dynamical systems: foundations of harmony theory. Parallel Distrib. Process. Explor. Microstruct. Cogn. **1**(1), 194–281 (1986)
14. Srivastava, N., Salakhutdinov, R., Hinton, G.: Modeling documents with a deep Boltzmann machine. In: Proceedings of the Twenty-Ninth Conference on Uncertainty in Artificial Intelligence, UAI 2013, Bellevue, WA, pp. 616–624. AUAI Press, Arlington (2013). http://dl.acm.org/citation.cfm?id=3023638.3023701
15. Teh, Y.W., Jordan, M.I., Beal, M.J., Blei, D.M.: Sharing clusters among related groups: Hierarchical Dirichlet processes. In: Advances in Neural Information Processing Systems, pp. 1385–1392 (2005)

Estimating Optimal Values for Intentional-Value-Substitution Learning

Takuya Fukushima[1(✉)], Taku Hasegawa[1,2P], and Tomoharu Nakashima[1]

[1] Graduate School of Humanities and Sustainable System Sciences, Osaka Prefecture University, Gakuen-cho 1-1, Naka-ku, Sakai, Osaka 599-8531, Japan
{takuya.fukushima,tomoharu.nakashima}@kis.osakafu-u.ac.jp,
hasegawa@ss.cs.osakafu-u.ac.jp
[2] NTT Media Intelligence Laboratory, NTT Corporation, Tokyo, Japan

Abstract. Intentional-Value-Substitution (IVS) learning was shown to be effective for missing data. This paper gives further insight into the optimal value for the substitution during the learning of a regression model. Function fitting is focused on as the task of the machine learning model. Theoretical analysis on the optimal substitution value for IVS learning is presented before a series of experiments with neural networks are conducted in order to confirm the validity of the theoretical analysis. This paper also proposes a method for estimating the optimal substitution values without using the information of the target function. Another series of computational experiments are conducted to evaluate the accuracy performance of the estimation method.

Keywords: Machine learning · Incomplete data · Regression · Neural network

1 Introduction

Many machine learning techniques assume that there are no missing values in the features of a given training dataset. An ideal situation is that the sample size is large enough to build a model as well as each data point is complete without any missing values. However, this does not always happen in the real world. For example in a medical diagnosis, some measurements might not be available due to the failure in the measuring equipment or patient's personal reasons.

There are several ways to overcome the issue of handling missing value [1]. One way is to impute the missing value by a certain value (e.g., zero, the average value of the feature value, or the output of an imputation model constructed from the training dataset). Another way is to omit the deficit features and construct a model without those features that include missing values. Some papers have presented how to handle the incomplete data with missing values by using a statistical modeling. Methods for the parameter estimation were also proposed [2,3]. Furthermore, the way of handling of missing values have been discussed as multiple imputation [4,5] and maximum likelihood estimation [6].

© Springer Nature Switzerland AG 2019
V. Torra et al. (Eds.): MDAI 2019, LNAI 11676, pp. 319–329, 2019.
https://doi.org/10.1007/978-3-030-26773-5_28

Volker et al. [7] provided a way to incorporate missing/uncertain value during the training of neural networks and it was shown that heuristic ways such as substitution can be harmful in the training of neural networks. Alan [8] discussed substitution strategies for missing values because non-optimum strategies for missing values could produce biased estimates, distorted statistical power, and invalid conclusions.

It should be noted that the above-mentioned methods consider the case where both training and test datasets have the missing values. On the other hand, we consider the case where the training dataset is complete without any missing values while there are missing values in the test dataset. This situation happens in a real world such as emergency medical cares and sports where more than enough information is available in the training/learning phase, but in the practical situation one must make a decision in a short time with a limited amount of information.

2   Intentional-Value-Substitution (IVS) Learning

Hasegawa et al. [9] proposed a method for training a data-driven model for the case where there are missing values only in the test data. In this paper, we call this method Intentional-Value-Substitution (IVS) learning. It is assumed in the IVS learning that the training dataset has no missing values while some values are missing in the test dataset. IVS learning substitutes a non-missing value in the training dataset with some value. In other words, they model the target function using a modified training dataset where some feature values are substituted with a certain value.

In this section, we first introduce the IVS method for training a model in the following subsections. Next, we also introduce a mathematical analysis by Hasegawa et al. [9] where the optimal substitution value for IVS learning is obtained in the case where we assume the target function is known. Finally, we propose a method that for estimating the optimal value even though the target function is unknown.

It should be noted that a feature value in the training dataset is substituted even if its true value is available (since the training dataset is complete with no missing values). We also note that the substitution does not always occur but a feature value is substituted with a pre-specified probability.

2.1   Method

In this subsection, we introduce the procedure of the IVS method for training a robust machine learning model. The following procedure assumes that a single input vector is used to train a model and we also know that which features of D_{test} will contain missing values beforehand. We can easily expand the procedure for a mini-batch training by iterating the process as many as the number of input vectors in the batch set.

Step 1: Draw an input vector with its associated target value from the training dataset D_{train}.

Step 2: With a pre-specified probability, substitute those feature values that are missing in D_{test} with certain values.

Step 3: Train a prediction model with the modified input vector and the target value.

In the IVS during training, we need to consider two things: Which value is used for substitution, and what is the best probability of the substitution. These settings are used in Step 2 in the above procedure. In the following subsection, we show a mathematical analysis on these two questions under the condition that the other experimental settings are ideal (e.g., the target function is known, a prediction model has a sufficient accuracy in approximating the target function). Then, we will validate the mathematical analysis through a series of computational experiments where synthetic regression problems are considered.

2.2 Analysis on the Optimal Values with the Target Function

Mathematical Formulation. This section first presents the formulation of a model that is trained using the IVS method. Next, we discuss the values for substitution in place of the missing values. The expected error of the trained model for test data is mathematically investigated. The mathematical investigation reveals that naive substitutions such as an average and a zero do not lead to a good trained model with a high prediction performance for unseen data. We also introduce the appropriate substitution value that is mathematically given under the above conditions. It should be noted that the mathematically appropriate substitution value can be obtained only for a situation where the target function is known and which feature will be missing in the prediction phase.

First of all, we mathematically describe our situation where a feature value is missing. It is generally assumed that such deficiency happens in any features. Let us consider an n-dimensional regression problem.

Let us denote the n feature variables as an n-dimensional random variable vector $\boldsymbol{X} = (X_1, X_2, \ldots, X_n)$. We also consider an n-dimensional random variable vector $\boldsymbol{R} = (R_1, R_2, \ldots, R_n)$, where each element of the vector represents whether the corresponding feature is observed or missing as follows:

$$R_i = \begin{cases} 1, & \text{if } x_i \text{ is observed,} \\ 0, & \text{otherwise (i.e., } x_i \text{ is missing).} \end{cases} \tag{1}$$

Now let us define a new random variable as follows:

$$X_i' = \begin{cases} X_i, & \text{if the } i\text{-th feature value is observed,} \\ ?, & \text{if it is missing.} \end{cases} \tag{2}$$

Then, we can define $\phi : X \times R \longrightarrow X'$, where ϕ is a bijective function.

When we consider the modeling problem with missing data using a joint probability distribution on the universe of discourse $(X_1, \ldots, X_n, R_1, \ldots, R_n)$, the joint probability function $p(\boldsymbol{x}, \boldsymbol{r})$ is defined as follows:

$$p(\boldsymbol{x}, \boldsymbol{r}) = p(\boldsymbol{x}|\boldsymbol{r})p(\boldsymbol{r}) = p(\boldsymbol{r}|\boldsymbol{x})p(\boldsymbol{x}), \tag{3}$$

where $p(\boldsymbol{x}) = p(x_1, \ldots, x_n)$ is the marginal probability density function of $X_1,$ $\ldots, X_n$, and $p(\boldsymbol{r}|\boldsymbol{x})$ is the probability function which represents whether x_i is observed or not for $X = \boldsymbol{x}$.

Secondly, we define a substituting operation for missing elements of the feature values. Let a mapping be $\psi : \mathbb{R}_?^n \to \mathbb{R}^n$ where $\mathbb{R}_?^n$ is $\{\boldsymbol{x}' = (x_1', \cdots, x_n')|x_i' \in \mathbb{R} \cup \{?\}\}$. Then the substituted data follow $\boldsymbol{x}^* = \psi(\boldsymbol{x}') = \psi(\phi(\boldsymbol{x}, \boldsymbol{r}))(\triangleq \psi^r(\boldsymbol{x}))$. Furthermore, when we put $\psi^r(\boldsymbol{x}) = \psi_1^r \times \cdots \times \psi_n^r(\boldsymbol{x}) = (\psi_1^r(\boldsymbol{x}), \ldots, \psi_n^r(\boldsymbol{x}))$, we can obtain

$$\psi_i(\phi(\boldsymbol{x}, \boldsymbol{r})) = \begin{cases} x_i, & \text{if } r_i = 1, \\ \psi_i'^r(\boldsymbol{x}_{\text{obs}}), & \text{if } r_i = 0, \end{cases} \tag{4}$$

where $\boldsymbol{x}_{\text{obs}}$ is a vector that consists of those observed features.

Next, we discuss the machine learning model and its loss function for a task. For simplicity, let the target function be f, and the prediction model be g. In this paper, for the sake of simplicity, we suppose $f : \mathbb{R}^n \to \mathbb{R}$ and $g : \mathbb{R}^n \to \mathbb{R}$. Moreover, let us define the distance (i.e., error) between f and g for an input vector $\boldsymbol{x}$ as $\delta(f(\boldsymbol{x}), g(\boldsymbol{x}))$, and also let us define a possible vector set for $\boldsymbol{r}$ as $S = \{s_1, \cdots, s_n | \forall i \in \mathbb{N}, s_i \in \{0, 1\}\}$. Then, the expectation of the error δ between f and g is represented as follows:

$$\mathbb{E}[\delta(f, g)] = \sum_{s \in S} \int \cdots \int_{D_X} p(\boldsymbol{x}, \boldsymbol{r} = s)\delta(f(\boldsymbol{x}), g(\psi^s(\boldsymbol{x})))dX \tag{5}$$

$$= \int \cdots \int_{D_X} p(\boldsymbol{x}, \boldsymbol{r} = 1)\delta(f(\boldsymbol{x}), g(\boldsymbol{x}))dX$$

$$+ \sum_{s \in S \setminus \{1\}} \int \cdots \int_{D_X} p(\boldsymbol{x}, \boldsymbol{r} = s)\delta(f(\boldsymbol{x}), g(\psi^s(\boldsymbol{x})))dX \tag{6}$$

Unless otherwise noted, we denote $\int \cdots \int_{D_X} = \int_{D_X}$ for simplifying equations hereafter. In Eq. (6), the first term is the expectation for those input vectors with no missing values, and the second term means the one for those input vectors with missing feature values. Here, when we suppose that the loss is evaluated by $\delta(f, g) = \{f - g\}^2$, then we have the following equation for obtaining the expected loss:

$$\mathbb{E}[\delta(f, g)] = \int_{D_X} p(\boldsymbol{x}, \boldsymbol{r} = 1)\{f(\boldsymbol{x}) - g(\boldsymbol{x})\}^2 dX$$

$$+ \sum_{s \in S \setminus \{1\}} \int_{D_X} p(\boldsymbol{x}, \boldsymbol{r} = s)\{f(\boldsymbol{x}) - g(\psi^s(\boldsymbol{x}))\}^2 dX \tag{7}$$

Now, we focus only on a single term in the latter part of Eq. (7) (i.e., the summation term starting with $\sum_{s \in S \setminus \{1\}}$). In this discussion, we assume that the s is $s_1 = s_2 = \cdots = s_k = 1$, $s_{k+1} = s_{k+2} = \cdots = s_n = 0$, without loss of generality. However, please note that the following discussion is same even if the assumption is not satisfied. Let us denote $X_{\text{obs}} = X_1, X_2, \cdots, X_k$, and

$X_{\mathrm{mis}} = X_{k+1}, X_{k+2}, \cdots, X_n$. When we suppose $p(\boldsymbol{x}, \boldsymbol{r} = \boldsymbol{s}) = p_s(\boldsymbol{x})$, the latter term that satisfies $\boldsymbol{r} = \boldsymbol{s}$ in Eq. (7) is written as follows:

$$\int_{D_X} p_s(\boldsymbol{x})\{f(\boldsymbol{x}) - g(\psi^s(\boldsymbol{x}))\}^2 dX$$

$$= \int_{D_X} p_s(\boldsymbol{x}) f^2(\boldsymbol{x}) dX$$

$$-2 \int_{D_{X_{\mathrm{obs}}} \times D_{X_{\mathrm{mis}}}} p_s(\boldsymbol{x}) f(\boldsymbol{x}) g(\boldsymbol{x}_{\mathrm{obs}}, \psi_{k+1}'^s(\boldsymbol{x}_{\mathrm{obs}}), \ldots, \psi_n'^s(\boldsymbol{x}_{\mathrm{obs}})) \, dX_{\mathrm{mis}} \, dX_{\mathrm{obs}}$$

$$+ \int_{D_{X_{\mathrm{obs}}} \times D_{X_{\mathrm{mis}}}} p_s(\boldsymbol{x}) g^2(\boldsymbol{x}_{\mathrm{obs}}, \psi_{k+1}'^s(\boldsymbol{x}_{\mathrm{obs}}), \ldots, \psi_n'^s(\boldsymbol{x}_{\mathrm{obs}})) \, dX_{\mathrm{mis}} \, dX_{\mathrm{obs}}$$

$$= \int_{D_X} p_s(\boldsymbol{x}) f^2(\boldsymbol{x}) dX$$

$$-2 \int_{D_{X_{\mathrm{obs}}}} g(\boldsymbol{x}_{\mathrm{obs}}, \psi_{k+1}'^s(\boldsymbol{x}_{\mathrm{obs}}), \ldots, \psi_n'^s(\boldsymbol{x}_{\mathrm{obs}})) \left[\int_{D_{X_{\mathrm{mis}}}} p_s(\boldsymbol{x}) f(\boldsymbol{x}) \, dX_{\mathrm{mis}} \right] dX_{\mathrm{obs}}$$

$$+ \int_{D_{X_{\mathrm{obs}}}} g^2(\boldsymbol{x}_{\mathrm{obs}}, \psi_{k+1}'^s(\boldsymbol{x}_{\mathrm{obs}}), \ldots, \psi_n'^s(\boldsymbol{x}_{\mathrm{obs}})) \left[\int_{D_{X_{\mathrm{mis}}}} p_s(\boldsymbol{x}) \, dX_{\mathrm{mis}} \right] dX_{\mathrm{obs}}, \qquad (8)$$

where $\int_{D_{X_{\mathrm{mis}}}} p_s(\boldsymbol{x}) \, dX_{\mathrm{mis}} = p_s(\boldsymbol{x}_{\mathrm{obs}}) \int_{D_{X_{\mathrm{mis}}}} p_s(\boldsymbol{x}_{\mathrm{mis}}|\boldsymbol{x}_{\mathrm{obs}}) \, dX_{\mathrm{mis}}$ in Eq. (8) denotes a marginal distribution by X_{mis}. By rearranging $\int_{D_{X_{\mathrm{mis}}}} p_s(\boldsymbol{x}) f(\boldsymbol{x}) dX_{\mathrm{mis}}$, the following equation is obtained:

$$\int_{D_{X_{\mathrm{mis}}}} p_s(\boldsymbol{x}) f(\boldsymbol{x}) \, dX_{\mathrm{mis}} = p_s(\boldsymbol{x}_{\mathrm{obs}}) \int_{D_{X_{\mathrm{mis}}}} p_s(\boldsymbol{x}_{\mathrm{mis}}|\boldsymbol{x}_{\mathrm{obs}}) f(\boldsymbol{x}) \, dX_{\mathrm{mis}}$$

$$= p_s(\boldsymbol{x}_{\mathrm{obs}}) \mathbb{E}_{X_{\mathrm{mis}}}[f(\boldsymbol{x})]. \qquad (9)$$

This represents the expected output value for the missed value. For simplicity, we denote this as $\mathbb{E}_{X_{mis}}[f(\boldsymbol{x})] = e_s(\boldsymbol{x}_{\mathrm{obs}})$ in the following equations. Based on these discussions, by setting $g'(\boldsymbol{x}_{\mathrm{obs}}) \triangleq g(\boldsymbol{x}_{\mathrm{obs}}, \psi_{k+1}'^s(\boldsymbol{x}_{\mathrm{obs}}), \ldots, \psi_n'^s(\boldsymbol{x}_{\mathrm{obs}}))$, then we have the following:

$$\int_{D_X} p_s(\boldsymbol{x})\{f(\boldsymbol{x}) - g(\psi^s(\boldsymbol{x}))\}^2 dX$$

$$= \int_{D_X} p_s(\boldsymbol{x}) f^2(\boldsymbol{x}) dX$$

$$+ \int_{D_{X_{\mathrm{obs}}}} -2g'(\boldsymbol{x}_{\mathrm{obs}}) e_s(\boldsymbol{x}_{\mathrm{obs}}) p_s(\boldsymbol{x}_{\mathrm{obs}}) + g'^2(\boldsymbol{x}_{\mathrm{obs}}) p_s(\boldsymbol{x}_{\mathrm{obs}}) \, dX_{\mathrm{obs}}$$

$$= \int_{D_{X_{\mathrm{obs}}}} p_s(\boldsymbol{x}_{\mathrm{obs}}) \{g'(\boldsymbol{x}_{\mathrm{obs}}) - e_s(\boldsymbol{x}_{\mathrm{obs}})\}^2 - p_s(\boldsymbol{x}_{\mathrm{obs}}) e_s^2(\boldsymbol{x}_{\mathrm{obs}}) \, dX_{\mathrm{obs}}$$

$$+ \int_{D_X} p_s(\boldsymbol{x}) f^2(\boldsymbol{x}) dX. \qquad (10)$$

If there is only one combination of the observed features and the missing features, that is, if $r = s$, the optimal model g' that minimizes Eq. (10) can be trained from the training using IVS method. Equation (10) is minimized when $g'(\boldsymbol{x}_{\text{obs}}) = g(\boldsymbol{x}_{\text{obs}}, \psi'^s_{k+1}(\boldsymbol{x}_{\text{obs}}), \dots, \psi'^s_n(\boldsymbol{x}_{\text{obs}})) = e_s(\boldsymbol{x}_{\text{obs}})$. As it is possible that there is no missing value in the input vector, it is necessary to minimize Eq. (7), which Eq. (10) is substituted with. Now, we suppose that f can be approximated by g, and $g \simeq f$. Then, we obtain the function $\psi'^s_{k+1}, \dots, \psi'^s_n$ satisfying $g(\boldsymbol{x}_{\text{obs}}, \psi'^s_{k+1}(\boldsymbol{x}_{\text{obs}}), \dots, \psi'^s_n(\boldsymbol{x}_{\text{obs}})) = e_s(\boldsymbol{x}_{\text{obs}})$. This leads to the substitution value for minimizing the expectation of error in the case where the above missing observations could occur.

Finally, we discuss this mathematical analysis in more detail by tackling to a concrete example. Let us assume that $R_1, R_2, \cdots, R_n$, and X are independent, that is, $p(\boldsymbol{x}, \boldsymbol{r}) = p(\boldsymbol{x})p(r_1)p(r_2)\dots p(r_n)$. Moreover, we also suppose $p(r_1 = 1) = \cdots = p(r_{n-1} = 1) = 1.0$, $p(r_n = 0) = p_{\text{mis}}$. This means that the value missing happens with a probability p_{mis} only at the last dimension. The expected error under these settings is obtained from Eq. (7) as follows:

$$\mathbb{E}[\delta(f, g)] = (1 - p_{\text{mis}}) \int_{D_X} p(\boldsymbol{x})\{f(\boldsymbol{x}) - g(\boldsymbol{x})\}^2 dX$$
$$+ p_{\text{mis}} \int_{D_X} p(\boldsymbol{x})\{f(\boldsymbol{x}) - g(\psi^s(\boldsymbol{x}))\}^2 dX \qquad (11)$$

and from Eq. (10), we obtain

$$\mathbb{E}[\delta(f, g)] = (1 - p_{\text{mis}}) \int_{D_X} p(\boldsymbol{x})\{f(\boldsymbol{x}) - g(\boldsymbol{x})\}^2 dX$$
$$+ p_{\text{mis}} \int_{D_{X_{\text{obs}}}} p(\boldsymbol{x}_{\text{obs}}) \{g'(\boldsymbol{x}_{\text{obs}}) - e(\boldsymbol{x}_{\text{obs}})\}^2 - p(\boldsymbol{x}_{\text{obs}})e^2(\boldsymbol{x}_{\text{obs}}) \, dX_{\text{obs}}$$
$$+ p_{\text{mis}} \int_{D_X} p_s(\boldsymbol{x})f^2(\boldsymbol{x})dX,$$

where $e(\boldsymbol{x}_{\text{obs}}) = \mathbb{E}_{X_n}[f(\boldsymbol{x})] = \int_{-\infty}^{\infty} p(x_n|\boldsymbol{x}_{\text{obs}})f(\boldsymbol{x})dX_n$ (see Eq. (9)). Now, when $g \simeq f$, the expected error is minimized if ψ satisfies the following equation:

$$\psi'_n(\boldsymbol{x}_{\text{obs}}) = \arg\min_{x_n} \left\{ \int_{-\infty}^{\infty} p(x_n|\boldsymbol{x}_{\text{obs}})f(\boldsymbol{x}_{\text{obs}}, x_n) \, dX_n - f(\boldsymbol{x}_{\text{obs}}, x_n) \right\}^2 \quad (12)$$

Examples. We use the following benchmark functions in the computational experiments:

f_1 (Sphere function): $f(\boldsymbol{x}) = \sum_{n}^{2} x_k^2, \qquad (-5 < x_k < 5)$

If we suppose that $n = 2$ and also if we suppose $p(x_1, x_2) = p(x_1)p(x_2)$ and $p(x_1) = p(x_2) = \frac{1}{10}$ (i.e., a uniform distribution), then the optimal substitution value in the ideal situation is obtained from Eq. (12) as follows:

$$\psi_2'(x_1) = \arg\min_{x_2} \left\{ \int_{-5}^{5} \frac{1}{10}(x_1^2 + x_2^2)\, dx_2 - (x_1^2 + x_2^2) \right\}^2 = \pm\frac{5}{\sqrt{3}}. \qquad (13)$$

f_2 **(for 2D variables):** $f(x) = (x_1 - x_2)^2, \qquad (-5 < x_1, x_2 < 5)$

If we suppose that $p(x_1, x_2) = p(x_1)p(x_2)$ and $p(x_1) = p(x_2) = \frac{1}{10}$ (i.e., a uniform distribution), the optimal substitution value in the ideal situation is obtained from Eq. (12) as follows:

$$\psi_2'(x_1) = \arg\min_{x_2} \left\{ \int_{-5}^{5} \frac{1}{10}(x_1 - x_2)^2\, dx_2 - (x_1 - x_2)^2 \right\}^2$$

$$= x_1 \pm \sqrt{x_1^2 + \frac{25}{3}}. \qquad (14)$$

3   Estimation of Optimal Substitution Values Without the Target Function

In the previous section, we obtained the function $\psi(\cdot)$ by assuming that we know the target function f beforehand. However, of course, the target function f is always unknown in many problem settings. On the other hand, according to the previous chapter, it is clear that the optimal substitution value has a very important meaning in the IVS learning. Therefore, in this section, we propose a method to estimate the optimal value without the target function.

For simplicity, we assume that the dimensionality of the problem is 2 and the missing occurs only at the second dimension in the following explanation. Our proposed method consists of the following four steps to calculate the function $\psi(\cdot)$ to estimate optimal substitution values.

Step 1: Divide the domain of the dimension that is not missing (i.e., first dimension) into the predefined size d.

Step 2: Calculate the average y_{avg} of the objective variable y^t for T training data points that exist in the interval $(x_{1i}, x_{1(i+1)}]$.

Step 3: Let the estimated substitution value in the interval $(x_{1i}, x_{1(i+1)}]$ be $x_2 = x_2^{t'}$ when $t' = \arg\min(y_{\text{avg}} - y^t)^2$.

Step 4: Repeat Step 2. and 3. for all intervals.

We also describe our proposed method as pseudo-code in Algorithm 1. Note that although we assume the dimensionality of data is 2 and the missing element is limited to be the last one for simplicity, we could apply this method regardless the missing elements.

Algorithm 1. Estimate $\psi(\cdot)$. Assume that the dimensionality of data is 2 and the value missing happens only at the last dimension.

Require: $D_{train} = \{(\boldsymbol{x}, y) | \boldsymbol{x} = (x_1, x_2) \text{ s.t., } a < x_1, x_2 < b\}$
Require: The number of division d
  bandwidth $\leftarrow (b - a)/d$
  for $i \leftarrow 0$ to bandwidth $- 1$ **do**
    count $\leftarrow 0$
    $y_{\text{sum}} \leftarrow 0$
    for $\boldsymbol{x}^t \in D_{train}$ **do**
      if $\boldsymbol{x}_1^t$ in $(x_{1i}, x_{1(i+1)}]$ **then**
        count $\leftarrow$ count $+ 1$
        $y_{\text{sum}} \leftarrow y_{\text{sum}} + y^t$
      end if
      $y_{\text{avg}} \leftarrow y_{\text{sum}}/$count
    end for
    $t' \leftarrow \arg \min(y_{\text{avg}} - y^t)^2$
    $x_2 \leftarrow x_2^{t'}$ at $(x_{1i}, x_{1(i+1)}]$
  end for

---

4 Computational Experiments

4.1 Experimental Setup

In this experiment, we assume that dimensionality of the problem is 2, and a value missing happens with a probability of p_{mis} only at the last dimension. The IVS method is performed with probability p_{sub}. We assume that the data follow a uniform distribution for all dimensions.

A neural network is employed to model benchmark functions. The neural network is trained with the following settings: The number of training data N_{train} is 500, and the number of epoch is 1000. The size of a mini-batch is 32. The number of layers in the neural network is set to three and the number of hidden units is specified as 50. The sigmoid function is used as an activation function for each layer and each unit. Adam algorithm is used as the optimizer that computes adaptive learning rates for updating the weights of the networks.

4.2 Estimated Values

Figure 1 depicts the result of the estimation method when x_1 is divided into $d = 22$. The horizontal axis is the value of the first element of the data, and the vertical axis means the substitution value at the second dimensional data. The thin line is the optimal substitution value obtained by Eqs. (13) and (14), and the bold line represent the value estimated by our proposed method. As shown in Fig. 1, the value around the optimal substitution value can be obtained. Thus, we found that the proposed method has the effectiveness for estimating the optimal substitution values.

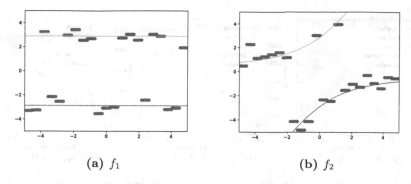

(a) f_1 **(b)** f_2

Fig. 1. Results of the proposed estimation method. The thin lines are the optimal substitution values and the bold lines represent the estimated value.

4.3 Generalization Performance

The results of the prediction errors when training with probability $p_{sub} = [0.00, 0.25, 0.50, 0.75, 0.90, 1.00]$ by using the substitution values are shown in Figs. 2, and 3. The test error in f_1 and f_2 are compared among five types of the substitution value. "Fixed best" substitutes the value that minimizes the test error and "Fixed avg" does so with the average in each setting of test probability, respectively. "Theory" and "Theory random" are set to the substitution value as shown in Eqs. (13), (14). The difference between them is that "Theory" indicates the two substitution values (positive and negative) obtained by Eqs. (13), (14) that give preference to the positive one. On the other hand, "Theory random" substitutes the two values randomly. It is fixed to the value when there is only one optimal substitution value as in the case of Eq. (14) (i.e. $-5 < x_1 < -2$, $2 < x_1 < 5$).

"Estimation" substitutes the value estimated by the our proposed method Algorithm 1. The colored area represents the variance of the test error. The optimal substitution value can be obtained by our proposed methods. Thus, the performance of "Estimation" is as good as "Theory" and "Theory Random" for all settings. Therefore, it is shown that the estimated substitution values work effectively for those data that include missing values even though the target function f is unknown. Moreover, the test error was not increased even in the case where the intentional substitution was performed at the learning phase but the unknown data had no missing data.

On the other hand, the models with "Fixed best" performed well only for the setting f_1. The substitution value that theoretically minimizes the expected error is fixed for f_1. In other words, the substitution value is independent from the x_1, as shown in Fig. 1(a). Thus, the optimal substitution value can be also obtained with the setting "Fixed best". However, as shown in Fig. 1(b) in the case of f_2, the optimal substitution value changes with the value of x_1. Therefore, "Fixed best" makes the models robust and performs well for f_1, but not for f_2.

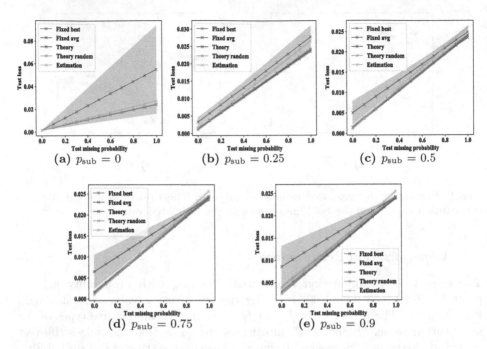

Fig. 2. Test error maps on f_1 (Sphere)

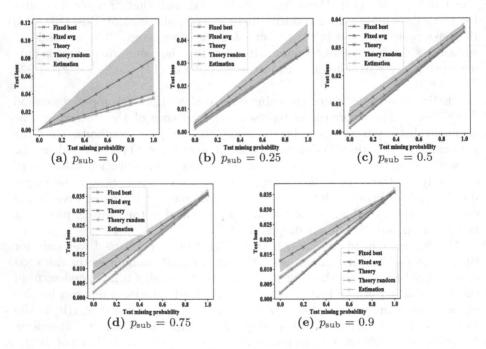

Fig. 3. Test error maps on f_2

5   Conclusions

In this research, we proposed the estimation method of the optimal substitution value in the IVS learning. As the results of numerical experiments, it was shown that the validity of the robust model against the loss for unknown data by estimating the optimal substitution values. For future work, we will conduct experiments with a biased-distribution data, and make use of the findings of this research for handling missing values in other noisy experimental settings.

References

1. Baraldi, A.N., Enders, C.K.: An introduction to modern missing data analyses. J. Sch. Psychol. **48**(1), 5–37 (2010)
2. Little, R.J.A., Rubin, D.B.: Statistical Analysis with Missing Data. Wiley, New York (1986)
3. Little, R.J.A., Rubin, D.B.: Statistical Analysis with Missing Data. Probability and Mathematical Statistics. Wiley, New York (2002)
4. Rubin, D.B.: Multiple imputation for nonresponse in surveys. SERBIULA (sistema Librum 2.0), 137, 11 (1989)
5. Rubin, D.B.: Multiple imputation after 18+ years. J. Am. Stat. Assoc. **91**(434), 473–489 (1996)
6. Schafer, J.L., Graham, J.W.: Missing data: our view of the state of the art. Psychol. Methods **7**(2), 147–77 (2002)
7. Tresp, V., Ahmad, S., Neuneier, R.: Training neural networks with deficient data. In: Proceedings of the 6th International Conference on Neural Information Processing Systems, NIPS 1993, pp. 128–135. Morgan Kaufmann Publishers Inc, San Francisco (1993)
8. Acock, A.C.: Working with missing values. J. Marriage Family 67(4), 1012–1028
9. Hasegawa, T., Fukushima, T., Nakashima, T.: Robust prediction against missing values by intentional value substitution. In: Proceedings of 7th International Symposium on Integrated Uncertainty in Knowledge Modelling and Decision Making, IUKM 2019, pp. 69–80 (2019)

Data Privacy and Security

Efficient Near-Optimal Variable-Size Microaggregation

Jordi Soria-Comas, Josep Domingo-Ferrer[⊠], and Rafael Mulero

Department of Computer Science and Mathematics, CYBERCAT-Center for
Cybersecurity Research of Catalonia, UNESCO Chair in Data Privacy,
Universitat Rovira i Virgili, Av. Països Catalans 26, 43007 Tarragona, Catalonia
{jordi.soria,josep.domingo,rafael.mulero}@urv.cat

Abstract. Microaggregation is a well-known family of statistical disclosure control methods, that can also be used to achieve the k-anonymity privacy model and some of its extensions. Microaggregation can be viewed as a clustering problem where clusters must include at least k elements. In this paper, we present a new microaggregation heuristic based on Lloyd's clustering algorithm that causes much less information loss than the other microaggregation heuristics in the literature. Our empirical work consistently observes this superior performance for all minimum cluster sizes k and data sets tried.

Keywords: Anonymization · Statistical disclosure control · Microaggregation · Lloyd's algorithm

1 Introduction

Collecting data and sharing them for secondary analysis is increasingly widespread and brings undoubted social and economic benefits. Yet, when data are personally identifiable information (PII), sharing them may be a threat to people's privacy. As a consequence, administrations have strengthened privacy regulation to protect the citizens. In a nutshell, these new privacy regulations, epitomized by the EU General Data Protection Regulation, require consent from data subjects for any PII collection, sharing or analysis. In the many situations in which obtaining consent is not feasible, anonymization is the only way to go. After anonymization, data no longer qualify as PII and, thus, are no longer subject to data protection regulations.

Anonymizing data involves not only suppressing any identifiers, but altering other attributes as well. The original data are first stripped from identifiers and then a statistical disclosure control method is used to mask the remaining attributes so that they no longer reveal information about original data subjects. Masking is not straightforward because, to keep the masked data statistically valid, the information loss must be minimized. Among the available statistical disclosure control techniques, in this paper we focus on microaggregation. Microaggregation replaces records in the original data set by (aggregated)

© Springer Nature Switzerland AG 2019
V. Torra et al. (Eds.): MDAI 2019, LNAI 11676, pp. 333–345, 2019.
https://doi.org/10.1007/978-3-030-26773-5_29

records that refer to groups of data subjects. The greater the groups, the stronger the protection. To guarantee at least a certain level of protection, microaggregation algorithms take a parameter k that determines the minimum required group size.

In recent years, the research on data anonymization performed by the computer science community has focused on privacy models. A privacy model describes the condition that data must satisfy for disclosure risk to be at an acceptable level, but it does not describe how this condition should be attained. k-Anonymity [15] is among the most popular privacy models. It seeks to limit the probability of successful record re-identification by altering the value of quasi-identifier attributes. Quasi-identifiers are attributes that are not re-identifying when separately considered (e.g. in general Age, Profession and Zipcode do not identify anyone separately), but such that their combination may identify the subject to whom a record corresponds (there may be a single 95-year old doctor in a certain zipcode, and it may be easy to find her name in an electoral roll). Interestingly, running microaggregation on the quasi-identifiers yields k-anonymity [8]. Microaggregation is also useful to enforce l-diversity and t-closeness, two extensions of k-anonymity [7,19], as well as a building block of ε-differentially private algorithms [17,18].

To minimize the information loss incurred by microaggregation, we need to carefully choose the groups of records to be aggregated. A common approach in numerical microaggregation is to attempt to minimize the sum of squared distances between original records and their corresponding aggregated records, which will be called SSE. Unfortunately, finding a microaggregation that minimizes SSE is an NP-hard problem. For this reason, existent approaches are heuristic. Most current microaggregation algorithms generate clusters with a fixed size (the minimum required cluster size). This cardinality constraint reduces the complexity of the microaggregation algorithm but it may result in large information loss. To reduce information loss, heuristic variable-size microaggregation algorithms have been proposed, but their computational complexity is greater than that of their fixed-size counterparts. Also, in some cases they need additional parameters whose optimal values are hard to determine.

Contribution and Plan of this Work

Microaggregation is closely related to clustering: in fact, it is clustering with a minimum cardinality constraint on clusters. In this work, we take advantage of the information loss minimization capabilities of Lloyd's clustering algorithm [12] to achieve near-optimal variable-size microaggregation. First, we embed a minimum cluster size constraint in the algorithm. Second, given that Lloyd's algorithm requires the number of clusters to be fixed beforehand, we modify it to allow a variable number of clusters. We call the resulting heuristic ONA (Near-Optimal microaggregation Algorithm). We then present empirical results on the information loss and the computing time of variable-size microaggregation with ONA.

In Sect. 2, we give some background on microaggregation and Lloyd's algorithm. In Sect. 3, we describe some limitations of current microaggregation algorithms. In Sect. 4 we present the ONA algorithm to deal with these limitations. In Sect. 5, we experimentally compare ONA with existing methods. We finalize with conclusions and future work directions in Sect. 6.

2 Background

2.1 Microaggregation

Microaggregation is a perturbative method for statistical disclosure control of microdata releases. It is based on the following two steps:

- *Partition:* The records in the original data set are partitioned into several clusters, each of them containing at least k records (the minimum cluster size). To minimize information loss in the following step, records in each cluster should be as close to one another as possible.
- *Aggregation:* An aggregation operator is used to compute the centroid of all the records in the cluster. If all attributes are numerical, the centroid record is the mean record. Finally, every record in the cluster is replaced with the cluster centroid record.

When replacing records by cluster centroids in the aggregation step of microaggregation, some information is lost. The ensuing loss of variability is a measure of information loss. A microaggregation algorithm is optimal if it minimizes information loss.

Let SST be the total sum of squares, that is, the sum of squared distances between each record r in an original data set D and the centroid record $c(D)$ of the entire data set:

$$SST = \sum_{r \in D} \|r - c(D)\|^2 .$$

Clearly, SST represents the total variability of D. Then compute the sum of squared records errors SSE, that is, the sum of squared distances between each record r and the centroid $c(r)$ of the cluster r belongs to:

$$SSE = \sum_{r \in D} \|r - c(r)\|^2 .$$

SSE represents the loss of variability incurred when replacing records with centroids. We can normalize SSE by dividing it by SST, so that SSE/SST accounts for the proportion of the total variability lost due to the microaggregation. With numerical attributes, the mean is a sensible choice as the aggregation operator, because for any given cluster partition it minimizes SSE in the aggregation step; the challenge thus is to come up with a partition that minimizes the overall SSE.

Finding an optimal algorithm is feasible for univariate microaggregation of a numerical attribute. There are two well-known necessary optimality conditions in this case [4]: clusters must contain consecutive records and the size of the clusters must be between k and $2k - 1$. Given these two conditions, a shortest-path algorithm can find the optimal univariate microaggregation with cost $O(n \log n)$ for n records [9].

Since realistic data sets contain multiple attributes, univariate microaggregation is not enough. Multivariate microaggregation is more complex: the first optimality condition above does not apply for want of a total order in the data domain. As a result, the search space for the optimal multivariate microaggregation remains too large and finding the optimal solution is NP-hard [14]. Therefore, heuristics are employed to obtain an approximation with reasonable cost. An example heuristic for the partition step of microaggregation is MDAV [8], which generates fixed-size clusters. Alternatively, VMDAV [16] is an adaptation of the MDAV heuristic that allows variable-size clusters.

2.2   MDAV

The MDAV algorithm aims at satisfying the optimality conditions of numerical univariate microaggregation:

1. *Optimal clusters must contain consecutive elements.* Since a total order is lacking in a multivariate domain, the meaning of consecutive elements is not well-defined. However, the intuition remains valid: it makes no sense to include a record r' in a cluster if a record r closer to the records of the cluster is not in the cluster.
2. *The size of optimal clusters ranges between k and $2k - 1$.* This condition remains valid in the multivariate case.

Thus, rather than minimizing the overall information loss, the MDAV heuristic proceeds by selecting specific records at the boundary of the set of records not yet assigned to any cluster and generating clusters of k elements around them: given a record r, a cluster is formed with r and the $k - 1$ records closest to r among those not clustered yet. See Algorithm 1.

2.3   VMDAV

VMDAV is an adaptation of MDAV that can yield variable-size clusters. The underlying idea is that variable-size clusters can be more adapted to the distribution of the records and, thus, reduce the information loss.

Essentially, VMDAV takes two steps: (i) generate a cluster of size k that contains the record that is farthest from the average record and its closest $k - 1$ records, and (ii) expand the cluster with neighboring records. These steps are repeated until all the records have been assigned to a cluster.

The first step is similar to MDAV. So we only describe the second step. Once we have a cluster with k records, we look for r_u, the unclustered record that minimizes the distance to the records in the cluster. Let d_{in} be such minimum distance. The we compute d_{out}, the minimum distance between r_u and

---

Algorithm 1. MDAV microaggregation algorithm with minimal cluster size k

1	**Let** D be a data set
2	**Let** k be the minimum cluster size
3	$Clusters = \emptyset$
4	**While** $|D| \geq 3k$
5	x_a=average record of D
6	x_r=record of D that is most distant from x_a
7	C=cluster containing x_r and the $k-1$ records of D closest to x_r
8	$Clusters = Clusters \cup C$
9	$D = D \setminus C$
10	x_s=record of D that is most distant from x_r
11	C=cluster containing x_s and the $k-1$ records of D closest to x_s
12	$Clusters = Clusters \cup C$
13	$D = D \setminus C$
14	**End while**
15	**If** $2k \leq |D| \leq 3k-1$ **Then**
16	x_a=average record of D
17	x_r=record of D that is most distant from x_a
18	C=cluster containing x_r and the $k-1$ records of D closest to x_r
19	$Clusters = Clusters \cup C$
20	$D = D \setminus C$
21	**End if**
22	$Clusters = Clusters \cup D$
23	**Return** $Clusters$

---

the remaining unclustered records. The cluster expansion procedure is based on these two distances. If d_{in} is smaller than d_{out}, then r_u is closer to the records in the cluster than to the other unclustered records. In that case, adding r_u to the current cluster is a sensible choice. To allow tuning cluster expansion, VMDAV introduces a threshold parameter γ, so that the current cluster is expanded with r_u if $d_{in} < \gamma d_{out}$.

2.4   Clustering and Lloyd's Algorithm

There are several approaches to generate clusters. In this work, we are interested in centroid-based clustering (a.k.a. c-means clustering). The purpose of c-means is to split the records in a fixed set of c clusters in a way that SSE is minimized.

Lloyd's algorithm is designed for c-means clustering. Starting from an arbitrary set of c centroids, the algorithm proceeds by iteratively assigning each record to the closest centroid and recomputing the centroids, until a convergence criterion is met. See Algorithm 2.

The runtime of Algorithm 2 is $O(ncdi)$, where n is the number of records, c is the number of clusters, d is the number of attributes per record and i the number of iterations needed until convergence. Lloyd's algorithm is thus often considered of linear complexity in practice, although in the worst case it can be superpolynomial.

Algorithm 2. Lloyd's online clustering of a data set D into c clusters

---
1   **Let** D be a data set
2   **Let** $Centroids = \{c_1, \ldots, c_c\}$ be the initial set of centroids
3   **Let** $C_i = \emptyset$ be the cluster associated with c_i for $i = 1, \ldots, c$
4   **Repeat**
5   **For each** $r \in D$
6   **If** r was assigned to a cluster C_j **Then** extract r from C_j
7   Compute the distance between r and $c_1, \ldots, c_c$
8   Assign r to the cluster around the closest centroid
9   **End for**
10   **Until** convergence condition
11   **Return** $\{C_1, \ldots, C_c\}$

---

3   Limitations of MDAV and VMDAV

MDAV is quite effective at generating clusters that are as compact as possible: it looks for the record that is farthest from the average record and then generates a cluster that contains it and the $k - 1$ records closest to it. In this way MDAV creates compact clusters and avoids the presence of intersecting clusters, which are undesirable because their records could be rearranged in non-intersecting clusters, thereby reducing information loss. The greatest limitation of MDAV is that all clusters (except perhaps the last one) have fixed size k. This is much more restrictive than the optimality condition according to which cluster cardinality must be between k and $2k - 1$, and it may have a significant negative impact on information loss. This limitation not only affects MDAV but all microaggregation methods that use fixed-size clusters.

VMDAV improves over MDAV by being more flexible about cluster sizes. However, the cluster expansion criterion is difficult to adjust. VMDAV uses an extra threshold parameter γ to decide between expanding the current cluster with an additional element (up to a maximum $2k - 1$ elements) or creating a new cluster. The difficulty comes from the fact that it is not known how to fix γ appropriately.

In [16], we find some vague recommendations, which suggest the use of large thresholds (*e.g.* $\gamma = 1.1$) when records are concentrated around specific areas of the data domain, whereas smaller thresholds (*e.g.* $\gamma = 0.2$) are preferable when records are scattered. The rationale for the rule that recommends the use of small γ for scattered records is clear: in this case, small clusters are preferable to avoid large SSE. However, we should keep in mind that by using small γ the cluster expansion mechanism is hampered, and VMDAV becomes closer to MDAV. The rationale for using large γ when records are concentrated around specific points is unclear to us. After all, regardless of the distribution of records, we should prefer smaller clusters to larger clusters. This is illustrated in Fig. 1, where two microaggregation partitions with minimum size $k = 3$ are displayed that could be obtained using VMDAV. On the left, all clusters have size 3, which is a result compatible with VMDAV for small γ (and also with MDAV). On the right, the size of the clusters is greater than 3, which is compatible with VMDAV for large γ. By looking at the distribution of the records, we observe that they are concentrated around two points; thus, according to the rules suggested in [16] we would select a large threshold, which would make the right-hand side partition likelier. However, SSE and hence the information loss is larger for this partition than for the left-hand side partition.

The issues of VMDAV that we have hinted are confirmed in the experimental section, where VMDAV and MDAV achieve comparable levels of information loss. That is, the cluster expansion procedure of VMDAV is not capable of offering noticeable reductions in the information loss.

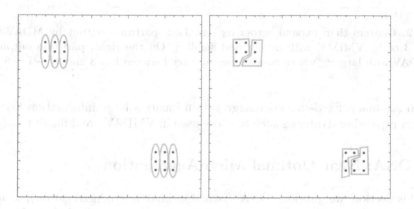

Fig. 1. Two microaggregation partitions with minimum size $k = 3$. Left, partition where all clusters have size 3. Right, partition where clusters have size greater than 3.

One justification for suggesting large γ when records are concentrated in different regions is to avoid obtaining clusters that expand across more than one region. On the left-hand side of Fig. 2, we show an example of this undesirable situation. This partition, where all clusters except one have size 3, could be the result of taking $k = 3$ in MDAV or in VMDAV with small γ. Taking a large threshold in VMDAV is expected to facilitate variable-size clusters, which might solve the problem. However, as shown on the right-hand side of Fig. 2, it is not guaranteed that variable-size clusters achieve the required result: there is still a cluster spread among two regions.

Even if the previous VMDAV threshold rules were effective for data sets that are clearly concentrated or scattered, we would still be at a loss for data sets that do not qualify as any of those two types. For example, consider a data set that has several small regions with concentrated records and a big region with scattered records.

Furthermore, in general it cannot be assumed that the data controller choosing anonymization parameters knows whether her data set is scattered, concentrated, etc. In fact, for large and high-dimensional data sets, it may be quite difficult to grasp how records are distributed in the domain of attributes.

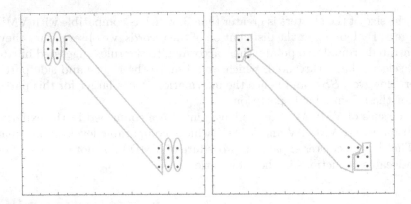

Fig. 2. Clusters than expand across regions. Left, partition output by MDAV with $k = 3$ or by VMDAV with $k = 3$ and small γ. On the right, partition output by VMDAV with large γ, where cluster size can vary between $k = 3$ and $2k - 1 = 5$.

In summary, fixed-size microaggregation incurs a large information loss and cluster expansion strategies such as those used in VMDAV are difficult to adjust.

4   ONA: Near-Optimal MicroAggregation

In this section we propose ONA (Near-Optimal microAggregation), a novel variable-size microaggregation method that is based on standard clustering algorithms. On the one hand, clustering algorithms adjust the size of each cluster automatically. We plan to take advantage of this property in ONA, while making sure that the size of the clusters stays within the known optimal bounds, that is, between k and $2k - 1$. On the other side, clustering algorithms usually take the number of clusters as a parameter. In microaggregation, we do not care about the number of clusters; we simply want a valid clustering that minimizes the information loss. Thus, the need to tell the microaggregation algorithm the number of clusters we want would be an artificial restriction that we prefer to avoid, both for the sake of algorithm clarity and to avoid unnecessary information loss.

ONA follows Lloyd's online algorithm (see Algorithm 2) but it makes several adjustments to guarantee that an appropriate number of clusters with an appropriate size is generated. Algorithm 3 formalizes ONA and its steps are explained next:

- We start (at line 3) by generating a random set of clusters whose cardinality is k or more. The minimum cardinality constraint of microaggregation is enforced by starting with a set of clusters that conforms to it and by making sure that any modification of the clusters does not violate it.
- The proposed algorithm is iterative. Each iteration (lines 4–29) is designed to reduce the SSE of the clustering, until convergence is reached. The convergence condition is not specified in the algorithm. To be strict, we should

require a completely stable set of clusters. However, as most of the reduction in SSE is attained in the first few iterations, it is usually safe to use less strict conditions to speed up the execution. We will describe alternative convergence conditions when reporting experiments in Sect. 5.

- Following Lloyd's online algorithm, loop through the records (lines 5–28) in the data set and reassign them (if needed) to the closest cluster so that SSE decreases.
- It is only possible to reassign a record if its current cluster contains more than k records (lines 7–11). Otherwise, there would remain less than k records in the cluster and the clustering would not satisfy the minimum cardinality constraint. If the cluster of the current record has more than k records, remove the record from the cluster (line 9) and assign it to the closest cluster (line 11).
- When the cluster of the current record has k records, the only way to reassign the current record to another cluster is to dissolve the cluster and reassign all its records to other clusters (lines 12–20). This is only done if it reduces SSE. In line 15 all reassignments are computed: $C_{j(s)}$ is the cluster to which record s is reassigned. The contribution to SSE of the original clusters (SSE_1, line 16) and the SSE of the reassigned clusters (SSE_2, line 17) are computed. If $SSE_2 < SSE_1$, the reassignments are applied; otherwise, the current clustering is kept unmodified.
- Finally, the algorithm checks that all clusters have at most $2k - 1$ records (as one of the optimality conditions requires). This condition must be checked because the reassignments can make clusters grow beyond $2k - 1$ records. If a cluster with $2k$ or more records is found, we apply the same Algorithm 3 to the cluster, which will split it into two clusters of size between k and $2k - 1$ thereby reducing SSE.

In spite of the distinction between the current cluster having more than k records or k records, the complexity of Algorithm 3 remains essentially the same as the one of Lloyd's algorithm (see Sect. 2.4).

5   Experimental Evaluation

5.1   Evaluated Methods

The motivation of our algorithm has been based on the limitations of MDAV and VMDAV. However, for completeness, the experimental section will not be limited to comparing with those two methods. We will compare the information loss using SEE and $100 \times SSE/SST$ (as described in Sect. 2.1) for the following methods: MDAV [4], VMDAV [16], MD-MHM [3], MDAV-MHM [3], CBFS-MHM [3], NPN-MHM [3], μ-Approx [6], M-d [10], TFRP-1 [2], TFRP-2 [2], DBA-1 [11], DBA-2 [11] and IMHM [13].

Algorithm 3. ONA algorithm for a data set D and minimal cluster size k.

1	**Let** D be a data set
2	**Let** k be the minimal cluster size
3	Randomly generate a set of clusters $\mathcal{C} = \{C_1, \ldots, C_{\lfloor |D|/k \rfloor}\}$ such that each cluster

contains at least k records

4	**Repeat**
5	**For each** $r \in D$
6	**Let** $C_{i(r)} \in \mathcal{C}$ be the cluster that contains r
7	**If** $|C_{i(r)}| > k$ **Then**
8	*// Should r be reassigned to another cluster?*
9	Extract r from $C_{i(r)}$
10	Compute the distance between r and the centroids of the clusters in $\mathcal{C}$
11	Add r to the cluster whose centroid is closest to r
12	**Else If** $|C_{i(r)}| = k$ **Then**
13	*// Should cluster $C_{i(r)}$ be dissolved?*
14	**Let** $C_{j(s)}$ be the cluster with the closest centroid to $s \in C_{i(r)}$ among

those in $\mathcal{C} \setminus C_{i(r)}$

15	**Let** $C'_k = C_k \cup \{s \in C_{i(r)} : j(s) = k\}$, for each $k \neq i(r)$
16	**Let** $SSE_1 = SSE(C_{i(r)}) + \sum_{k \in \{j(s): s \in C_{i(r)}\}} SSE(C_k)$
17	**Let** $SSE_2 = \sum_{k \in \{j(s): s \in C_{i(r)}\}} SSE(C'_k)$
18	**If** $SSE_1 > SSE_2$ **Then**
19	$\mathcal{C} = \{C'_k : k \neq i(r)\}$
20	**End if**
21	**End if**
22	*// Split clusters that have become too large*
23	**For each** $C \in \mathcal{C}$
24	**If** $|C| \geq 2k$ **Then**
25	Run Algorithm 3 on C with minimal cluster size k
26	**End if**
27	**End for**
28	**End for**
29	**Until** convergence_condition

5.2 Data Sets

The evaluation was performed on data sets [1] that have been used in the literature to evaluate microaggregation algorithms:

- *Census.* Data set with 1080 records and 13 numerical attributes.
- *Tarragona.* Data set with 834 records and 13 numerical attributes.
- *EIA.* Data set with 4092 records and 11 numerical attributes.

5.3 Evaluation Results

The evaluation results are shown in Table 1. We observe that, while there are only small differences in the information loss reported by other methods, our proposal achieves a significantly smaller information loss. This behavior is consistent across cluster sizes and data sets.

Table 1. Information loss $100 \times SSE/SST$ for several values of k and several data sets

Data set	Method	$k = 3$	$k = 5$	$k = 10$
Census	**ONA**	**1.59**	**2.33**	**3.88**
	MDAV	5.69	9.09	14.16
	VMDAV	5.69	8.98	14.07
	MD-MHM	5.69	8.99	14.40
	MDAV-MHM	5.65	9.08	14.22
	CBFS-MHM	5.67	8.89	13.89
	NPN-MHM	6.34	11.34	18.73
	μ-Approx	6.25	10.78	17.01
	M-d	6.11	10.30	17.17
	TFRP-1	5.93	9.36	14.44
	TFRP-2	5.80	8.98	13.96
	DBA-1	6.15	10.84	15.79
	DBA-2	5.58	9.04	13.52
	IMHM	5.37	8.42	12.23
Tarragona	**ONA**	**5.75**	**9.54**	**14.40**
	MDAV	16.93	22.46	33.19
	VMDAV	16.96	22.88	33.26
	MD-MHM	16.98	22.53	33.18
	MDAV-MHM	16.93	22.46	33.19
	CBFS-MHM	16.97	22.53	33.18
	NPN-MHM	17.39	27.02	40.18
	μ-Approx	17.10	26.04	38.80
	M-d	16.63	24.50	38.58
	TFRP-1	17.23	22.11	33.19
	TFRP-2	16.88	21.85	33.09
	DBA-1	20.70	26.00	35.39
	DBA-2	16.15	25.45	34.81
	IMHM	16.93	22.19	30.78
EIA	**ONA**	**0.23**	**0.41**	**1.02**
	MDAV	0.48	1.67	3.84
	VMDAV	0.53	1.30	2.88
	MD-MHM	0.44	1.26	3.64
	MDAV-MHM	0.41	1.26	3.77
	NPN-MHM	0.55	0.96	2.32
	μ-Approx	0.43	0.83	2.26
	TFRP-1	0.53	1.65	3.24
	TFRP-2	0.42	0.91	2.59
	DBA-1	1.09	1.89	4.26
	DBA-2	0.42	0.82	2.08
	IMHM	0.37	0.76	2.18

The algorithm has been implemented in Java and the experiments have been run on a AMD Ryzen 1700X machine under Ubuntu 17.04 x64. Table 2 shows the runtimes of ONA for the various test data sets and cluster sizes. To compute these runtimes, we have used the strictest convergence criterion: we keep iterating until no more record reassignments take place. We should remark that the steepest SSE decrease takes place during the first few iterations. Thus, a less strict convergence condition could offer significantly shorter runtimes without a substantial difference in the SSE. Indeed, we have observed that the SSE reaches a stationary value long before the number of reassignments reaches 0.

Table 2. ONA runtimes in seconds for the test data sets and the tested cluster sizes.

Time (s)	$k = 3$	$k = 5$	$k = 10$
Census	0.295	0.376	0.196
Tarragona	0.254	0.485	0.212
EIA	1.751	1.430	1.607

6   Conclusions and Future Research

We have proposed ONA, a novel microaggregation algorithm that significantly reduces the information loss with respect to existent algorithms. ONA operates iteratively and is based on Lloyd's clustering algorithm. Each iteration of ONA decreases the information loss until it converges to a (possibly local) minimum.

In the design of ONA, we have tried to match the two necessary conditions for optimal microaggregation as closely as possible. First, we make sure that each cluster contains only adjacent records. This is achieved by reassigning records to the cluster with the closest centroid. Second, we make sure that the size of clusters ranges between k and $2k - 1$. In record reassignments, we take care that a source cluster is never left with less than k records (otherwise we disband it) and that a destination cluster never increases to more than $2k - 1$ records (otherwise we split it into two clusters).

In the experimental section, we have presented an exhaustive comparison of the information loss with existent microaggregation algorithms. The results show that ONA offers a very significant reduction of the information loss. It is also important to remark that such a reduction is effected without resorting to complex procedures. Indeed, the internal operation of ONA is simpler than that of most of the microaggregation algorithms included in the comparison.

As future work, we plan to conduct a detailed analysis of the convergence conditions for ONA and also to extend it to categorical data. Currently, the range of microaggregation algorithms available for dealing with this kind of data is rather limited. The work in [5] provides a good starting point.

Acknowledgments and disclaimer. The following funding sources are gratefully acknowledged: European Commission (project H2020-700540 "CANVAS"), Government of Catalonia (ICREA Acadèmia Prize to J. Domingo-Ferrer and grant 2017 SGR 705) and Spanish Government (project RTI2018-095094-B-C21 "Consent"). The views in this paper are the authors' own and do not necessarily reflect the views of UNESCO or any of the funders.

References

1. Brand, R., Domingo-Ferrer, J., Mateo-Sanz, J.M.: Reference data sets to test and compare SDC methods for protection of numerical microdata. European Project IST-2000-25069 CASC (2002). http://neon.vb.cbs.nl/casc/CASCtestsets.htm
2. Chang, C.C., Li, Y.C., Huang, W.H.: TFRP: an efficient microaggregation algorithm for statistical disclosure control. J. Syst. Softw. **80**(11), 1866–1878 (2007)
3. Domingo-Ferrer, J., Martínez-Ballesté, A., Mateo-Sanz, J.M., Sebé, F.: Efficient multivariate data-oriented microaggregation. VLDB J. **15**(4), 355–369 (2006)
4. Domingo-Ferrer, J., Mateo-Sanz, J.M.: Practical data-oriented microaggregation for statistical disclosure control. IEEE Trans. Knowl. Data Eng. **14**, 189–201 (2002)
5. Domingo-Ferrer, J., Sánchez, D., Rufian-Torrell, G.: Anonymization of nominal data based on semantic marginality. Inf. Sci. (Ny) **242**, 35–48 (2013)
6. Domingo-Ferrer, J., Sebé, F., Solanas, A.: A polynomial-time approximation to optimal multivariate microaggregation. Comput. Math. Appl. **55**, 714–732 (2008)
7. Domingo-Ferrer, J., Soria-Comas, J.: Steered microaggregation: a unified primitive for anonymization of data sets and data streams. In: IEEE International Conference on Data Mining Workshops, ICDMW, pp. 995–1002. New Orleans (2017)
8. Domingo-Ferrer, J., Torra, V.: Ordinal, continuous and heterogeneous k-anonymity through microaggregation. Data Min. Knowl. Discov. **11**, 195–212 (2005)
9. Hansen, S.L., Mukherjee, S.: A polynomial algorithm for optimal univariate microaggregation. IEEE Trans. Knowl. Data Eng. **15**(4), 1043–1044 (2003)
10. Laszlo, M., Mukherjee, S.: Minimum spanning tree partitioning algorithm for microaggregation. IEEE Trans. Knowl. Data Eng. **17**(7), 902–911 (2005)
11. Lin, J.L., Wen, T.H., Hsieh, J.C., Chang, P.C.: Density-based microaggregation for statistical disclosure control. Expert Syst. Appl. **37**(4), 3256–3263 (2010)
12. Lloyd, S.P.: Least squares quantization in PCM. IEEE Trans. Inf. Theory **28**(2), 129–137 (1982)
13. Mortazavi, R., Jalili, S., Gohargazi, H.: Multivariate microaggregation by iterative optimization. Appl. Intell. **39**, 529–544 (2013)
14. Oganian, A., Domingo-Ferrer, J.: On the complexity of optimal microaggregation for statistical disclosure control. Stat. J. UN Econ. Comm. Eur. **18**, 345–354 (2001)
15. Samarati, P., Sweeney, L.: Protecting privacy when disclosing information: k-anonymity and its enforcement through generalization and suppression. Technical report, SRI International (1998)
16. Solanas, A., Martínez-Ballesté, A.: V-MDAV: a multivariate microaggregation with variable group size. In: Proceedings in Computational Statistics, pp. 917–926 (2006)
17. Soria-Comas, J., Domingo-Ferrer, J.: Differentially private data publishing via optimal univariate microaggregation and record perturbation. Knowl.-Based Syst. **153**, 78–90 (2018)
18. Soria-Comas, J., Domingo-Ferrer, J., Sánchez, D., Martínez, S.: Enhancing data utility in differential privacy via microaggregation-based k-anonymity. VLDB J. **23**, 771–794 (2014)
19. Soria-Comas, J., Domingo-Ferrer, J., Sánchez, D., Martínez, S.: t-Closeness through microaggregation: strict privacy with enhanced utility preservation. In: 2016 IEEE 32nd International Conference on Data Engineering, ICDE 2016. pp. 1464–1465 (2016)

Mitigating the Curse of Dimensionality in Data Anonymization

Jordi Soria-Comas and Josep Domingo-Ferrer[✉]

Department of Computer Science and Mathematics, CYBERCAT-Center for
Cybersecurity Research of Catalonia, UNESCO Chair in Data Privacy, Universitat
Rovira i Virgili, Av. Països Catalans 26, 43007 Tarragona, Catalonia
{jordi.soria,josep.domingo}@urv.cat

Abstract. In general, just suppressing identifiers from released microdata is insufficient for privacy protection. It has been shown that
the risk of re-identification increases with the dimensionality of the
released records. Hence, sound anonymization procedures are needed to
anonymize high-dimensional records. Unfortunately, most privacy models yield very poor utility if enforced on data sets with many attributes.
In this paper, we propose a method based on principal component analysis (PCA) to mitigate the curse of dimensionality in anonymization. Our
aim is to reduce dimensionality without incurring large utility losses. We
instantiate our approach with anonymization based on differential privacy. Empirical work shows that using differential privacy on the PCA-
transformed and dimensionality-reduced data set yields less information
loss than directly using differential privacy on the original data set.

Keywords: Privacy preserving data publishing ·
Curse of dimensionality · Differential privacy

1 Introduction

Under the EU General Data Protection Regulation (GDPR), which is becoming
a *de facto* global standard, personally identifiable information (PII) cannot be
accumulated for secondary use. Hence, to stay GDPR-compliant, exploratory
analytics must take place on anonymized data.

As repeatedly aired in the media [1,15,19], just suppressing direct identifiers
(names, passport numbers, etc.), let alone replacing them by pseudonyms, is not
enough to anonymize a data set. In the case of high-dimensional data even adding
a small amount of noise may not be enough to prevent re-identification [11].

In this work we deal with anonymization of microdata with a moderate to
large number of attributes. This type of data are more and more common in
the big data landscape, in which data sets are constructed by merging several
sources and may end up having a great number of attributes.

Direct application of statistical disclosure control (SDC) methods [8] to high-
dimensional data is likely to result in high information loss and hence in poor
utility, due to the sparsity of such data.

© Springer Nature Switzerland AG 2019
V. Torra et al. (Eds.): MDAI 2019, LNAI 11676, pp. 346–355, 2019.
https://doi.org/10.1007/978-3-030-26773-5_30

In this work we tackle the problem by investigating a pre-processing transformation aimed at reducing dimensionality before anonymization is carried out. Specifically, we propose principal component analysis (PCA), which could also be replaced by another dimensionality reduction technique, e.g. if data are not numerical. Subjects send their original records to the controller, who gathers the entire original data set and can compute the principal components of it. The dimensionality reduction comes from the controller choosing only those components that carry most information. Anonymization is carried out on this smaller set of components and the result if mapped back to the domain of the original attributes to obtain the anonymized data set.

For the sake of concreteness, we exemplify dimensionality reduction to obtain differentially private data sets. According to our experiments, *using differential privacy on the PCA-transformed and dimensionality-reduced data incurs substially less information loss than directly applying differential privacy on the original data.*

The rest of this paper is organized as follows. In Sect. 2, we recall some background concepts. In Sect. 3 we deal with dimensionality reduction. Section 4 exemplifies our approach to generate differentially private data sets. Empirical work on the presented approach is reported in Sect. 5. Finally, in Sect. 6 we summarize conclusions and indicate future research lines.

2 Background

2.1 Principal Component Analysis

PCA is a dimensionality-reduction technique that uses an orthogonal transformation to convert a set of observations of possibly correlated attributes into a set of values of linearly uncorrelated attributes called *principal components*. This transformation is defined in such a way that the first principal component has the largest possible variance (that is, accounts for as much of the variability in the data as possible), and each succeeding component in turn has the highest possible variance under the constraint that it is orthogonal to the preceding components.

More formally, consider a data set X containing n records with m attributes, such that the attribute means are zero. The covariance matrix of X is

$$C_X = \frac{1}{n} X^T X.$$

We want to find a new basis $\{p_1, \ldots, p_m\}$ such that: (i) $Var(P_i) \geq Var(P_j)$, for all $i < j$, (ii) $Cov(P_i, P_j) = 0$, for all $i \neq j$ (low redundancy) and (iii) the vectors p_i are orthonormal.

The PCA transformation is attained by means of the change of basis matrix $P = [p_1 \ldots p_m]$, where the p_i's are eigenvectors of C_X sorted by descending order of the corresponding eigenvalues. That is, the resulting data set is XP.

Since the principal components are sorted in descending order of the information they contain (the first component has the largest amount of information), dimensionality reduction is very simple. We just have to drop some components starting from the last one (which is the least informative one).

2.2 Differentially Private Data Sets

Definition 1 (ϵ-differential privacy). *A randomized function κ gives ϵ-differential privacy (ϵ-DP, [6]) if, for all data sets X_1 and X_2 that differ in one record (a.k.a. neighbor data sets), and all $S \subset Range(\kappa)$, we have*

$$\Pr(\kappa(X_1) \in S) \leq \exp(\epsilon) \Pr(\kappa(X_2) \in S).$$

Given a query function f, the goal in DP is to find a randomized function κ_f that satisfies ϵ-DP and approximates f as closely as possible. For the case of numerical queries κ_f can be obtained via noise addition; that is $\kappa_f() = f() + N$, where N has been properly adjusted. Adding Laplace distributed noise whose scale has been adjusted to the global sensitivity of the query f is, probably, the most common approach (other alternatives include [10,12,16]). The global sensitivity or L_1-sensitivity, Δf, of a function $f : \mathcal{D}^n \to \mathbb{R}^d$ is the maximum variation of f between neighbor data sets:

$$\Delta f = \max_{d(X_1,X_2)=1} \|f(X_1) - f(X_2)\|_1.$$

Proposition 2. *Let $f : \mathcal{D}^n \to \mathbb{R}^d$ be a function. The mechanism $\kappa_f(X) = f(X) + (N_1, \ldots, N_d)$, where N_i are drawn i.i.d. from a Laplace$(0, \Delta f/\epsilon)$ distribution, is ϵ-DP.*

The above original definition of DP was intended for interactive queries to a database. If we want to obtain a DP data set, a straightforward procedure would be to add noise to the results of queries about individual attribute or record values. However, this is not feasible, because the sensitivity of individual records is large, and according to Proposition 2 a lot of noise would be needed and the results would be very inaccurate. Better methods to produce DP microdata sets can be found in the literature, based on histograms [7], microaggregation [17], Bayesian networks [20] or synthetic data [9,14]. Although using histograms is especially problematic for high-dimensional data (due to sparsity), the rest of the above approaches also yield increasingly inaccurate results as the dimensionality of the original data set increases. In the rest of this paper, we will use the microaggregation-based procedure [17], that we next recall.

Microaggregation [4,5] is a well-known SDC technique that works in two steps: (i) split the original data set into clusters of at least k records following a criterion of maximum within-cluster similarity; (ii) replace the records in each cluster by the cluster centroid. In [18], the focus is moved from the original data set X to its microaggregated version $\bar{X}$. After $\bar{X}$ is generated, X is dropped and the goal becomes to protect $\bar{X}$ via DP. Since $\bar{X}$ contains less information than X, this change of focus does not increase the risk of disclosure: that is, $\bar{X}_\epsilon$ (an ϵ-DP version of $\bar{X}$) entails less disclosure risk than X_ϵ (an ϵ-DP version of X). Additionally, $\bar{X}$ is less sensitive to changes in one individual than X. In particular, when centroids are computed as arithmetic averages, the sensitivity of the centroid of a cluster C is a factor $1/|C|$ of the sensitivity of the original records.

Let X be the original data set (with attributes $A_1, \ldots, A_m$) and $\bar{X}$ be a microaggregated version of X (with attributes $\bar{A}_1, \ldots, \bar{A}_m$). Among the possibilities described in [17], we use the individual ranking microaggregation approach: each attribute is microaggregated separately from the rest. The DP data set $\bar{X}_\epsilon$ is generated by combining ϵ_i-DP versions of each of the attributes: by the sequential composition property of DP, if we have an ϵ_i-DP version of each attribute $\bar{A}_i$, the combination is $(\epsilon_1 + \ldots + \epsilon_m)$-DP.

To obtain the ϵ_i-DP version of attribute $\bar{A}_i$ we proceed as illustrated in Fig. 1. Each centroid is masked with an appropriate amount of noise that depends on the cardinality $|C_j|$ of the corresponding cluster, on the sensitivity ΔA_i of the original attribute, A_i, and on the ϵ_i assigned to the attribute. For each cluster C_j of A_i values, we sample a noise n_j from a $Laplace(0, \Delta A_i/(|C_j| \times \epsilon_i))$ distribution and add it to the corresponding centroid c_j, to obtain $c_j + n_j$. It is important to remark that each occurrence of c_j is replaced by the same masked value $c_j + n_j$.

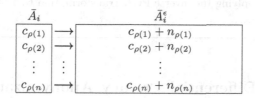

where $\rho(r)$ =number of record r's cluster
$$n_j = Laplace(0, \Delta A_i/(|C_j| \times \epsilon_i))$$

Fig. 1. Procedure to generate an $(\epsilon_1 + \ldots + \epsilon_m)$-DP data set by combining an ϵ_i-DP version of the attribute A_i, for $i = 1, \ldots, m$.

3   PCA-Based Dimensionality Reduction

Given a data set X, PCA applies a change of basis to X. Let X^{PC} be the resulting data set and $\{p_1, \ldots, p_m\}$ be the new basis (the principal components). A change of basis is a bijective transformation: it alters the representation of the information but does not alter the amount of information. Thus, both X and X^{PC} have the same risk of disclosure.

PCA concentrates the information in the first components. That is, the amount of information contained in p_i is greater than in p_j for any $i < j$. As explained in Sect. 2, we can take advantage of this property to reduce the dimensionality of the data set. This is done by dropping some of the last principal components (the ones with less information). Let $X_{PC'}$ be the data set generated from X_{PC} by dropping components $p_{l+1}, \ldots, p_m$. This transformation is not injective: as the result of dropping components, we reduce the amount of information in the data set. Thus, $X^{PC'}$ incurs no more risk of disclosure than X.

After generating an anonymized data set $Y^{PC'}$ by applying an appropriate statistical disclosure control method $\mathcal{G}$ to $X^{PC'}$, we would like to release the protected data set in terms of the original basis. That is, we want to generate an anonymized data set Y in the same basis as X. To do so, we need to apply the inverse PCA transformation. In other words, we need to change the basis from $\{p_1, \ldots, p_l\}$ to $\{x_1, \ldots, x_m\}$.

Algorithm 1 presents a formal description of the previously described steps.

Algorithm 1. Reducing data dimensionality via PCA in anonymization

---

Require:
  X : original data set
  $\mathcal{G}$: Anonymization algorithm

Let $X^{PC} :=$ PCA representation of X
Let $X^{PC'} :=$ projection of X^{PC} over the first l principal components
Let $Y^{PC'} := \mathcal{G}(X^{PC'})$
Let $Y :=$ result of applying the inverse PCA transformation to $Y^{PC'}$
Return Y

---

4 Example: Differential Privacy Anonymization on PCA-Reduced Dimensionality Data

We illustrate the PCA-based dimensionality reduction when the anonymization algorithm $\mathcal{G}$ in Algorithm 1 intends to reach differential privacy. Algorithm 1 entails a pre-processing of the original data set (PCA representation and projection over the first l principal components) and a post-processing of the DP data set (inverse PCA transformation).

It is well established that post-processing DP results preserves the same level of DP: if an output satisfies ϵ-DP, any transformation of it that adds no new information offers at least the same level of privacy. However, the PCA pre-processing step should be performed in a DP way. In [3] it is described how to attain DP in PCA.

In this section, we slightly depart from strict DP. We claim that, in our specific case, we can use a non-DP PCA transformation without degrading the privacy of subjects with respect to DP. If using a non-DP PCA transformation is safe, it should be preferred over DP PCA because it is simpler and yields more accurate data.

Our claim is based on the following observation: data sets X and X^{PC} are equivalent (they contain exactly the same information) and, hence, the privacy risk of X_ϵ (an ϵ-DP version of X) is equivalent to the privacy risk of X_ϵ^{PC}

(an ϵ-DP version of X^{PC}). After generating X^{PC}, we drop some of the components to reduce the dimensionality of the data and obtain $X^{PC'}$. As the amount of information in $X^{PC'}$ is smaller than in X^{PC}, the privacy risk of $X^{PC'}$ is smaller than the privacy risk of X^{PC}, and thus than the privacy risk of X.

Even though publishing the principal components would entail privacy leakage for subjects (as principal components are based on the original data set X), we assert that the way PCA is used in Algorithm 1 does not increase the privacy leakage associated with the anonymization algorithm $\mathcal{G}$, because:

- Algorithm 1 does not disclose the data in their principal components representation: the PCA transformation is done in the first step and it is undone in the last step.
- The anonymized data Y returned for publication are computed based on an anonymized version $Y^{PC'}$ of the principal components coefficients. In the case of using DP as $\mathcal{G}$, Y is computed based on coefficients $Y^{PC'} := X_{\epsilon}^{PC'}$ that are DP; thus Y does not leak more private information than the one DP leaks about the coefficients.

5   Empirical Work

In this section we present experimental work on the example given in Sect. 4 above. To evaluate the DP anonymization on PCA-reduced dimensionality data, we compared the accuracy of the DP data set generated using the microaggregation-based approach described in Sect. 2.2, with and without the proposed PCA-based dimensionality reduction methodology. In particular, we compared against the accuracy of DP data sets obtained in [13].

The empirical evaluation was performed on the Census data set [2], which is a widely used reference numerical data set and was also employed in [13]. Census contains 13 numerical attributes and 1080 records. For the sake of comparability with [13], we defined the domain of each attribute to range between 0 and 1.5 times the maximum attribute value in the data set. The difference between the bounds of the domain of each attribute A_i determines the sensitivity of that attribute (ΔA_i) and, as detailed above, determines the amount of Laplace noise to be added to microaggregated outputs. Since the Laplace distribution is unbounded, for consistency we bounded noise-added outputs to the domain ranges define above.

We used the same information loss measure as in [13], namely the relative error (RE). RE is measured as the absolute difference between each original attribute value a and its masked version a' divided by the original value. A sanity bound was included in the denominator to mitigate the effect of very small values. We used the same sanity bound as in [13]: for an attribute A it is $|\max(Dom(A)) - \min(Dom(A))|/100$. Thus,

$$RE_A(a, a') = \frac{|a - a'|}{\max\{a,\ sanity\_bound_A\}}.$$

When using PCA, the amount of dimensionality reduction is the number of discarded principal components. First, we examined the accuracy of the generated DP data set in terms of RE as we changed the amount of dimensionality reduction. In these experiments, we took $\epsilon = 1$ as the target DP level, which is smaller than the ϵ values usually chosen when generating DP data sets —in spite of DP not offering any effective privacy guarantee for large ϵ, it is not uncommon to see $\epsilon = 10$ or larger in this context. In Fig. 2 we report the average RE (y-axis) over the remaining l principal components for different microaggregation cluster sizes (x-axis); each curve represents a different number l of remaining principal components.

RE is the result of three errors: the dimensionality reduction error, the microaggregation error, and the noise introduced by DP. We focused on analyzing the interplay between the error due to dimensionality reduction and the other types of errors.

The effect of dimensionality reduction is twofold. On the one side, the fewer the components that remain, the greater the error due to dimensionality reduction. On the other side, the fewer the components, the larger the share of ϵ we can assign to each component, which reduces the error due to DP. This is clear by looking at the curves with one and two components in Fig. 2. For small values of k, the total error in the one-component curve is smaller. The reason is that,

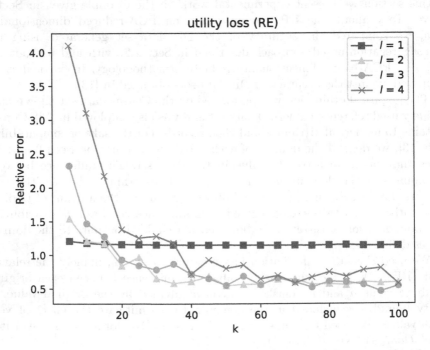

Fig. 2. Average RE (y-axis) of the DP data set for $\epsilon = 1$ in terms the cluster size (x-axis). Each curve corresponds to a different number l of remaining principal components.

for a small k, the error due to DP dominates and having multiple components magnifies it. In contrast, for larger k, the error due to DP becomes less significant and it is preferable to decrease the error due to dimensionality reduction, even if that implies increasing the error due to DP.

Even if the total RE is a combination of opposing factors, it becomes apparent in Fig. 2 that preserving a large number of components is, in general, not a good compromise. To emphasize this fact and show the real value of the proposed method we compared against the results in [13], which do not use dimensionality reduction. Figure 3 shows the RE of the method described in [13] when dealing with the 13 attributes of the Census data set. We observe in Fig. 3 that for DP with $\epsilon = 1$ the best RE is substantially greater than 1 (actually around 4) because the scale is logarithmic. Using PCA dimensionality reduction with 3 principal components we reached an RE of 0.5. Even though RE $= 0.5$ may still be too much, the improvement is very significant. To reduce the RE to more acceptable rates, we could slightly increase ϵ (e.g. $\epsilon = 2$). Indeed, despite DP becomes meaningless if a large ϵ is used, greater values are common in DP data set generation. In contrast, increasing ϵ is not very useful in [13]: even using $\epsilon = 10$ (too large to offer any privacy guarantees) yields results worse than those obtained for $\epsilon = 1$ with our approach.

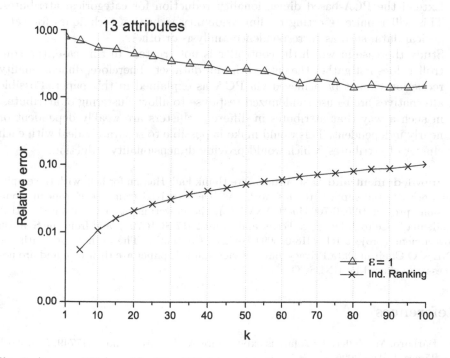

Fig. 3. Average RE (y-axis) of the DP data set in terms the cluster size (x-axis) of the DP data set generation method described in [13] and of the baseline microaggregation algorithm.

6   Conclusions and Future Research

In the context of privacy-preserving data publishing, attaining privacy guarantees at the individual level while preserving a reasonable data utility is only feasible for low-dimensional data sets. We have presented an approach for dimensionality reduction that is aimed at improving the loss of information caused by anonymization.

We leverage principal components decomposition: subjects send their original records to the controller, who gathers the entire original data set and computes the principal components of it. The dimensionality reduction comes from the controller choosing only those components that carry most information.

We have exemplified our approach for the specific case of DP data set generation. We have shown that pre-processing data (to transform the original data set into its principal components) does not increase the risk of disclosure. Experimental results are based on the DP data set generation mechanism of [17], and *they show a very significant information loss reduction in the DP data set.*

Future research lines include the following:

- Amplify empirical work by working on data sets with more attributes and more records.
- Extend the PCA-based dimensionality reduction for categorical attributes. This will require resorting to dimensionality reduction techniques for categorical data, such as correspondence analysis or others.
- Study the case in which the controller is not trusted. In this case, the controller does not gather the entire original data set. Therefore, dimensionality reduction cannot be achieved via PCA as explained in this paper. Possible alternatives are to use randomized response to allow clustering of attributes, in such a way that attributes in different clusters are weakly dependent or nearly independent. This would make it possible to separately deal with each cluster of attributes, which would provide dimensionality reduction.

Acknowledgment and Disclaimer. We thank Fadi Hassan for help with the empirical work. Partial support to this work has been received from the European Commission (project H2020-700540 "CANVAS"), the Government of Catalonia (ICREA Acadèmia Prize to J. Domingo-Ferrer and grant 2017 SGR 705), and from the Spanish Government (project RTI2018-095094-B-C21 "Consent"). The authors are with the UNESCO Chair in Data Privacy, but the views in this paper are their own and are not necessarily shared by UNESCO.

References

1. Barbaro, M., Zeller, T.: A face is exposed for AOL searcher no. 4417749. New York Times, 9 Aug 2006
2. Brand, R., Domingo-Ferrer, J., Mateo-Sanz, J.M.: Reference data sets to test and compare SDC methods for the protection of numerical microdata. Deliverable of the CASC project (IST-2000-25069) (2002)

3. Chaudhuri, K., Sarwate, A.D., Sinha, K.: A near-optimal algorithm for differentially-private principal components. J. Mach. Learn. Res. **14**(1), 2905–2943 (2013)
4. Domingo-Ferrer, J., Mateo-Sanz, J.M.: Practical data-oriented microaggregation for statistical disclosure control. IEEE Trans. Knowl. Data Eng. **14**(1), 189–201 (2002)
5. Domingo-Ferrer, J., Torra, V.: Ordinal, continuous and heterogeneous k-anonymity through microaggregation. Data Min. Knowl. Discov. **11**(2), 195–212 (2005)
6. Dwork, C., McSherry, F., Nissim, K., Smith, A.: Calibrating noise to sensitivity in private data analysis. In: Halevi, S., Rabin, T. (eds.) TCC 2006. LNCS, vol. 3876, pp. 265–284. Springer, Heidelberg (2006). https://doi.org/10.1007/11681878_14
7. Hardt, M., Ligett, K., McSherry, F.: A simple and practical algorithm for differentially private data release. Adv. Neural Inf. Process. Syst.-NIPS **2012**, 2339–2347 (2012)
8. Hundepool, A., et al.: Statistical Disclosure Control. Wiley, New Jersey (2012)
9. Machanavajjhala, A., Kifer, D., Abowd, J., Gehrke, J., Vilhuber, L.: Privacy: theory meets practice on the map. In: IEEE 24th International Conference on Data Engineering-ICDE 2008, pp. 277–286. IEEE (2008)
10. McSherry, F., Talwar, K.: Mechanism design via differential privacy. In: 48th Annual IEEE Symposium on Foundations of Computer Science- FOCS 2007, pp. 94–103. IEEE Computer Society (2007)
11. Narayanan, A., Shmatikov, V.: Robust de-anonymization of large sparse datasets. In: 2008 IEEE Symposium on Security and Privacy, pp. 111–125. IEEE (2008)
12. Nissim, K., Raskhodnikova, S., Smith, A.: Smooth sensitivity and sampling in private data analysis. In: 39th Annual ACM Symposium on Theory of Computing-STOC 2007, pp. 75–84. ACM (2007)
13. Sánchez, D., Domingo-Ferrer, J., Martínez, S., Soria-Comas, J.: Utility-preserving differentially private data releases via individual ranking microaggregation. Inf. Fusion **30**, 1–14 (2016)
14. Snoke, J., Slavković, A.: *pMSE* mechanism: differentially private synthetic data with maximal distributional similarity. In: Domingo-Ferrer, J., Montes, F. (eds.) PSD 2018. LNCS, vol. 11126, pp. 138–159. Springer, Cham (2018). https://doi.org/10.1007/978-3-319-99771-1_10
15. Solon, O.: 'Data is a fingerprint': why you aren't as anonymous as you think online. The Guardian, 13 Jul 2018
16. Soria-Comas, J., Domingo-Ferrer, J.: Optimal data-independent noise for differential privacy. Inf. Sci. **250**, 200–214 (2013)
17. Soria-Comas, J., Domingo-Ferrer, J.: Differentially private data publishing via optimal univariate microaggregation and record perturbation. Knowl.-Based Syst. **125**, 13–23 (2018)
18. Soria-Comas, J., Domingo-Ferrer, J.: Differentially private data sets based on microaggregation and record perturbation. In: Torra, V., Narukawa, Y., Honda, A., Inoue, S. (eds.) MDAI 2017. LNCS (LNAI), vol. 10571, pp. 119–131. Springer, Cham (2017). https://doi.org/10.1007/978-3-319-67422-3_11
19. Sweeney, L.: Simple Demographics Often Identify People Uniquely. Carnegie Mellon University, Data Privacy Working Paper 3, Pittsburgh (2000)
20. Zhang, J., Cormode, G., Procopiuc, C.M., Srivastava, D., Xiao, X.: PrivBayes: private data release via bayesian networks. ACM Trans. Database Syst. **42**(4), 25 (2017)

Author Index